Durability of Materials and Structures in Building and Civil Engineering

Edited by

the late **C. W. Yu**
Consulting Civil and Structural Engineer

and

John W. Bull
Senior Lecturer in Structural Engineering, School of Civil Engineering and Geosciences, The University, Newcastle upon Tyne, UK

Whittles Publishing

CRC Press
Taylor & Francis Group

Published by
Whittles Publishing,
Dunbeath Mains Cottages,
Dunbeath,
Caithness KW6 6EY,
Scotland, UK
www.whittlespublishing.com

Distributed in North America by
CRC Press LLC,
Taylor and Francis Group,
6000 Broken Sound Parkway NW, Suite 300,
Boca Raton, FL 33487, USA

ISBN 1-870325-58-3
USA ISBN 08493-9239-X

Typeset by Compuscript Ltd, Shannon, Ireland
Printed and bound in Poland, EU
Produced by Polskabook

Contents

Preface

When I was approached by Joe Yu to provide a chapter on the durability of aluminium structures, I was very pleased to be associated with the impressive array of world experts Joe had assembled for this book. It was, therefore, with great sadness, that during Joe's editing of the book I learnt of his death following his serious illness. When later I was approached by the publishers and Joe's wife, Beryl, to take over the editorship of the book I felt honoured to accept the invitation. I hope I have been able to maintain Joe's high standards.

Practices in the building and civil engineering construction industry vary around the world. Local conditions, conventions and work practices mean that design methods, design loadings, construction and maintenance have evolved using a range of life expectancies for the structures so designed. Today a client wants a value-managed construction that will fulfil the design requirement, which usually requires high quality, low cost, low maintenance and a defined working life. Each of these requirements can be met singly by relaxing one or more of the other requirements. However, Eurocode EN 1990 *Basis of structural design*, establishes for all the other structural Eurocodes, not only requirements for safety and serviceability but also guidelines for reliability and durability. Emphasis is therefore given to the importance of durability.

The durability of a structure can be defined as the structure's ability to remain fit for purpose during its design working life. This means that the design engineer must take into account many of the following factors: detailing, environment, future use of the structure, maintenance, performance and properties of materials, member shape, performance specifications, structural system, use of the structure and workmanship. This book discusses these aspects in relation to the durability of materials used in the construction industry, and includes aluminium, composites, concrete, masonry, steel and timber among others. It further covers standards/codes and the legal aspects of durability

The state of the art is presented for each material and the effect of the environment on their durability is covered. The book includes numerous examples illustrating the durability of the materials, their performance and how the materials react with their environment. Increasing the materials' durability is an integral part of the book.

The book draws upon an international team of authors with extensive cumulative experience making this wide-ranging comprehensive book a necessary guide for academics, architects, design engineers, civil engineers, construction technologists, maintenance engineers, materials technologists, postgraduate students, structural engineers and undergraduates. Readers will note that chapters differ in some aspects of presentation, for example regarding references. The conscious decision was taken to retain as much as possible of each author's style and since this in no way detracts from, or devalues, the content, I hope this will be considered the correct decision.

My thanks go to Joe Yu, whose idea started this book, to the publishers for their invaluable help and guidance and especially to the chapter authors who have spent so much of their energy and time in making this book such a memorable achievement.

John Bull

Authors

Professor G Baronio, Politecnico di Milano, Dipartmento di Ingegneria Strutturale, Milano, Italy

Professor Giulia Baronio is lecturer in the postgraduate school of Restoration of Monuments at the Polytechnic of Milan and a member of the Board of Professors of the PhD course in Conservation of Politecnico di Milano. She has over 100 publications focusing on the behaviour of construction materials and their service life under different conditions. Giulia Baronia has ongoing research with Laboratoire Central des ponts et Chaussées and the National Technical University of Athens and she is responsible for the Chemical Section of the Chemical Technological Area of the Materials Testing Laboratory of the Structural Engineering Department of the Polytechnic. Professor Baronio is a member of several commissions and RILEM committees dealing with masonry and construction materials.

Professor B Benmokrane, NSERC Research Professor in Innovative FRP Composite Materials for Infrastructure, Department of Civil Engineering, University of Sherbrooke, Sherbrooke, Québec, Canada J1K 2R1

Professor Benmokrane is a project leader in ISIS (Canadian Network of Centers of Excellence on Intelligent Sensing for Innovative Structures) with research interests that include the development and application of advanced FRP composite materials in civil engineering structures. He is an active member on Canadian Standard Association and ACI committees where he has contributed to the development of new codes. He has published more than 200 technical papers and received numerous awards or distinctions. During the last ten years, Professor Benmokrane has trained more than 50 graduate students and postdoctoral fellows on various aspects of FRP materials and concrete structures reinforced with FRP.

Professor Luigi Binda, Politecnico di Milano, Dipartmento di Ingegneria Strutturale, Milano, Italy

Luigia Binda has been teaching at the Polytechnic of Milan since 1963, becoming Professor in 1990. She teaches courses on the diagnosis and strengthening of historic buildings, and on restoration. Professor Binda has numerous research interests concerned with masonry including long-term behaviour, durability, modelling and experimental investigations. Luigia Binda is a member of several international societies and chairs or convenes several RILEM committees. She has undertaken *in situ* experimental investigation for the evaluation of structural and physical decay of masonry materials and structures with particular interest in the physical deterioration and mechanical damage of brick and stone masonries. She has also been involved in the damage evaluation and rehabilitation of brick masonry facades and plasters of ancient monuments, and on the diagnosis of damaged masonry structures.

Dr John W Bull, Senior Lecturer in Structural Engineering, School of Civil Engineering and Geosciences, The University, Newcastle upon Tyne, UK

Dr Bull worked in the construction industry between 1974 and 1979 before moving to Newcastle University. His research areas include computational engineering, structural optimization, finite element development and analysis, life cycle costing, soil-structure interaction, precast concrete runways, airfield damage repair and structural design. Dr. Bull has successfully completed six research grants and has 147 publications, including 12 books. Further he has two edited books and three book chapters awaiting publication. He has been external examiner for many PhDs, a visiting professor in Australia and in Japan, an advisor on book proposals to a number of publishers, refereed papers for many journals and has been a member of the Editorial Board for over 25 conferences and journals. He has been a member of BSI Standards Sub-Committees on the Structural use of Aluminium and the BSI Committee on Timber Testing Methods. He was also a member of the Aluminium Federation's Aluminium Foresight Programme Construction Committee.

Dr WWL Chan, Consulting Civil and Structural Engineer, Middlesex, UK

Dr Bill Chan is a consulting civil and structural engineer with a special interest in timber, having been responsible for many timber frame and civil engineering projects. He is a Fellow of the Institute of Wood Science and member of British Standards technical committees on timber codes of practice, Eurocodes and product standards.

David H Deacon, Director, The Steel Protection Consultancy Ltd., Leighton Buzzard, UK

The author has many years experience, of which more than thirty have been as an independent consultant. He has been retained by a number of clients both in the UK and overseas on a range of major projects. As Director of the UK's Steel Protection Consultancy he is currently advising on numerous projects including both the Forth Road and Rail Bridges, and the Thames Barrier.

David K Doran, Consulting Civil and Structural Engineer, London, UK

David Doran has been a consulting civil/structural engineer for over 20 years and previously was for 20 yearsChief Structural Engineer for George Wimpey plc where he was a Director of Wimpey Laboratories, Wimpey Group Services and Wimpey Construction UK. He has served on a number of technical committees including BSI Committees on Durabilty from 1965 to 2004. He has chaired a number of Task Groups dealing with, inter alia, *Cladding and Alkali-silica reaction in concrete.*

Dr Graham Gedge, Associate Director, Arup Materials Consulting, London, UK

Graham Gedge specialises in corrosion, corrosion control and durability of materials in the built environment. His work spans all areas of building, civil and structural engineering where durability of steel (and other metals) is a concern. He has a particular interest in bridge structures and the development of appropriate solutions to corrosion related problems in civil engineering applications and has been responsible for developing innovative solutions to long established corrosion problems.

Professor Odd E Gjørv, Norwegian University of Science and Technology, Trondheim, Norway

Dr Odd E. Gjørv is Emeritus Professor of Structural Engineering at the Norwegian University of Science and Technology at Trondheim, has conducted research in various aspects of concrete technology since 1959, with major interest in durability and construction of reinforced and prestressed concrete in severe environments and he has over 350 scientific and technical publications and some patents to his name. In addition to working with several international technical organizations on various committees and commissions, he has contributed to many conferences and meetings around the world. He has advised companies and governmental agencies on the subject f concrete technology and his contributions to engineering have been recognized in several honours.

Emeritus Professor AW Hendry, Edinburgh, UK

Loh Yan Hui, Project Director (Reclamation), Surbana Consultants Pte Ltd, Singapore

Loh Yan Hui has more than 25 years experience in civil engineering and reclamation. He is currently the Division Head (Reclamation & Infrastructure) of Surbana International Consultants Pte Ltd, Singapore.

Dr John Morton, Masonry Consultant, Surrey, UK

Dr Morton is the author of many papers, technical notes and design guides, including some that are considered 'industry standards' such as *Designing for movement in brickwork*. With 40 years experience in masonry design and "material properties", such as durability issues, he now runs his own practice specialising in masonry in its broadest sense. He represents the Institution of Structural Engineers on the main technical drafting committees for masonry at BSI and is the National Technical Coordinator for Masonry for the UK.

Dr LS Norwood, former Group Technical Manager, Scott Bader Company Ltd., Wellingborough, Northamptonshire, UK

Timbak Pavate, visiting Research Professor at the University of Sherbrooke, Sherbrooke, Québec, Canada J1K 2R1

Timbak Pavate was formerly Chief Research Engineer at the Indian Institute of Technology in Bombay, India. He is a multidisciplinary researcher specialising in electrokinetics as applied to the processes of solid-liquid systems. Presently he is engaged in research in concrete technology and material development.

Mathieu Robert, postgraduate student at the University of Sherbrooke, Sherbrooke Québec, Canada J1K 2R1

Mathieu Robert received his BS and MS from the Laval University and University of Sherbrooke, respectively. His research interests include fibre-reinforced polymer for infrastructure applications, mechanical and durability characterisation and micro-structural analysis.

Dr Steen Rostam, COWI A/S, Kgs., Lyngby, Denmark

Steen Rostam has been with COWI A/S, Denmark since 1972 and has overall responsibility for concrete durability technology and was, with others, responsible for the 100 years service life design concept for the Great Belt Link in Denmark. He specialises in durability and service life concepts in the design, analysis, detailing, operation, assessment, maintenance and repair of reinforced and prestressed concrete structures combining structural and non-structural aspects into an operational durability technology. He is currently the concrete durability expert for numerous international projects and chairman of Commission 5 (Structural Service Life Aspects) of the ***fib*** (International Federation for Structural Concrete). He is a member of several code committees and has gained international recognition for his work; in 2003 receiving the ***fib*** Medal of Merit for his international achievements within the field of durability and service life of concrete structures and in 2004 being awarded the ANIFER, International Calavera Award. He has over 100 publications including *CEB Guide to Durable Concrete Structures* and the ***fib*** book *Structural Concrete*.

Neil le Roux, Solicitor, formerly with Reynolds Porter Chamberlain, London, UK

Dr C T Tam, Honorary Researcher, Department of Civil Engineering, National University of Singapore, Singapore

Tam Chat Tim has more than 40 years of experience in teaching, research and providing consultancy services in concrete construction; he is currently retired but has an honorary appointment to continue in research at the Department of Civil Engineering, National University of Singapore.

Tee Choon Peng, Civil Engineer, Surbana Consultants Pte Ltd., Singapore

In 1997 Tee Choon Peng graduated joined the Housing and Development Board as a civil engineer and is now with Surbana International Consultants Pte Ltd. In the last nine years, he has worked on major reclamation projects and has been greatly involved in the construction of marine structures.

Dr R C de Vekey, Associate Technical Director, BRE Masonry, Construction Division, Building Research Establishment, Watford, UK

After an initial degree in chemistry, Bob de Vekey undertook training in materials science related to building materials and gained a PhD from Imperial College London. Although semi-retired from the Building Research Establishment, he still consults and writes technical material on his main specialism of masonry design and performance.

Dr Peng Wang, Postdoctoral Researcher, University of Sherbrooke, Sherbrooke, Québec, Canada J1K 2R1

Peng Wang received his PhD from the University of Sherbrooke. His research interests include fibre-reinforced polymer for infrastructure applications, and durability characterisation.

1 Durability of Concrete Structures – The State of the Art

Dr Steen Rostam

1.1 The uniqueness of structural concrete

A pleasant visual appearance and attractive ageing properties have contributed substantially to the reputation of concrete. One of the most versatile and robust construction materials available, it has obtained a dominant position in the construction industry and can expect to remain in this position for the foreseeable future.

In the vast majority of cases concrete structures work well (see Fig. 1.1). Nevertheless, there are situations where premature deterioration has rendered concrete structures unfit for use. It is an economic disaster when a huge urban dwelling, a large bridge, or a major marine structure deteriorates rapidly soon after coming into service, and since the 1970s such examples have increasingly been reported. The collapsed deck of a multi-storey car park in Fig. 1.2 is just one example. The reasons are complex but fortunately the main causes of premature deterioration have been identified. It is essential to have these causes highlighted

Figure 1.1 *The Sydney Opera House reflects the potential of close co-operation between architect and engineer.*

Figure 1.2 *Consequences of a structural failure in an aggressive environment caused by de-icing salts carried in by the vehicles. Deterioration had developed unattended until failure, indicating a complete lack of maintenance. (Photo: Arminox).*

so that design methods, construction procedures, material compositions, as well as maintenance and repair procedures, can be rectified to ensure reliable new structures in the future.

For the large stock of existing structures which do not live up to expectations, rational assessment and rehabilitation procedures to prolong their service life are a major challenge. Concrete structures have characteristic properties which, with respect to deterioration, differ fundamentally from structures made from other materials (Rostam 1998). These characteristic properties are:

- The quality and the performance of the concrete is assumed at the design stage of a specific structure. Nevertheless, these specific quality and performance requirements have to be specified.
- The true quality and performance characteristics of the concrete are determined during the construction process. Hence, this very short time period is the most important factor in ensuring the required durability of the finished structure.
- If the durability performance turns out to be sub-standard, it is often not apparent nor detectable until some considerable time has passed – which may often be longer than the contracted liability period of the contractor and the designer, but very much shorter than the service life expected by the owner.

1.2 Ageing and deterioration of concrete structures

The durability and service life of a concrete structure is determined by the interaction between the structure and its environment. Depending upon the type and importance of the structure, its service life may be between 10 and 200 years. For example 100–120 years is the design service life of large bridges.

Figure 1.3 *Marine structure. Synergy effect between chloride corrosion and freeze–thaw action.*

If the aggressiveness of the structure's environment has not been adequately identified and dealt with during the design and construction process, premature deterioration may render the structure unfit to serve its intended purpose. It has been recognised that some environments are more aggressive than others, e.g. cyclic wetting and drying at high temperatures with water that contains chloride is one of the most severe environments for reinforced concrete.

Marine structures which are exposed to freeze–thaw action, in addition to the saline environment, are highly vulnerable (Fig. 1.3).

Apart from chlorides entering the concrete from seawater spray or de-icing salts, two other causes of excessive amounts of chloride in concrete need to be mentioned. Due to general chloride contamination of fine aggregates and water resources in several regions, such as the Gulf countries, numerous concrete structures have been built – and, due to lack of other resources, are still being built – with contaminated sand and water in the initial mix. This has led to very serious early corrosion damage (Fig.1.4).

During the construction boom of the 1960s and early 1970s calcium chloride based accelerators were used in many industrialised countries, later leading to serious corrosion damage (Fig. 1.5). Reinforcement corrosion is probably the most serious and widespread type of deterioration in concrete structures. Costs related to the repair of corrosion damage dominate the maintenance and repair budgets in many countries.

It has become evident that all deterioration mechanisms depend upon an aggressive substance penetrating from the surrounding environment into the outer layer of concrete – the cover (Fig. 1.6).

Knowing and understanding the transport mechanisms of liquid and gaseous substances into and within concrete structures is, therefore, the most important element in ensuring sufficiently durable concrete structures. This is also the fundamental basis for quantifying durability in the form of service life performance (CEB 1992).

The development, over time, of nearly all types of deterioration mechanism can be modelled by a two-phase curve as illustrated in Fig. 1.7 (Tuutti 1982). This is the basis

Figure 1.4 *The cracking of a column due to chloride corrosion caused by contaminated aggregates and mixing water. The age of the structure at first cracking was 3 years (Gulf climate).*

for defining the service life of a structure. The technical service life is the time in service until a defined unacceptable state of deterioration is reached.

The two phases of deterioration are as follows:

1. The initiation phase. During this phase no noticeable weakening of the material or reduction in the function of the structure occurs, but some of the protective barrier is broken down or overcome by the aggressive media. Carbonation, chloride penetration and sulphate accumulation (the latter two are accelerated by cyclic wetting and drying) are examples of mechanisms that determine the duration of the initiation period.
2. The propagation phase. During this phase an active deterioration develops and loss of function is observed. A number of deterioration mechanisms develop with time at an increasing rate. Reinforcement corrosion is one important example of propagating deterioration.

Because concrete structures are composed of two materials, concrete and steel reinforcement, it is necessary to differentiate between the deterioration of the concrete and the corrosion of the reinforcement. The deterioration of reinforced concrete structures includes mechanical, chemical, physical as well as electrochemical mechanisms (Fig. 1.8).

It is the possible interaction between the deterioration mechanisms of each material when interacting with the environment which determines the durability of a

Figure 1.5 *Delamination of a precast balcony partition wall due to chloride corrosion caused by calcium chloride added as an accelerator during casting. The age of structure was 15 years (North European climate).*

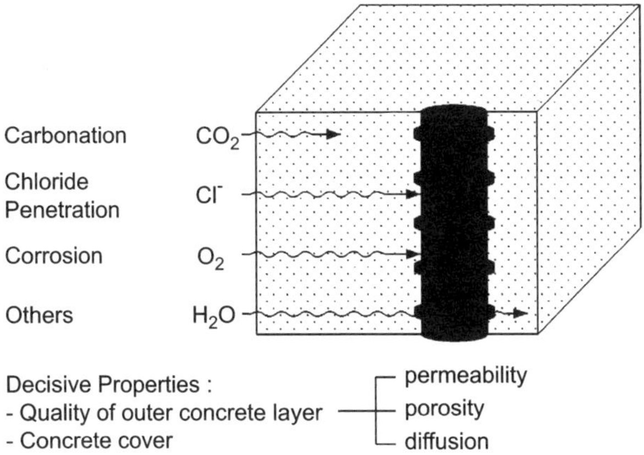

Figure 1.6 *The importance of the outer concrete layer – or the skin of concrete – to protect the structure against deterioration.*

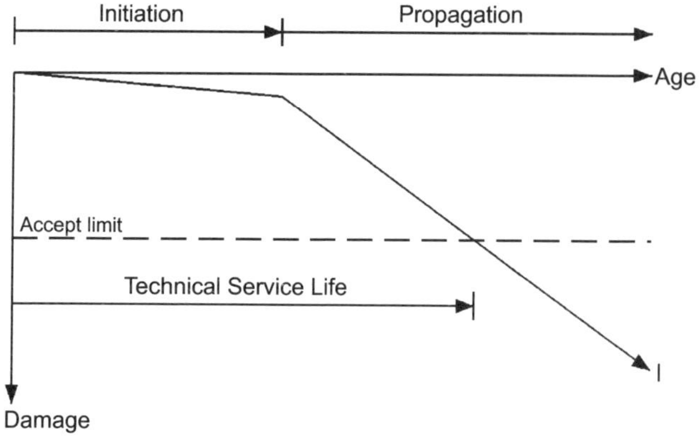

Figure 1.7 *The service life of concrete structures. A two-phase modelling of deterioration.*

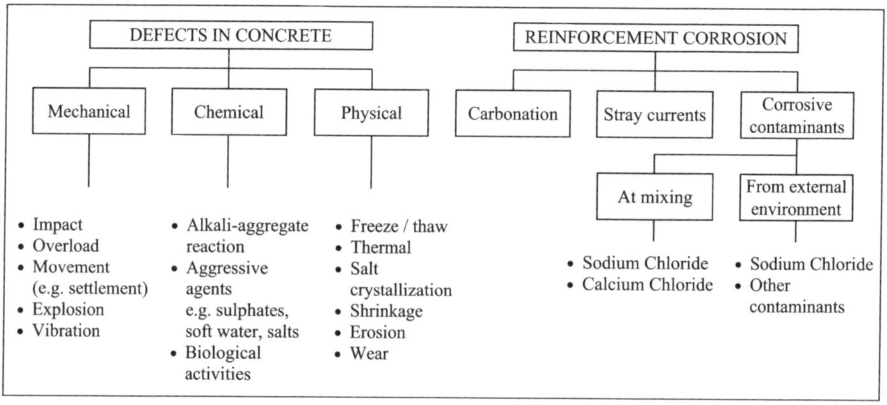

Figure 1.8 *Common causes of deterioration of concrete structures [ENV 1504-9 (1997)].*

concrete structure. It is important to keep this issue in mind, as it is the performance and durability of the structure in its environment which is important, and not the potential durability of each individual material.

1.3 Transport and deterioration mechanisms in concrete

1.3.1 Transport mechanisms

Particular transport mechanisms which are decisive factors for durability include:

- permeation – such as water under a hydrostatic head;
- diffusion – such as the penetration of carbon dioxide into the concrete or of water vapour out of the concrete;
- capillary suction or absorption of water or of solutions; as well as
- combinations – such as the transport of ions by moving water.

The rate of transport is different for the various mechanisms. In particular the take-up of water or solutions by capillary suction is considerably more effective than the loss of moisture due to drying. The presence of water or moisture is the single most important factor controlling the various deterioration processes, apart from physical deterioration.

The transport of water within the concrete is determined by the pore type, pore size and distribution, and by cracks. Controlling the nature and distribution of pores and cracks is essential when designing for durability. In turn, the type and rate of the deterioration processes for concrete (and for normal reinforcement and prestressing reinforcement) determine the resistance and the rigidity of the materials, sections and structural elements making up the structure.

1.3.2 Mechanical actions

Mechanical defects are defined as damage caused by impact, overload, movement, vibration and explosion (Fig. 1.8). The damage which occurs will normally be cracking, spalling, failure of the concrete due to overstressing, yield or rupture of the reinforcement, or fatigue loading of these materials. These defects can be investigated by traditional structural analysis.

Cracks will influence the type and rate of transport of substances into the concrete. However, structural and load-induced cracks are inevitable in concrete structures and do not necessarily indicate undue lack of serviceability or durability provided the crack type, extent, width and orientation are controlled through the design and construction processes.

The different types of mechanical actions and their consequences for the performance of concrete structures are described in CEB (1992).

1.3.3 Chemical processes

The chemical reactions which lead to a decrease in quality and to increased deterioration of the concrete are well established. The most important are:

- reduction of the pH value of the concrete due to carbonation;
- reaction of alkalis with reactive aggregates in the concrete, as shown in Fig. 1.9;
- reaction of sulphates with the aluminates in the cement;
- reaction of acids, ammonium salts, magnesium salts and soft water with hardened cement;
- effects of biological activities.

Chemical and biological reactions which result in concrete deteriorating are given a comprehensive treatment in Biczók (1972), CEB (1992) and FIB (1999). A few of the most important mechanisms are discussed below.

The alkalinity of concrete, represented by the high pH value of the pore water in the concrete, provides a very efficient protection to the reinforcement against corrosion because of the passivation of the steel. The alkalinity of the concrete can be reduced by carbonation (one of the chemical processes in concrete) which may cause depassivation of the embedded steel and trigger corrosion. Carbonation is a chemical reaction between the calcium hydroxide in the concrete due to the hydration of the cement and the carbon dioxide of the surrounding atmosphere. Carbonation itself does not generally harm concrete, but the lowering of the pH value may lead to corrosion of the reinforcement.

In recent years, increased air pollution has acidified rain and very low pH values have been measured in some areas. This has proved detrimental to plants and many organic materials. However, acid rain is not a problem for concrete so long as the concrete quality is good. The alkalinity of concrete will spontaneously neutralise acid rain because of the reserve calcium hydroxide, even in carbonated concrete, as rain has no acid buffer capacity.

If concrete contains aggregates with a critical amount of amorphous silicates a deleterious reaction may take place with alkalis originating mainly from the cement. Such alkali–silica reactions can in adverse conditions cause serious cracking of the concrete (Fig. 1.9). Depending upon the type of aggregate and the concrete mix such cracks may appear as fine map cracking, delamination of thin concrete layers, or large single cracks.

In general, alkali–silica reactions develop during warm and humid weather. If a structure which is prone to such reactions is located in an area with cold winters then

Figure 1.9 *Cracking due to alkali–silica reaction in a column. The cracks form in the direction of least resistance thus indicating the path of the compression trajectories in this column.*

cracking which appears during the summer may increase the take-up of water and consequently raise the risk of further cracking due to frost damage during winter. This leads to a detrimental synergistic effect between summer and winter conditions.

In some parts of the world deleterious expansions may occur in the concrete when exposed to sulphates contained in groundwater or soils. Also, in sewage installations sulphate-reducing bacteria may produce sulphuric acid which dissolves the concrete. Sulphates in combination with chlorides, as in seawater, will usually not produce sufficient expansions to crack the concrete. Therefore seawater is not considered as aggressive to concrete as the content of sulphates would usually warrant.

1.3.4 Physical processes

The main physical defects in concrete are caused by:

- freezing and thawing, aggravated by de-icing salts;
- early age cracking due to shrinkage and thermal strains;
- salt re-crystallisation within the concrete;
- drying shrinkage strains;
- erosion and wear.

Owing to the fact that the volume of water increases by 9% during freezing, a sufficient volume of pore space not filled with water should be available to allow the water to expand, thus preventing frost damage. The limit value of the water content where damage occurs is defined by the critical degree of saturation.

The application of de-icing agents to a concrete surface covered with ice will cause a substantial drop in temperature at the surface of the concrete (temperature shock) during thawing of the ice. The difference in temperature between the surface and the interior of the concrete gives rise to internal stresses which are likely to induce cracking in the outer layer of the concrete. For the reasons outlined above, any frost attack should be regarded as more severe when de-icing agents are present. Consequently, to ensure frost resistance under these circumstances a higher content of entrained fine air voids in the concrete will be required. The corresponding loss in concrete strength must be accounted for in the mix design. Aggregates, which are not frost-resistant will, as a rule, absorb water that will expand during freezing and destroy the cement paste. Typical indications of such processes are local spallings above larger-sized aggregates called pop-outs.

In concrete the expansive forces due to salt re-crystallisation near to the surface will usually cause only minor problems provided the concrete is not weak and porous. Of more importance is the chemical effect of the increased concentration near to the surface of aggressive substances. Nevertheless, there are examples of severe scaling of concrete due to salt crystallisation (Fig. 1.10).

Young concrete is prone to cracking. During the transition phase from fresh to hardening concrete, a critical period with low tensile strength and low deformability starts a few hours (at the earliest 2 hours) after casting and lasts for 4 to 16 hours. During this early plastic phase of concrete, plastic shrinkage cracking and plastic settlement or slump cracking can occur. The former type of cracking is due to early drying (particularly in slabs) while the latter type occurs typically in deep members.

Figure 1.10 *Salt scaling of concrete in a pile foundation carrying a pipeline. An in-situ cast top part of the pile is of better quality than the pile itself and is thus undamaged.*

The heat set free during the hydration of concrete creates a condition of self-equilibrating stresses, with tensile stresses in the outer layers and compressive stresses in the core. If the tensile stresses exceed the low tensile strength of the hardening concrete, thermo-cracks are formed. The cracks are surface cracks, mostly in the form of map cracking. More serious are thermo-cracks developed due to the tensile strains caused by temperature differences across construction joints between a previously cast and a new cast section. Such cracks can be 1–3 mm wide and will remain open permanently.

The hydration of cement has a feature, which can increase the tendency of early age cracking; this is autogenous shrinkage, sometimes called chemical shrinkage. This shrinkage strain coincides with the thermal contraction when the concrete cools after the peak temperature from the heat of hydration has passed. The consequence is that the available tensile strain capacity of the young concrete during cooling is reduced, thus reducing the thermal differences which can be resisted by the concrete without cracking. The lower the water/cement ratio and the higher the absolute cement content, the stronger is the tendency for micro- and macro-cracking due to chemical shrinkage.

Drying shrinkage occurs in all concrete structures exposed to the atmosphere. The rate depends upon the concrete mix, the average relative humidity of the air, the temperature and the geometric form and dimensions of the members. Shrinkage is an imposed deformation, which can generate large forces in restrained members. Shrinkage cracking often develops in long walls, retaining walls, edge beams on bridges, etc. The right amount and layout of the reinforcement can control the location and width of such shrinkage cracks but cannot avoid fully their occurrence. Shrinkage cracks may be avoided by prestressing the concrete before shrinkage has developed.

Erosion and wear are physical/mechanical processes, which cause a reduction in concrete dimensions. Apart from the obvious effects of abrasive machinery and traffic on concrete pavements, flowing water and possibly ice may cause abrasion of the concrete. Fast flowing water passing abrupt changes in geometric section can cause cavitation, which has a strong percussive effect on concrete. This phenomenon occurs in spillways from dams and in canals and pipelines that carry fast flowing water.

1.4 Reinforcement corrosion

1.4.1 Depassivation of reinforcement

Normal reinforcement is very efficiently protected against corrosion when cast into a good quality alkaline and chloride-free concrete with a pH value in the range of 12.5–13.5. This is the well-known unique benefit of using reinforced concrete structures in building and construction. This inherent corrosion protection may be eliminated and corrosion initiated due to:

- carbonation of concrete reaching the level of the reinforcement (1.3.3);
- corrosive contaminants either mixed into the concrete at construction or penetrating from the external environment (1.4.3);
- stray current corrosion (1.4.4).

A further depassivating mechanism, but of considerably less importance, is a reduction of alkalinity due to leaching out of lime by flowing water. In practice, this may happen in the region of weak points in the structure (e.g. leaky construction joints and wide cracks) in combination with inferior quality concrete.

1.4.2 Corrosion process

As a simplified model, the corrosion process can be separated into two processes: the cathodic process and the anodic process, as illustrated in Fig. 1.11. Under practical conditions, rust products are more or less water-containing compounds and are therefore usually expansive with a volume increase of up to 600–700% of the volume of the dissolved iron. Consequently, they are able to exert very high expansion pressures (Fig. 1.12).

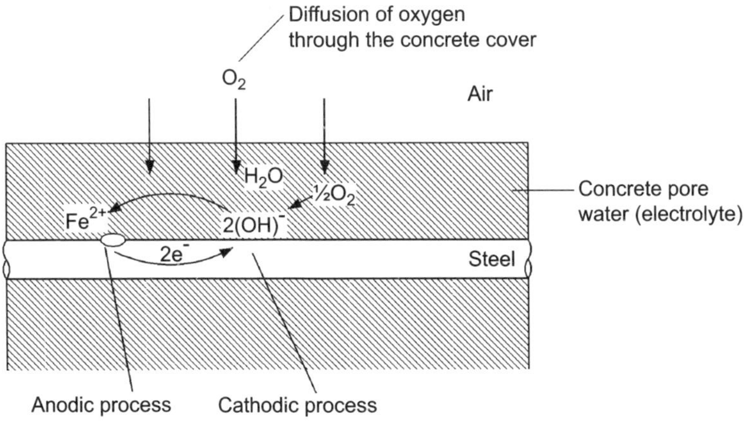

Figure 1.11 *Model of the corrosion mechanisms for steel reinforcement.*

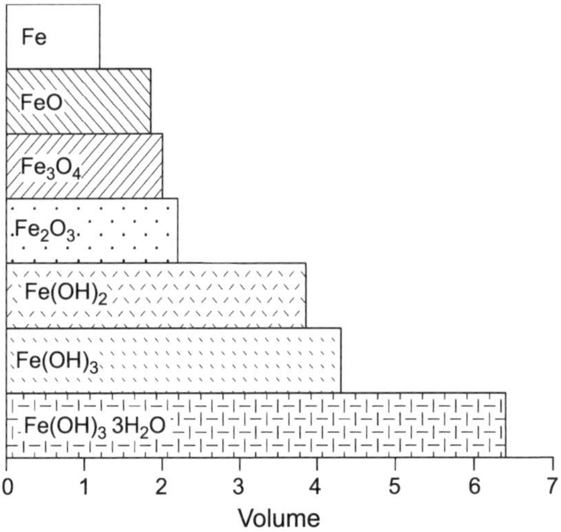

Figure 1.12 *Relative volume of the different types of rust products.*

This means that only oxygen is consumed to form rust products. This oxygen must normally diffuse through the concrete cover towards the reinforcement. Water is only necessary to enable the electrolytic process to take place, and combines with the rust depending on availability. A comprehensive treatment of reinforcement corrosion is given in Nürnberger (1995).

When corrosion develops in environments with low availability of oxygen, the volume of the rust products may only be 50–200% greater than the volume of the steel. Such corrosion processes proceed slowly, and in special cases the rust products may diffuse into the voids and pores of the porous concrete without causing cracking and spalling. In such rare cases, serious corrosion may develop on the reinforcement without any visible warning and a sudden failure may occur.

From the above it follows that the preconditions for corrosion to occur are as follows:

- Depassivation of the reinforcement must have taken place creating a location for an anodic process to develop.
- The depassivated area must be in metallic contact with a neighbouring or nearby area of the reinforcement which is passivated, in order for a cathodic process to take place.
- The cathodic area and the anodic area must also be connected electrolytically in order for ions to move between the two areas.
- There must be sufficient moisture available in the concrete surrounding the reinforcement in order for the ions to migrate between the cathode and the anode. This means that the electric resistivity of the concrete must be sufficiently low for this migration to take place at a rate which sustains corrosion as an active deterioration mechanism.

- There must be oxygen available, primarily at the cathode in order to sustain the cathodic process creating hydroxyl ions, and also at the anode, where the availability of oxygen determines the type and volume of the rust products produced.

Both CO_2 and chlorides may penetrate to the steel surface through cracks at some order of magnitudes faster than through uncracked concrete. This may influence the initiation period but with respect to carbonation it has been shown convincingly that with well distributed cracks, and with relatively large cover, a corrosion of the reinforcement at the crack will usually self-heal. For chloride penetration at a crack, the initiation period may be reduced but increased cover will considerably enhance the durability of the concrete.

1.4.3 Chloride-induced corrosion

Chloride-induced corrosion is currently considered to be the most important and most serious deterioration mechanism for concrete structures. Therefore, although chloride penetration into concrete may be considered a physical process, chloride penetration is described as an essential part of reinforcement corrosion.

Chloride ions (originating from seawater or de-icing salt) may penetrate through the pores to the interior of the concrete. Chloride intrusion is due to either diffusion, taking place in totally or partially water-filled pores, or capillary suction of chloride-containing water. Cement has a certain chemical and physical binding capacity for chloride ions (forming Fridell salt), depending upon the chloride concentration in the pore water. However, not all the chlorides can be bound. Equilibrium will always exist between bound chlorides and free chloride ions in the pore water.

Only the free chloride ions are relevant to the corrosion of the reinforcement. It is important to note, therefore, that after carbonation of concrete, bound chlorides are released again, so that the chloride content in the pore water, and consequently the risk of corrosion due to chlorides, will increase considerably. The critical chloride concentration or threshold value at which corrosion will occur depends upon many parameters and cannot generally be fixed. An indication of the threshold value is presented in Table 1.1 which is based on experience in temperate climates. In hot, humid regions risk levels may be increased by one level.

Chlorides may also be mixed into the concrete through contaminated fine aggregates and mixing water. In some cases seawater has been used as mixing water due to ignorance. This has caused serious corrosion damage in marine concrete structures and must be avoided as seawater is not allowed by almost all standards and codes.

Table 1.1 Corrosion risk classification at different chloride threshold levels

Risk of corrosion	Chloride (% wt of cement)	Chloride (% wt of concrete)*
Negligible	<0.4	<0.07
Possible	0.4–1.0	0.07–0.17
Probable	1.0–2.0	0.17–0.35
Certain	>2.0	>0.35

*Based on 400 kg cement/m^3

Previously, calcium chloride has been used as an admixture in concrete to accelerate the hardening process. The accelerating effect is pronounced, illustrated by the possibility of the precasting industry being able to use concrete forms twice in a day with this type of admixture (Fig. 1.5). This admixture has also caused very extensive and costly damage to a large number of concrete structures, and as such its use is not allowed in reinforced concrete and it is banned in all of today's codes of practice.

1.4.4 Stray current corrosion

Electric currents may flow through the soil due to leaking electric current sources, or due to general systems of cathodic protection such as those in the neighbourhood of pipelines, oil and gas tanks and marine structures. In addition, moving trains and trams develop a magnetic field, which can generate an electric current in neighbouring metallic installations. If concrete structures have reinforcement parallel to the direction of passing trains they may generate high electric currents, proportional to their lengths, which trigger corrosion at the ends of the bars where the current leaves the bar. Preventing such stray current corrosion requires earthing of the reinforcement or an active potential equilibration in the system.

1.4.5 Effect of temperature

It is generally recognised that increasing temperatures result in increased rates of deterioration. This is valid for both chemical deterioration and for the corrosion of reinforcement. The rate-increasing effect of increasing temperature is due mainly to the effect on the transport rate as higher temperature results in the higher mobility of ions and molecules. Depending upon the type of reaction, the accessibility will be determined by the permeability of the still sound concrete or by the passivating layer of the reaction products. A simple rule-of-thumb says that a temperature increase of 10°C doubles the rate of deleterious reactions. This factor alone makes hot and tropical environments considerably more aggressive than temperate climates. This effect of temperature can be clearly demonstrated by comparing Figs 1.13 and 1.14. In the former case the average yearly temperature is approximately 30°C higher than in the latter case, which would lead to a $2 \times 2 \times 2 = 8$ times faster deterioration in the Gulf compared to the rate of deterioration in the Nordic countries. The pictures are clear documentation of this dependency on temperature.

1.5 Durability-enhancing measures

For the large majority of ordinary structures placed in aggressive environments a conscious choice of cement type, concrete mix (especially w/c-ratio), concrete cover, curing (moisture and temperature control) and geometry of the exposed parts of the structure, will normally result in a satisfactory service life. The design procedures, concrete mix and detailing that enhance durability and service life are described in more detail in a later chapter.

It has been recognised that in some selected or special cases the usual choice of design parameters for durability will not provide an adequate service life, often

Figure 1.13 *Reinforced concrete jetty in the Gulf which exhibited extensive chloride-induced corrosion damage after only 2–3 years, and reached a stage of failure and collapse after 7.5 years, when this picture was taken.*

Figure 1.14 *Bridge piers exposed for 18 years to a temperate marine environment showing the extensive damage in the splash zone due to chloride-induced reinforcement corrosion.*

because a very small or local part of the structure is particularly prone to premature deterioration. This would typically be in the following situations:

- Where the environment is particularly harsh and aggressive. Typically this would be a hot, humid and saline environment, such as the splash zones of

Figure 1.15 *Permanent thin stainless steel linings used as lost formwork for the concrete columns. The structure is nearly 70 years old and has been exposed to heavy de-icing salts each winter for 30 years without any trace of deterioration.*

marine structures, or structures exposed to the extensive use of chloride-based de-icing chemical. Further aggravation would be caused by cyclic wetting and drying or by other types of evaporative effects causing increased local concentrations of chlorides, sulphates, magnesium salts, etc.

- The available ingredients of the concrete mix being unavoidably polluted, particularly with chlorides in the fine aggregates and in the mixing water.
- The available workforces not having adequate experience, knowledge or understanding of the importance of the construction process in providing a long-term durable structure.
- The specified service life being very much longer than for normal structures, say 100 or 200 years, or more.

The most serious conditions occur when the first two or three of the above situations occur together. In such situations it is now recognised that concrete alone should not always be relied upon to provide the service life needed and that a durability-enhancement system should be applied. Fig. 1.15 illustrates a special case where a thin stainless steel lining was used as lost formwork for a reinforced concrete column exposed to many years of salt spray. This has proved to be a highly reliable and life cycle cost-efficient solution.

1.6 Coatings and surface protection

A few of the reasons for applying surface protection to new concrete are as follows:

- resistance to aggressive chemicals (e.g. in a chimney, sewage treatment works, chemical plant, etc.)
- resistance to abrasion (e.g. industrial floors, etc.)

- decreased contamination of the environment/contents (e.g. dust from walls, ceilings and floors, tanks, etc.)
- cleaning conditions (e.g. food factories, hospitals, etc.), or appearance (e.g. roof tiles, crash barriers, etc.).

Concrete structures designed and built in accordance with the current codes of practice with regard to materials characteristics, architectural and structural design, processes of execution, inspection and maintenance procedures, normally do not deteriorate prematurely due to alkali–silica reactions, freeze–thaw action or reinforcement corrosion. However, some older and some new concrete structures show (or will show) decreased durability due to insufficient cover to the reinforcement, porous concrete, reactive aggregates or faulty design. Surface treatments may successfully be used to diminish the rate of degradation of these structures, by reducing the rate of penetration of water and aggressive substances into the concrete.

1.7 Corrosion protection of reinforcement

1.7.1 Non-corrodible reinforcement

Recently the development of non-corrodible reinforcement for concrete has opened promising new doors in the fight against reinforcement corrosion. The following products are presently available:

- stainless steel reinforcement;
- non-metallic reinforcing bars made from glass fibres, aramid fibres or carbon fibres.

Stainless steel reinforcement

During the past few years stainless steel reinforcement has become commercially available in dimensions, strengths and alloy types which are fully compatible with normal structural requirements for reinforced concrete structures, and at competitive prices – relatively speaking! Stainless steel reinforcement may be corrosion resistant even in highly chloride-contaminated environments. Used selectively in the most exposed zones of the structure the increased costs per kilogram of stainless steel compared with the costs of normal steel will have a marginal or negligible effect on the overall initial construction costs. In addition, the service life costs will be reduced due to the savings in future repair and maintenance (Gedge et al. 2001; Abbott 1999; Concrete Society 1998).

From a practical point of view this technology is particularly interesting because it solves ‘only’ the corrosion problem. All other techniques and technologies within design, production and execution remain unchanged – a fact that is very attractive to the traditionally very conservative construction industry (Markeset et al. 2000).

It is recognised that the most serious durability problem for concrete structures is that of reinforcement corrosion. Therefore, the reliable and ready availability of stainless steel reinforcement at reasonable and foreseeable prices may change – or revolutionise – major parts of the building sector in aggressive environments, simply by solving the corrosion problem (FIB 1999).

Another reason is the added value which follows from the possibility of accepting the use of locally available materials (even with chloride contamination) and also accepting a poorly qualified local workforce, yet still being able to produce highly-durable long-lasting reinforced concrete structures. Examples where this could be of particular value are the extremely aggressive environments in hot, arid and saline regions such as the Middle East, particularly in the Gulf and in the Red Sea, and in similar regions of the Far East and Central America.

A very valuable and convincing documentation of the performance of stainless steel reinforcement in highly chloride-contaminated concrete is presented by the 70-year-old 2 km long concrete pier out into the Mexican Gulf at Progreso in Mexico (Fig. 1.16). No corrosion has taken place within the structure, despite the harsh environment and poor quality materials used in the construction. The chloride levels at the surface of the reinforcement were between 10 and 20 times the traditionally assumed corrosion threshold level (Knudsen et al. 1998).

The selective use of stainless steel can also be extremely beneficial in the industrialised regions that suffer heavy winters where extensive quantities of de-icing salts are used, such as in Northern, Central and Eastern Europe, North America and Northern Japan.

Stainless steel reinforcement can be combined with black steel cast into concrete without risks of galvanic corrosion due to bi-metal action. The reason is that the two types of steels reach nearly the same electrochemical potentials when cast into concrete. This is an important observation, and is the precondition for the general economical application of stainless steel reinforcement used selectively in parts of the structure where this protection is needed (Markeset et al. 2000; Klinghoffer 1999; Knudsen et al. 1998; Pedeferri et al. 1998; Concrete Society 1998; Nürnberger et al. 1994).

Figure 1.16 *The stainless steel reinforced pier at Progreso in Mexico (constructed 1937–43) still fully intact without any maintenance whereas in the foreground the remains of what is left of a 30-year-old pier reinforced with ordinary steel.*

Figure 1.17 *Two photographs of the reinforced concrete cable stayed foot bridge in Herning, Denmark. The main span is 80 m, being a record with carbon fibre cable stays, carbon fibre reinforcing bars in the deck and carbon fibre prestressing strands. The side spans are reinforced with stainless steel reinforcement.*

Fibre reinforcing bars

Fibre reinforcement bars have also been developed and it is recognised that corrosion resistance is equally achieved with such reinforcing bars. At present such non-metallic bars are based on glass, aramid or carbon fibres. However, the different mechanical characteristics between fibre reinforcement and steel reinforcement, together with the different conditions for their practical use on site, seem to indicate their rather limited application in concrete structures. The precast concrete industry seems to present the largest potential use of fibre reinforcing bars. Fig. 1.17 shows a recent application of carbon fibre reinforcement and stainless steel reinforcement used in a cable-stayed concrete footbridge which has an 80 m central span, the largest bridge using carbon fibres as cables and main reinforcement in the deck (Christoffersen et al. 1999).

1.7.2 Epoxy coating and hot-dip galvanising of reinforcement

Numerous types of protective coatings to reinforcement have been tested. The most used are:

- epoxy-coated reinforcement, which has been commercially available for several decades;
- hot-dip galvanised (HDG) steel reinforcement, which has been tested in several specific cases.

Epoxy coating of reinforcement has been used in North America since the mid-1970s but serious doubts have been expressed regarding the traditional North American approach (Clear 1992; Manning 1996). This is supported by recent experience from the Gulf (Fig. 1.18). The conventional procedure has been, first to coat straight bars individually, then to cut them to length and bend them to the

Figure 1.18 *Heavily corroded epoxy-coated reinforcement of a marine structure in the Arabian or Persian Gulf.*

required shape – which damages the epoxy coating. The fine cracking that occurs during bending, together with the pinholes that appear in the coating, and the inferior protective ability of the patched repaired zones have made epoxy-coated reinforcement a less attractive solution. There is also the side-effect that this technology will prevent the future use of cathodic protection, leaving no alternative to the replacement of damaged members if corrosion develops. Currently, the coatings industry is working on enhancing the technology.

Using the fluidised-bed dipping technique for 3-D fully welded, grit blasted and pre-heated reinforcement cages will eliminate the need for cutting and bending coated bars (Fig. 1.19). In addition, the welding of the cages prior to coating will ensure

Figure 1.19 *The fluidised-bed dipping technique used to coat 62 000 fully welded 3-dimensional reinforcement cages for the precast tunnel segments of the East Tunnel of the Great Belt Link, Denmark.*

electrical continuity, thus maintaining the possibility of installing cathodic protection in the future. This is a novel technique and damage to the coating is reduced to almost zero due to the avoidance of the cutting and bending of the coated bars. More durable coating formulations can be used when there are no strict requirements for flexibility to allow for bending.

With respect to galvanised reinforcement, little practical experience is available regarding its performance in very aggressive environments. However, it has been shown that HDG steels may have a slightly higher threshold value for chloride-induced corrosion (initiation) compared to black steel. Their protective ability depends upon the reactions at the iron–zinc interface. The risk of pit corrosion is considered to increase for HDG steel if it is exposed to chloride attack. HDG-reinforcement should not be mixed with ordinary black steel reinforcement. Hence, HDG is not a realistic alternative to SSR.

1.7.3 Cathodic prevention

Cathodic prevention has become an interesting option to protect the parts of structures exposed to high chloride concentrations. They can readily be installed within new structures.

When an impressed current system is foreseen, the system can either be energised from the start or remain dormant until corrosion sensors indicate that the initiation period is nearing its end and corrosion may begin. In marine structures a sacrificial anode system may be installed by placing anodes in the water near to or on the structure and linking them to the reinforcement. This can provide a very valuable initial protection of the reinforcement in the air-containing part of the water, the tidal zone and part of the splash zone (Fig. 1.20 and Storebælt 1998).

Figure 1.20 *Sacrificial anodes placed on the bottom slab of prefabricated caissons to be placed permanently submerged in the sea. (The Great Belt Link, West Bridge).*

1.8 Execution

1.8.1 General

The construction stage constitutes the most critical step in achieving the required quality in a concrete structure. From a durability point of view the essential part of the construction process is to ensure the correct concrete quality and thickness of the concrete cover.

The following two issues relate to the resistance of the concrete cover to ingress of aggressive substance from the environment:

- compaction must ensure uniform, dense and strong concrete cover;
- during curing, the moisture levels and temperature differences must be controlled to limit or stop early age cracking.

In addition, the following two recent developments are expected to have a considerable influence on the quality and service life of exposed structures:

- permeability controlled formwork liners (PFL);
- self-compacting concrete (SCC).

The means of providing the required thickness of the concrete cover using spacers is an integral, but often neglected, part of ensuring the durability of structures. Because the construction phase has a dominating influence on the final quality and performance of a structure, the on-site supervision and quality control become essential parts of the construction process.

1.8.2 Compaction of concrete

Adequate compaction of the concrete in the cover may be difficult to achieve due to the limited space and the need for the cover concrete to be moved through the outer layer of reinforcement. This movement of the concrete may cause 'sieving' of the concrete if the spacing of the reinforcing bars is small or the concrete is stiff. When designing the reinforcement layout the ability to compact the concrete on site must be realistically considered. Fig. 1.21 illustrates a situation where this is not the case. The compaction by vibrators can be achieved through vibrating the concrete inside the reinforcement cage. Experience has shown that vibrating directly in the cover may lead to inferior compaction of the concrete, e.g. by leaving porous traces from the vibrator. The use of external form vibrators is also a questionable procedure due to accumulation of air voids where the form vibrator was placed.

1.8.3 Curing

Curing of the concrete is part of the hardening process, which ensures an optimal development of the fresh, newly cast concrete into a strong, impermeable and durable hardened concrete in the cover zone, free from plastic shrinkage and thermal cracks. The increased sensitivity to the premature drying out of some types of cement and concrete (composite and blended cements; chemical and mineral admixtures) has accentuated the need to develop simple and rational heat and moisture curing

Figure 1.21 *An example of reinforcement detailing which does not respect the need for the reliable casting and compaction of concrete. The quality of the concrete cover is particularly in danger.*

procedures. Several comprehensive tools are now available to design the appropriate types of moisture and heat curing.

In practice, temperature profiles can be calculated and the whole curing process can be designed prior to casting the concrete (see for example RILEM 1995; HETEK 4 1997; HETEK 5 1997).

In short, good curing is needed to benefit from a good concrete mix. Bad curing destroys an otherwise good concrete mix and good curing cannot compensate for a bad concrete mix. All efforts to ensure optimal heat and moisture curing may be in vain if the initial quality of the concrete mix is inferior.

1.8.4 High-performance concrete (HPC)

The continuous demand for increased strength and improved durability and performance of concrete structures has led to the development of HPC. This development has three main objectives (Rostam 2000):

1. To protect the reinforcement against corrosion; in particular to provide protection against the ingress of chlorides by creating a dense, impermeable concrete in the cover zone which has very low penetrability of aggressive ions (such as chlorides and sulphates) and CO_2.

2. To resist the deterioration of the concrete when it is exposed to the aggressiveness of an environment which might include sulphates, seawater and other chemical attacks, as well as resisting freeze–thaw attack.
3. To provide an adequate high strength to satisfy the structural requirements.

The development of HPC has been very successful in that the advanced HPC products that are available have met the very complex and demanding structural challenges of today.

HPC for normal structures has a relatively high cementitious binder content (blended cements), a low water/cement ratio (say in the range of 0.35–0.40), and a high content of plasticiser and superplasticiser. Such concrete can conveniently be used for bridges, marine works, offshore structures, high-rise buildings, etc. where strength requirements usually remain within the range of, say, 50–80 MPa.

One drawback has been that these more refined concrete mixes become more sensitive to their handling during construction. They set high demands on the workforce regarding competence, experience and workmanship. To varying degrees these concretes have different long-term characteristics to the more usual types of structural concrete.

The dosing and mixing of HPC may need special precautions and due to a possible thixotropic nature of such concrete much more vibration may be needed to ensure proper compaction. Curing demands are higher as the risk of plastic shrinkage cracking is very high, and for some mixes, thermal cracking likely due to the large autogenous shrinkage. HPC is, in general, sensitive to freezing, particularly in combination with de-icing salts, and may require several times more air entrainer to produce sufficient air voids to ensure frost resistance (Rostam 2000).

The sensitive elements of HPC mentioned above are not necessarily valid for all types and uses of such concretes. It is, however, necessary to have these potential problems in mind when designing structures based on the application of HPC, taking the realistically available local workforce into account when making the selections. One new issue has evolved during the last few years, which may influence the broad – but not always successful – application of HPC. This is related to the increased availability and competitiveness of non-corrodible reinforcement bars to prevent reinforcement corrosion in heavily chloride-containing environments (see Section 1.7.1).

The HPC is usually only needed in the thin cover zone but not automatically needed in the bulk of the structural section. The measures introduced to create the HPC are thus to a large extent wasted. Considering that the cost of HPC is approximately twice that of traditional good quality concrete, and the execution is made more difficult, this approach should be carefully evaluated in each case (Rostam 2004). The problem is not solved in practice merely by writing strict specifications and enforcing strict quality control.

1.8.5 Permeability controlled formwork liner (PFL)

Recognising the dominant influence of the quality of the outer concrete skin to protect the concrete (as well as the reinforcement) against the penetration of aggressive substances has led to the development of a special permeable formwork liner, which can be either flexible and tissue-like, or stiff like a plastic board. Such liners are able to improve the quality of the outer few centimetres or millimetres of the concrete cover by controlling the water/cement content and enhancing the curing in this thin outer zone.

The liner leaves a characteristic tissue-like pattern on the surface of the concrete and requires that considerable care be taken when placing the liner on the formwork. Numerous tests and practical experience have demonstrated the beneficial effects of PFL. From a technical point of view, PFL technology could be considered one of the most durability-enhancing achievements of the last two decades of the 20th century. The main drawbacks are the difficulties in fixing the lining to the formwork so that an attractive appearance of the final concrete surface is ensured, and the relatively high costs of the products.

PFL cannot compensate for local bad compaction and honeycombing. However, new developments within the field of self-compacting concrete could be a realistic solution to this problem, and thus a combination of PFL and self-compacting concrete would have a valuable synergistic effect in improving the durability of concrete structures.

1.8.6 Self-compacting concrete (SCC)

The need for durable concrete in an aggressive environment leads automatically to concrete with optimised mix composition and a low water/cement ratio. Such concrete may be difficult to compact and the risk of honeycombing, particularly in the cover concrete, increases. In addition, the quality and efficiency of compaction is extremely dependent on the person handling the vibrator. Hence, the better the concrete mix, from a durability point of view, the greater the risk of having inferior or bad compaction leading to reduced quality in the final structure.

With the aid of a range of chemical admixtures and the optimal grading of the aggregates, concrete with low water/cement ratio can be made to flow through complicated form geometry and around complex reinforcement layout without the use of vibrators. Flowing concrete will exert an increased pressure on the formwork, which must be taken carefully into account when designing the formwork.

The use of SCC is also an environmentally friendly technology as the noise level from vibrators is almost eliminated because the concrete workers rarely need to work with vibrators, thus also eliminating the adverse effects the vibration of concrete has on the body ('white fingers').

The main current drawback with this technology is the sensitivity of such concrete to precise dosing, mixing and transporting – and the dependence on the weather conditions while casting the concrete. Under adverse conditions the concrete's ability to flow may suddenly be lost. In addition, the increased cost of such concrete is noticeable, and the demand for expertise and experience is moved almost fully to the mixing plant.

1.8.7 Spacers

The minimum concrete cover specified in design is usually the value used to design the expected service life – based on assumptions regarding the penetration of depassivating and corrosive substance to the reinforcement. Therefore this minimum value must be ensured in the final structure by taking the relevant tolerances into account in the selection of type, dimension and spacing of the spacers. Spacers must be designed according to c_{nom} and not to comply only with c_{min}. This fact is often overlooked. When controlling concrete cover after the placing and hardening of concrete the measured values must not be less than c_{min}.

The concrete cover in aggressive environments is usually a minimum of 50 mm, according to several codes and standards. However, the special aggressive conditions found in tidal and splash zones would lead to greater cover in the above-water zone. Values of 75 mm have become normal for long-life marine structures requiring 100–120 years service life (Rostam 1996).

The spacer material must have a good bond to the concrete and must have similar hygro-thermal deformation characteristics as the concrete. In this respect plastic spacers are not compatible with the surrounding concrete, in the sense that they have no direct adhesion and they have different temperature coefficients (up to a factor of ten different) to concrete, and furthermore they age under exposure to air, sun and marine environments. In aggressive environments high-quality concrete spacers shall be the preferred option and it is important to ensure that the spacers are of the same high quality as required for the structural concrete itself.

1.9 Quality assurance

1.9.1 On-site supervision: QA versus QC

A quality assurance procedure (QA) must be followed in order to document the real quality obtained in the structure throughout the actual construction process. This shall also include a factual documentation of durability-related parameters obtained – based on selected tests made on the finished structure and included in the 'birth certificate' (see 1.9.2), as part of the operation and maintenance manual to be handed over to the owner.

The main value of a quality assurance system enforced on the site operations is to motivate the contractor's workers to exert their true skills when doing their professional work. Bad quality is not produced wilfully but is the result of a lack of awareness of what is good and what is not good, and a lack of knowledge by the people on site. Hence, a formalised system provides one part of a motivating factor in improving quality and workmanship. On-site guidance and supervision by competent personnel are also important quality-improving measures. Finally, quality control (QC) and the physical presence on site of competent supervision personnel is a further quality enhancing measure, which shall be adapted to the real nature and possibilities of the construction industry.

Modern quality assurance systems following the ISO 9000 methodology, and based mainly on self-control, are not uncritically adaptable to the large-scale concrete construction industry. It is, in part, incompatible with on-site concrete construction due to the fact that the true, final concrete quality is determined by the actions – and lack of actions – taking place during the first few hours and days after mixing and placing the concrete. Current state-of-the-art regarding the QA of concrete construction has not yet found a satisfactory level.

1.9.2 Handing-over from the contractor to the owner: the 'birth certificate'

In order to document the fulfilment of the design specifications, and verify the subsequent performance, the quality assurance documentation for design and construction should be enlarged to include information gathered during the operation and use of the structure. Developing an operation and maintenance manual specific for each structure can do this. This manual should be prepared by the designer and must include all information from the structural design and the construction relevant to future inspections and maintenance. The manual must include recommendations regarding type and frequency of future inspections and should highlight possible sensitive or critical parts of the structure which are assumed beforehand to need particular attention during use.

When the structure is handed over to the owner, the initial operation and maintenance manual will constitute a 'birth certificate' of the structure (Rostam 1999). Information from future inspections and all other relevant events such as accidental impacts are then filled in as they occur. Depending on the nature and contents of such future information the type, frequency and selected special areas of concern shall be revised or updated by the owner.

1.10 Maintenance of concrete structures

1.10.1 Scope of maintenance

Maintenance is an action to cope with the effects of minor deterioration, degradation or mechanical wear of the structure or its components. In practice it often turns out that the owner considers the distinction between maintenance and repair to depend on whether the intervention has been foreseen or not foreseen within the long-term operation and maintenance budget. The costs for maintenance have thus, in due time, been included in the long-term budget and the timing has been adjusted to fit into the long-term activity plans of the personnel involved. Repair is when the intervention has become necessary following a normal inspection and possibly special testing has revealed the need for interventions within a short period of time and requiring costs outside the normally allocated funds.

1.10.2 The 'Law-of-Fives'

This relates to the approximation of the true relative costs to ensure the required service life of a concrete structure, depending on which approach was chosen to ensure this service life. The simple curve in Fig. 1.7 representing the two-phase service life modelling of concrete structures is valuable when illustrating these alternative approaches as shown in Fig. 1.22.

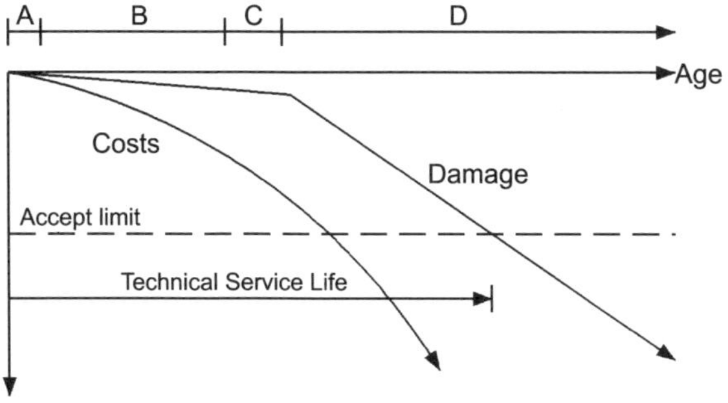

Figure 1.22 *The 'Law-of-Fives' representing the costs of alternative approaches to ensure the required service life of concrete structures.*

According to de Sitter (1983) the service life can be divided into four different phases:

- Phase A: Design, construction and curing period;
- Phase B: Initiation processes under way, no propagation of damage has begun yet;
- Phase C: Propagating deterioration has just begun;
- Phase D: Advanced state of propagation with extensive damage occurring.

These phases and the estimated relative costs are illustrated in Fig. 1.22. Based on experience the 'Law of Fives' was introduced, which states:

One dollar spent in Phase A equals five dollars in Phase B, equals twenty-five dollars in Phase C, equals one-hundred and twenty-five dollars in Phase D.

This means that the same lifetime may be achieved by a little extra attention to 'good engineering practice' during Phase A, compared to the very costly approach of relying on future maintenance, repair or rehabilitation to provide the required service life. Significant savings can be made in this way which can then be used for better purposes.

From the above it follows that interventions during the initiation period, so-called preventive maintenance, generally is much less costly than delaying the intervention until the deterioration has developed to a stage where major repair or rehabilitation becomes necessary. However, the timing of such interventions is also a factor strongly influenced by economy and non-technical issues such as reputation, aesthetics, expectations by the users and society, and politics.

All experience from practice confirms that a planned and budgeted regular upkeep of a structure leads to the lowest costs for operating a structure, provided the correct assessment and the correct interventions are made at the correct time. This latter condition is the most difficult part of the process.

1.10.3 Preventive maintenance

Maintenance of the concrete structure introduced during the initiation period with the specific aim of preventing a foreseen later occurrence of major deterioration developing during the initiation phase is considered preventive maintenance. In short, preventive maintenance intends to delay the transition from initiation to propagation. The possible economic consequences can be evaluated based on the 'Law of Fives' (Fig. 1.22).

1.10.4 Durability monitoring

Designing for long service life needs some form of verification during the use of the structure. Such a need for verifying the performance of the structure is just as relevant after a repair. This need for verification, and the benefits of obtaining more and more reliable forecasts of the service life to be expected, leads naturally to a need to monitor (non-destructively) the durability performance of the structure while in service.

Fig. 1.23 shows an example of corrosion sensors able to detect the depth of penetration of chlorides in quantities able to trigger corrosion. The sensors consist of

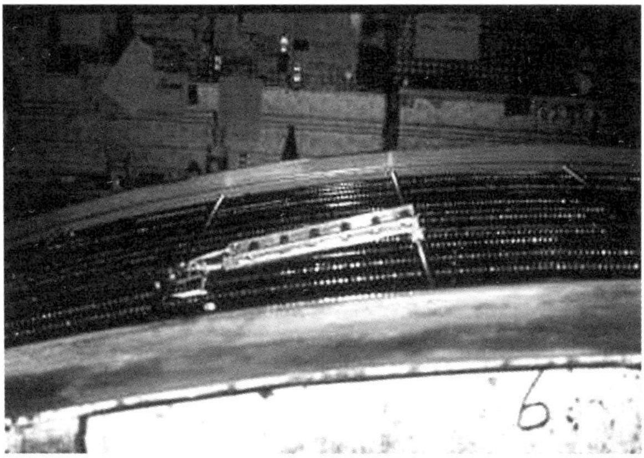

Figure 1.23 *Corrosion sensor to be cast into selected spots of a new structure.*

six individual black steel bars acting as anodes. For the cast-in sensors the anodes are fixed between two stainless steel bars forming a ladder.

Placing the sensors in the cover zone of the exposed part of the structure, and with the anodes at different depths from the surface, enables detection of the progression of the critical chloride concentration when the anodes one-by-one are depassivated and corrosion starts. A microammeter inserted between the anodes and cathode will record a jump in corrosion current once depassivation of an anode has taken place. Corrosion signals are recorded at time intervals and an increasingly precise prognosis can be made as to when the critical amounts of chlorides will have penetrated to the level of the reinforcement and reinforcement corrosion will start (Schiessl & Raupach 1992).

1.11 Repair principles and methods

European Standards EN 206 and EN 1504-1-10 have recently been adapted for the protection and repair of damage to reinforced concrete structures. A comprehensive package of standards, ranging from test methods for specific properties, through specifications for the key repair materials, including coatings, mortars, bonding agents and injection materials, to a statement of general principles for repair work has been written [ENV 1504-9 (1997)]. The principles and methods cover methods relevant both to maintenance and repair of concrete structures, i.e. related to defects in concrete and to corrosion in reinforcement.

A successful repair is very dependent on the technical competence and practical experience of the assessment and the repair engineer. In particular the engineer must have a good understanding of the mechanisms causing deterioration in order to take rational decisions regarding assessment of the structural and durability-related deterioration as well as making sound selections of maintenance and repair systems and materials. Understanding the mechanisms is also a prerequisite for choosing the optimal timing of making such interventions, i.e. as preventive maintenance during the initiation period, or as repair and rehabilitation some time into the propagation period.

1.11.1 Durability of patch repairs

Local patch repairs constitute the most usual means of repairs to concrete structures. Few repair methods have been misused as seriously as this approach, in particular when the cause of deterioration is chloride-induced reinforcement corrosion. The following description may illustrate how important it is to have a clear understanding of the causes of the deterioration that have led to the localised damage and which might be repaired using a patch repair.

Chloride corrosion of reinforcement has the effect that the local corrosion pit tends to depress the potentials in the near vicinity of the pit, even if the chloride concentration in this neighbouring zone is above the threshold value for corrosion.

A locally spalled concrete section suffering chloride-induced corrosion is illustrated in Fig. 1.24. If repaired by the traditional local patch repair, only the anodic area is repaired. The relatively high chlorides content in the neighbouring concrete may just shift the anode to these areas and corrosion continues (Fig. 1.25). In fact, if the repair is very well made, with good cleaning of the corroding reinforcement then the repair has, in fact, created an enlarged cathode, which increases the rate of the continued corrosion. This effect is verified in Fig. 1.26.

One of the problems with chloride-induced corrosion is that the chlorides are not consumed by the corrosion process, but are liberated again from the rust products. They act as a catalyst for the corrosion, and therefore remain available for further corrosion. This means that preventing additional chlorides from entering into the concrete after corrosion has occurred does not by itself stop further chloride corrosion. This effect makes it difficult to perform reliable long-lasting repairs of chloride-corrosion damaged concrete structures.

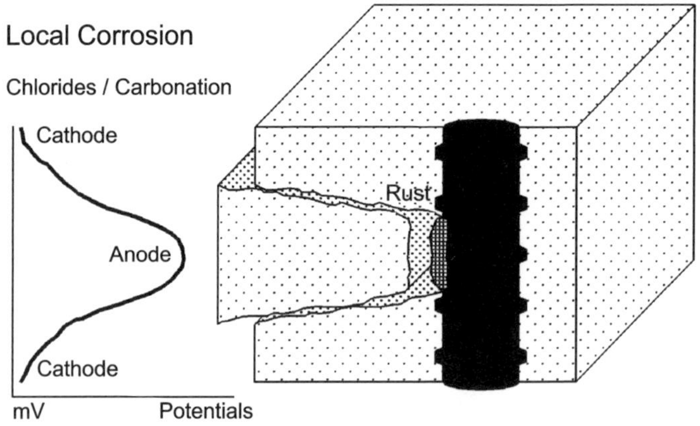

Figure 1.24 *Damage due to local chloride-induced corrosion. The electrochemical potentials of the reinforcement are also indicated, with low potentials indicating the location of the corrosion. The spalling is due to the expansive effect of the rust products.*

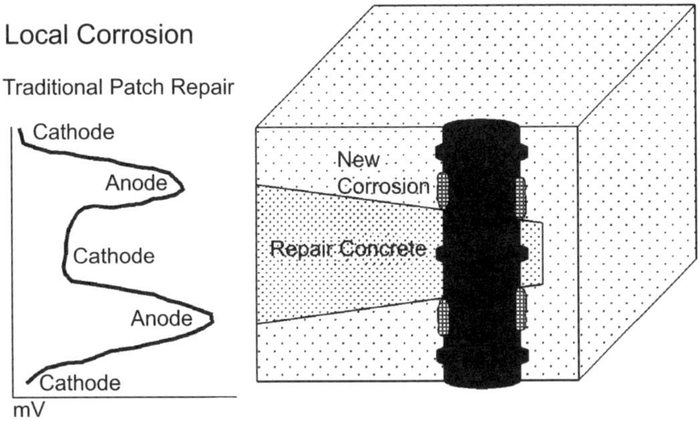

Figure 1.25 *Traditional patch repair of the damage illustrated in Fig. 1.26. Due to the corrosion in the neighbouring concrete the corrosion is just moved and continues to develop.*

Figure 1.26 *Cracks along a previous patch repair of chloride-induced corrosion indicating renewed corrosion due to inadequate understanding of the chloride corrosion mechanism.*

The most reliable way to repair chloride corrosion damage – or means of preventing chloride corrosion from developing – is to use cathodic protection or cathodic prevention.

1.11.2 Electrochemical repairs

Electrochemical repairs refer to three very closely related methods of solving the problems of reinforcement corrosion, either by preventing corrosion from developing or by stopping ongoing corrosion. The methods rely on three different effects of imposing an electric field between the cast-in reinforcement and an external electrode (Fig. 1.27).

1. By connecting the reinforcement to the negative connection of a direct current electric source and the external electrode to the positive connection, the reinforcement is forced to remain negative, i.e. to act as a cathode, and will thus not corrode: 'cathodic protection'.
2. The electric field will tend to lead negatively charged ions away from the reinforcement and towards the electrode, hence chloride ions (among others) will tend to move out of the concrete to the external electrode and they can be removed: 'chloride extraction'.
3. The electrolysis taking place at the surface of the reinforcement will produce hydroxyl ions which create an alkaline environment. Properly controlled this can re-alkalise carbonated concrete and restore the corrosion protection of the reinforcement: 're-alkalisation'.

Two of the major differences between the practical application of these methods are the current density needed and the time needed for the process, as highlighted below.

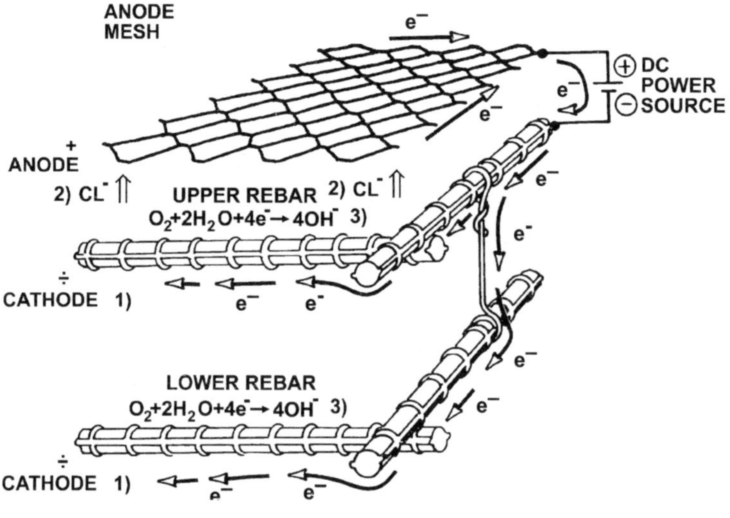

Figure 1.27 *Electrochemical processes of imposing an electric field between a cast-in reinforcement and an external electrode.*

Cathodic protection

Cathodic protection is a long-term, well known technique for protecting submerged or buried sheet piles, steel piles, pipelines, water mains, etc. against corrosion. From the early 1980s this technique was also developed for protecting concrete structures in the atmosphere, particularly when threatened by chloride-induced corrosion. For submerged or buried parts of such structures, classical methods can be used with either sacrificial anodes or with impressed current systems and inert anodes. For parts of the structure above water or above ground the system is more complicated. For structural parts in the atmosphere the problem is to achieve a sufficiently even distribution of the protective current to the reinforcement being covered by the concrete, which has a rather high electrical resistivity. The most reliable and robust system seems to be the use of a surface-mounted titanium mesh coated with a precious metal catalyst.

For new structures, preventive cathodic protection in very aggressive environments may be required, and in such cases the anode mesh can be placed in the formwork prior to casting the formwork, but making sure that there are no short circuits between this mesh and the reinforcement. In addition the reinforcement must be electrically continuous in selected zones.

For an existing structure, the customary approach is to make necessary repairs to the damaged concrete, then fix the anode mesh onto the surface and then cast a thin concrete overlay to ensure electrolytic contact between the anode mesh and the reinforcing bars (Fig. 1.28). The anodes are then linked up to the positive connection of a transformer rectifier and the reinforcement linked up to the negative connection.

Cathodic protection is generally a permanent installation needing regular inspection and maintenance. However, the current density needed to provide long-term reliable protection, even in heavily chloride-contaminated zones, is in the range of 5–20 mA/m² referring to the steel surface to be protected. This very low current

Figure 1.28 *Cathodic protection of a 55-year-old pier for a bascule bridge. A precious-metal coated anode mesh has been fixed to the surface and a sprayed concrete overlay is being applied.*

density has not given cause to any detected changes in concrete permeability or in reduced bond of the reinforcement.

The situation seems to be different when much higher current densities are used. The current densities generally used for chloride extraction and re-alkalisation, as discussed below, seem to cause some collapse of the pore walls in the concrete rendering the concrete considerably more porous after such a treatment. Hence a high-quality surface protection is needed following treatment with these short-term measures. In addition, there seems to be a loss of adhesion between the concrete and the steel, as the pull-out force of smooth round bars decreases proportionally to the amount of coulombs passed through the installation.

Special precautions are therefore needed when short-term treatments such as chloride extraction or re-alkalisation are to be used.

Chloride extraction

Chloride extraction is a temporary installation lasting one or two months and requiring a design current density in the range of 500–2000 mA/m^2.

Re-alkalisation

Re-alkalisation is also a temporary installation lasting a few weeks to a month and requiring a design current density in the same range of 500–2000 mA/m^2 as for chloride extraction.

References

Abbott, C.J. (1999) 'Steel Corrosion Solution', *Concrete Engineering International*, Vol. 3, No. 4, May 1999.

Biczók, I. (1972) *Concrete corrosion. Concrete protection*, Budapest: Akadémiai Kiadó, 1972.

CEB (1992) *Durable Concrete Structures*, Design Guide, London: Thomas Telford Services Ltd., (First draft published as CEB Bulletin d'Information 166, 1985.)

CEB-FIP (1992) *Model Code 1990.*

Christoffersen, J., Hauge, L. and Bjerrum, J. (1999) 'A footbridge with carbon-fibre-reinforced polymers, Denmark', *Structural Engineering International*, No. 4, 1999.

Clear, K.C. (1992). 'Effectiveness of Epoxy Coated Reinforcing Steel', Memorandum summarizing findings and experience with epoxy coated reinforcing steel sent to clients, associates, fellow researchers and engineers, Kenneth C. Clear Inc., January 10th, 1992, incl. replies by CRSI Concrete Reinforcing Institute by Theodore L. Neff, February 1992, pp. 17.

Concrete Society (1998) 'Guidance on the use of stainless steel reinforcement', Technical Report No. 51, Report of a Concrete Society Steering Committee, October 1998.

de Sitter, W.R. (1983) 'Costs for service life optimization: 'The Law of Fives", CEB–RILEM Workshop 18–20 May 1983, *Durability of Concrete Structures*, Workshop Report, pp. 131–134, Copenhagen 1984.

ENV 206 (1992) *Concrete: Performance, production, placing and compliance criteria.*

ENV 1504-9 (1997) *Products and Systems for the Protection and Repair of Concrete Structures*, Part 1–10. Specifically Part 9.

FIB (1999) *Structural Concrete*, Textbook on Behaviour, Design and Performance. Updated knowledge of the CEB/FIP Model Code 1990. Manual – Textbook Volume 3. Steen Rostam: Chapter 3: 'Durability', and Chapter 8: 'Assessment, Maintenance and Repair'. Fédéretion Internationale du Béton, 1999.

Gedge, G., Martin, J. and Cooke, R.S. (2001) 'An investigation of initial and life cycle costs of using stainless steel reinforcement in highways structures', *Paper presented at the Conference on Structural Faults and Repair*, June 2001, UK, pp. 18.

HETEK 4 (1997) 'Control of early age cracking in concrete', Guidelines. Report No. 120, 1997, Road Directorate Denmark, Ministry of Transport.

HETEK 5 (1997) 'Curing' Guideline. Report No. 121, 1997, Road Directorate Denmark, Ministry of Transport.

ISO 9000 (1994, 1997, 2000) *Quality management and quality assurance standards.*

Klinghoffer, O. (1999) 'Corrosion aspects of galvanic coupling between carbon steel and stainless steel in concrete', Test Report, FORCE Institute, Broendby, Denmark, May 21, 1999, pp. 15.

Knudsen, A., et al. (1998) 'Cost effective enhancement of durability of concrete structures by intelligent use of stainless steel reinforcement' *Conference on Corrosion and Rehabilitation of Reinforced Concrete Structures, 8–11 December 1998, Florida, USA.*

Manning, D. G. (1996). 'Corrosion performance of epoxy-coated reinforcing steel: North American experience', *Construction and Building Materials*, Vol. 10, No. 5, Elsevier Science Ltd., pp. 349–365.

Markeset, G., Rostam, S. and Skovsgaard, T. (2000) 'Stainless steel reinforcement – an owner's, a designer's and a producer's viewpoint', *Proceedings, 6th International Conference on Deterioration and Repair of Reinforced Concrete in the Arabian Gulf, 20–22 November 2000, Bahrain,* pp. 495–513.

Markeset, G., Rostam, S. and Klinghoffer, O. (2006) 'Guide for the use of stainless steel reinforcement in concrete structures', *Scandinavian Project sponsored by the Danish, Swedish and Norwegian Road Authorities, the Norwegian Defence Estates Agency, and by Nordic Innovation* (in print).

Nürnberger, U., et al. (1994) 'Stainless Steel in Concrete', European Federation of Corrosion, Working Party WP 11, *Corrosion in Concrete: State of the Art Report,* July 1994.

Nürnberger, U. (1995) *Korrosion und Korrosionsschutz im Bauwesen*, Band 1, Grundlagen, Betonbau. Bauverlag Gmbh, Wiesbaden und Berlin. pp. 643.

Pedeferri, P., et al. (1998) 'Behaviour of stainless steels in concrete' *Repair and rehabilitation of reinforced structures: the state of the art*, Ed. by Silva Araya, W.F., et al., American Society of Civil Engineers, 1998, pp. 192–206.

RILEM (1995) 'Thermal cracking in concrete at early ages' *Proceedings of the International Symposium held by RILEM, 10–12 October, 1994, Munich*, Ed. by R. Springenschmid, London: E. & F.N. Spon.

Rostam, S. (2004) 'Service life of concrete structures – Challenging the steel reinforcement industry. Shall concrete remain the dominating means of corrosion prevention?' *ANIFER, International Calavera Award paper 2004.*

Rostam, S. (2000) 'Does high performance concrete provide high performance concrete structures?', *Proceedings, PCI/FHWA/FIB International Symposium on High Performance Concrete, 25–27 September 2000, Orlando, Florida, USA*, pp. 64–73.

Rostam, S. (1999) 'Performance-based design of structures for the future', *Proceedings, IABSE Symposium, Structures for the Future – The Search for Quality, Rio de Janeiro, Brazil.*

Rostam, S. (1998) 'Aesthetics and service life performance of concrete bridges' *Bouwkroniek*, special edition, Belgian Concrete Day – 22 October 1998.

Rostam, S. (1996) 'High performance concrete cover – why it is needed, and how to achieve it in practice', *Construction and Building Materials*, Vol. 10, No. 5. Elsevier Science.

Schiessl, P. & Raupach, M. (1992) 'Monitoring system for the corrosion risk of steel in concrete structures', *Concrete International*, Vol. 14, No. 7, July 1992, pp. 52–55.
Storebælt (1998) *West Bridge*, The Storebælt Publications. Published by A/S Storebæltsforbindelsen.
Tuutti, K. (1982) *Corrosion of Steel in Concrete*, Cement och Betong Institutet, Stockholm.

2 The Durability of Concrete Structures in the Tropics

Tam Chat Tim
Loh Yan Hui
Tee Choon Peng

2.1 Introduction

The durability of concrete structures in the tropics involves many issues that are similar to those in other climates. Historically, and indeed at the present, much of the research and development activity in the field of durability has been carried out in countries with temperate climates. In adapting such experience and practices to the tropics, there is a need to recognise the different effects of the tropical climate on the behaviour of the concrete. The major difference between the performance of concrete in the tropical climate compared to that in other regions is that the temperature is very much higher throughout the year. This constant high temperature not only leads to a higher rate of reactivity but also the longer duration of high temperatures further promotes a sustained rate of reactivity. Chemical reactions take place rapidly when reactants are in the liquid state and the high humidity throughout the year provides adequate moisture to sustain faster rates of reaction within concrete. The ingress of chemicals and moisture is made easier by micro-cracks and cracks developed due to structural action and non-structural causes.

2.2 Tropical climatic conditions

The tropical climate may be subdivided into the hot–dry type and the hot–wet type. The difference lies only in the amount of rainfall throughout the year. In both cases rainfall is high, but in the case of the hot–wet climate the rainfall is at least 150 mm per month. As an example, the monthly rainfall data for Singapore for the period 1869–1980 compiled by Rahman[1] is shown in Table 2.1.

A typical hot–wet climate is the fifty-year (1929–1941, 1948–1984) average climatic conditions for Singapore published by the Singapore Meteorological Service as summarised by Rahman[1] and shown in Table 2.2.

Unlike high temperatures in temperate regions that are generally associated with low humidity, the tropical environment is always hot and humid. In the tropics, the ambient temperature remains nearly constant throughout the year. Of greater importance in concreting practice is the fact that in a temperate region, cooler temperatures may follow a particularly hot day, whereas in the tropics, the hot days are present all the year round. It may well be that everyday normal concreting practice in temperate regions is inadequate on a hot summer's day, when the temperature exceeds 30°C. However, in the tropics, everyday tropical concreting practice has already been adjusted to meet daily temperatures of this level.

Table 2.1 Monthly rainfall data for Singapore (1869–1980)

Month	Average (mm)	Highest (mm)	Lowest (mm)
January	243.1	818.6	21.6
February	173.5	566.7	8.4
March	189.2	528.3	18.5
April	185.4	454.9	16.5
May	171.5	386.6	59.2
June	170.4	378.7	56.9
July	160.5	527.3	24.9
August	182.6	526.8	18.0
September	172.7	426.7	39.6
October	204.3	497.1	31.7
November	257.1	521.5	91.2
December	278.4	681.0	62.5
Total annual	**2388.7**	**3452.4**	**1563.4**

Table 2.2 Fifty-year (1929–1941, 1948–1984) average climatic conditions for Singapore

	Mean (°C)	Range (°C)
Mean daily minimum temperature	30.7	29.7 to 31.4
Mean daily maximum temperature	23.7	22.9 to 24.4
Mean daily minimum relative humidity	64.7	60.5 to 67.7
Mean daily maximum relative humidity	96.4	95.3 to 97.2

2.3 Effects of a tropical climate on the properties of concrete

In order to produce durable concrete, a suitable degree of workability of the fresh concrete is required for placing the concrete and to adequately delay the setting to avoid potential cold joints. The higher ambient temperature in the tropics leads to a higher initial temperature of fresh concrete and a more rapid initial reaction as well as a faster rate of loss of moisture during transportation and placing. The advance in chemical admixtures, such as plasticisers and retarders, has provided technical solutions to overcome these issues.

The initial degree of workability of a given mix proportion is lower at higher temperatures due to a greater initial chemical reaction during mixing. The actual reduction in workability depends not only on the ambient temperature but also on the mix proportions, particularly the type and amount of cement and the method of mixing. The chemical reaction rate will double for an increase in temperature of 10°C. A drop of 20 mm in slump is associated with an increase of 10°C in the initial concrete temperature. This is translated into an increase of about 5–10 kg/m^3 in water content to achieve an equal slump in the case of plain concrete as reported by Klieger[2], Tam[3], and Yamamoto and Kobayashi[4]. In the case of high-strength concrete with a water/cement ratio of 0.4 or less, the concomitant increase in cement content is more than 10–20 kg/m^3. Much less information is available in cases of higher levels of workability as these are achieved generally with the addition of a water-reducing admixture (WRA) or a high range water-reducing admixture (HWRA). The use of such chemical admixtures to provide the increase in workability leads to a technically better concrete in terms of heat generation and volume changes than the traditional approach by an increase in both water and cement contents at a constant water/cement ratio.

Typically, a nominal dosage of WRA is equivalent to an increase of 10% to 15% and for HWRA it is equivalent to an increase of 20% to 30% in water content. The effectiveness of a water-reducing admixture varies with the chemical and physical properties of the cement, particularly its tricalcium aluminate content, alkali content, fineness and the temperature of the concrete at the end of mixing. The use of cement replacement materials such as pulverised fuel ash (PFA), ground granulated blastfurnace slag (GGBS) and condensed silica fume (CSF) calls for further consideration of their effects on workability. As the number of variables is increased and the possible interactions become less quantifiable, for any specific combination of materials, past experience or a trial mix is often the best guide to performance.

Unless a retarding admixture is added, the initial degree of workability drops off rapidly with time after remaining nearly constant for a short period (30 to 45 minutes). The length of this workability retention time and the subsequent rate of loss of workability are dependent upon the initial temperature of the concrete and the temperature and relative humidity of the environment. The period of workability retention for a given mix may be determined from the stiffening time test in BS 5075 or ASTM C 403[5,6]. The tests serve not only the purpose of quantifying retarders and plasticisers in product specification, but also provide a direct assessment of the stiffening times of a mix. Although there are minor differences in the details of the two standard test methods, in both cases, the penetration resistance of the wet-sieved mortar fraction of a concrete mix is measured over a period of time after mixing. The test methods included the interpretation of the time to reach various degrees of penetration resistance in relation to the observed behaviour of the concrete mix during the construction process. The three degrees of penetration resistance and their practical significance are as shown in Table 2.3.

The time to reach a strength of 0.5 MPa can be extended by the addition of retarders. However, the rate of increase in penetration resistance beyond this time is not changed significantly. In practice the delay to reach each of the three degrees of penetration resistance is almost the same. The use of water-reducing and set-retarding admixtures is a regular practice in the tropical climate in order to achieve a sufficient degree of workability and an adequate time after mixing for the transportation and the placing of the concrete.

The major effects of temperature on the strength development of hardened concrete are due to the high temperature during curing and setting. The influence of high temperature during setting on the strength of cement has been reported. In particular, Neville[7] explains that a rapid initial rate of hydration can lead to the formation of products of a poorer physical structure, probably more porous, so that a large part of these pores remain unfilled even with continuing hydration. Verbeck and Helmuth[8] state that the rapid rate of initial hydration at higher temperatures

Table 2.3 Penetration resistances and their significance in practice

Penetration resistance – MPa (psi)		Significance in practice
ASTM C403	BS 5075	
[not applicable]	0.5 (72)	Time up to which initial workability retained (workability retention)
3.5 (500)	3.5 (500)	Time beyond which cold joint may form (cold joint formation)
27.6 (4000)	[not applicable]	Time at which concrete begins to harden (start of hardening)

retards the subsequent hydration and further extend this concept to the non-uniform distribution of the products of hydration within the paste. Such products have a poorer physical structure and are probably more porous. These pores may remain unfilled and their resultant lower gel-to-space ratio leads to a lower strength according to Powers[9]. Price[10] and Klieger[2] have studied the influence of high temperature on the rate of strength development as well as on the long-term strength that can be achieved. In the majority of cases, the reported loss of strength for temperatures up to 40°C when compared to temperatures of 20°C is 10% to 15% for the 28-day strength levels of 30 to 40 MPa concrete. For setting temperatures at 50°C, a strength loss of 30% has been indicated by Klieger[2] and by Verbeck and Helmuth[8]. However, their work was based on a high setting temperature followed by a lower curing temperature. This is similar to concrete cast on a hot summer's day, followed by cooler days. However, the situation in the hot–wet tropical environment is quite different. Both the setting and curing temperatures remain high throughout. Under these conditions, studies by those in the tropics, as observed by Ackroyd and Rodes[11] in Nigeria, Quao[12] in Ghana and Tam[13] in Malaysia, showed that the 28-day strength of concrete of comparable mix proportions increases at curing temperatures of around 30°C. Tam[13] observed that it was only at the age of about 1 year when the strength of concrete cured in the tropical climate falls below that cured in a temperate climate. In fact, the results show that at the age of 28 days, similar strengths can be obtained even at an increase in water/cement ratio of about 0.08. Thus, on an equal 28-day strength basis, concrete in the tropics is of a lower quality and hence of poorer durability than corresponding mixes in temperate climates. Temperature-matched curing studies for cement containing partial replacement material with mineral admixtures such as pulverised fuel ash (PFA) by Mani et al.[14], or ground granulated blastfurnace slag (GGBS) by Bamforth[15], observed similar trends in strength development under temperature-matched curing at much higher early temperatures simulating conditions generated by the heat of hydration in thick sections.

2.4 Non-structural cracks in a tropical environment

The control of load-induced cracks forms part of the design consideration in reinforced concrete structures. The introduction of higher yield strength steel (460

MPa) for reinforcement bars compared to mild steel (250 MPa) in itself has resulted in higher degrees of micro-cracking even under the serviceability limit state. The higher steel strain induces a higher degree of micro-cracking in the tensile zone of the concrete. In addition, non-structural cracks are induced by environmental factors. Cracks induced by any cause, including micro-cracks, tend to reduce the resistance of concrete to the ingress of gases and fluids. The rate of deterioration of concrete structures is dependent upon the ease with which deleterious gases and fluids penetrate the concrete. These are factors for the initiation of corrosion as well as the depassivation of the concrete cover.

The need for retarding chemical admixtures in a tropical environment has been discussed. For the casting of large pours, such as raft foundations, the existing practice of building up the full thickness at one part of the section before moving on to the rest of the section calls for a high retardation time to prevent the formation of cold joints. After completing the build of the thickness, the concrete has a long period for plastic settlement cracking to potentially form. When such cracks form, they need to be closed by revibration of the top layer just prior to the setting of the concrete. Generally, a large horizontal surface area of the completed raft is exposed to rapid drying, in spite of the higher relative humidity of the tropical environment. Due to the availability of a ready-mixed concrete supply, large floor areas in buildings are cast in a single operation. In such situations, unless curing is started soon after the finishing off of the surface, rapid drying of the concrete in its plastic state can lead to plastic shrinkage cracking. ACI Committee 305[16] provides guidance on the expected rate of evaporation and suggests that a rate of 1 kg/m^2 may lead to the potential formation of plastic shrinkage cracking. The graphical method provided is limited to the range of air temperature up to 40°C. Uno[17] has proposed an equation for the same purpose, which enables any temperature to be used. Examples of evaporation rate under tropical conditions are presented by Tam et al.[18] in the recommended strategy for the casting of a raft foundation to minimise both potential plastic settlement and potential plastic shrinkage cracking. This calls for a change in the method of placing a thick slab. The usual practice of building up to the full depth of the raft and then moving the fresh concrete front forward from one end to the other of the area is changed. The new approach is to place concrete over the whole area in shallow (300 mm to 400 mm) layers. This ensures proper compaction and more definite control on the retardation time needed to prevent cold joints. In particular, the topmost layer requires just sufficient delay in setting, for the purpose of transportation and placing, as there is no concrete to be placed over it for which a cold joint can form. The reduced delay in setting provides less time for the development of potential settlement cracking. The change in retardation is achieved by adjusting the dosage of the WRA, which has a higher retardation effect, and that of the HRWRA, which has lower retardation effect, but maintaining the same degree of workability and strength of the mix. In addition, should the environmental conditions be critical, polypropylene fibres may be added to the concrete, for this last layer, to increase the concrete's resistance to plastic shrinkage cracking. The new approach has the advantage that the interior layers of a thick raft, which develop higher temperatures (due to the heat of hydration) than the exterior layers, are placed at a cooler initial fresh concrete temperature. The higher initial fresh concrete temperature at the exterior zones results in a temperature gradient opposite to that when the peak temperature is developed in the interior. This reduces the overall temperature differential and therefore potential early thermal cracking and has the added benefit of a cost saving in having to cool a smaller volume of concrete.

2.5 Durability of concrete structures in a tropical climate

Deterioration of concrete structures may arise from chemical attack directly onto the concrete or by corrosion induced on the embedded steel reinforcement due to the ingress of chemicals causing depassivation of the steel. Since these are chemical reactions, it is to be expected that the tropical hot–wet climate is conducive to such reactions. The current approach to designing for durability is one of prescription. This is an implicit design approach for which the expected more severe tropical environment cannot be evaluated directly. The move to provide an explicit design approach for durability design life is being developed by international groups (see the RILEM Report No. 12[19] and the Concrete Society Special Publication CS 109[20]). An approach parallel to that in structural design has to be developed whereby environmental loading is matched by the resistance of the concrete in a mathematically related formulation for each chemically induced deterioration process, such as alkali–silica reaction, sulphate attack and carbonation or chloride-induced corrosion of embedded steel reinforcement. Suitable partial safety factors for selected design life have yet to be agreed upon by consensus. Calibration of such durability design with performance under actual service conditions has to be conducted in order that the design equations lead to safe and economic solutions. The monitoring of existing concrete structures and their service environment, then relating their performance to the relevant properties of the *in-situ* concrete, provide the necessary information for the calibration and selection of appropriate partial safety factors for the design life subjected to each type of deterioration process.

The presence of sulphate in the soil, groundwater or in the marine environment is well known. These are external sources and the current recommendations in BS 5328[21] are commonly adopted in practice for such cases. The adequacy of these provisions may be verified in due time or when explicit durability design formulation becomes available. For the tropical climate, the phenomenon of internal sulphate attack (ISA) or delayed ettringite formation (DEF) is of special concern in the casting of thick raft foundations, deep transfer girders and large-size column sections. This phenomenon, first associated with steam-cured products, may be present when high cement content mixes are used. The heat generated by cement hydration may reach similar temperatures to those used in steam curing (70°C to 80°C). The formation of ettringite between the tricalcium aluminate and the added gypsum in the manufacture of ordinary Portland cement (OPC) may be retarded at high temperatures. This delayed formation of ettringite from tricalcium aluminate with internally available sulphate may cause disruption of the bond between aggregate and cement paste (particularly at the zones where bleed water has been trapped – called 'water gain'). Others believe that the formation is within the paste at the interfacial zone; which is referred to as the transition zone. Although the two explanations are different, the resultant expansion produces the interfacial cracks as well as within the paste structure causing loss of strength. Such formation has been observed by petrographic examination. Details on this new and controversial topic are provided by Thaulow et al.[22], Lewis and Scrivener[23], Lawrence[24], and St John et al.[25]. However, when pozzolanic materials such as PFA, or GGBS are added, the effect is not observed. The full explanation for this has yet to be found, but one possible cause is the dilution effect when part of the OPC is replaced by the pozzolans in blended cements.

The other common deterioration in reinforced concrete structures is the result of corrosion of embedded reinforcing steel. The initial stable state of a thin layer of oxide under the highly alkali environment of the pore fluid in concrete is destroyed by the neutralisation of the alkalinity by ingress of carbon dioxide and/or chloride. Chloride has the additional effect of causing the dissolution of the oxide layer even under a highly alkali environment when its concentration is high. The processes of gas diffusion and chemical reaction are both expected to be faster at high tropical temperatures compared to the experience of temperate regions where most studies have been conducted. Available published works on carbonation in tropical climate, for example, by Chin[26], and Roy et al.[27] are limited in scope and extent. Chin[26] based on accelerated testing in an exposure condition of 7% CO_2, 65% RH and for temperatures ranging from 20°C to 40°C found that the effect of temperature was not significant. This is similar to that reported by Papdakis et al.[28] who found it very weakly sensitive to temperature – based also on accelerated testing at 65% RH. However, based on the Arrhenius relationship, Roy[29] estimated that between 27°C and 37°C, the carbonation rate is raised by a ratio of 1.7 times, pointing to the higher rate of carbonation due to the effect of tropical temperatures on the diffusion coefficient. Some typical carbonation depths measured for natural exposure reported under the Japan–Asean Co-operation Programme on Materials Science and Technology, JICA[30] are shown in Table 2.4.

Roy et al.[27] reported similar observations. Chin[26] reported on natural exposure in relation to accelerated testing results. It was observed that 1 week of accelerated testing at 12% CO_2, 30°C and 65% RH is equivalent to 4.74 years of exposure under indoor conditions (average 28°C and 82% RH) or 7.29 years of exposure under outdoor unsheltered conditions. Although no detailed reports are available, it has been generally observed that many structures built in the period from 1960 to 1980 (when Grade 20 concrete was often used in buildings), carbonation-induced concrete spalling occurred 15 to 20 years later. The carbonation depth within the first year of exposure is often greater than 5 mm. Based on the simple model that carbonation depth is proportional to the square root of time, the depth after the first year of exposure is often called the carbonation constant. For a typical cover of 20 to 30 mm, the cover is expected to be fully carbonated by these ages. In recent years, the grade of concrete used for general construction in Singapore has moved up to Grade 30 and Grade 35. Given the same range of cover thickness and the specified curing time of seven days, such concrete may be expected to take about 50 years to be fully

Table 2.4 Depth of carbonation

Age of structure and element	Depth of carbonation (mm)	Carbonation constant, k ($mm/y^{1/2}$)	Concrete strength (MPa)
22 years – Beams	13 to 43	2.8 to 9.2	21
– Slabs	4 to 20	0.9 to 4.3	21
19 years – Beams	0 to 19	0 to 4.4	42
– Slabs	4 to 10	0.9 to 2.3	42
7 years – Beams	0 to 8	0 to 3.0	28
– Slabs	4 to 18	1.5 to 6.8	42
16 years – Beams	0 to 24	0 to 6.0	26
– Slabs	0 to 53	0 to 13.3	42
21 years – Beams	0 to 69	0 to 15.1	21
– Slabs	7 to 32	1.5 to 7.0	21

carbonated. However, as often reported, e.g. Concrete Society[20], the variability of cover thickness in an 'as built' structure is often in excess of the ±5 mm specified in BS 8110[31]. Another factor is the quality of the *in-situ* concrete cover which is often below that of the standard test specimens. The carbonation depth within a structure may well range from 2 mm to 7 mm due to construction variability. Hence, in terms of carbonation, the construction tolerance achieved is an important factor in service life formulation.

The number of publications on chloride-induced corrosion of actual structures in the tropical climate is limited. These are cases for the marine environment as this is the common chloride exposure condition in the region. Norzan and Clark[32] provided an overview of environmental loading in a hot–humid marine environment. Norzan and Clark[33] described the results of an investigation of a bridge on the east coast of the peninsula of Malaysia. The bridge was less than ten years old at the time of investigation. The structure consists of a reinforced concrete slab resting on five 31 m long prestressed concrete beams sitting on a pier supported by three 0.9 m diameter columns rising from pilecap. The results indicated that deterioration developed within the lower part of the columns within six years of construction and that initiation had started at other parts of the structure after ten years, even though the concrete strength in the structure was well above the specified characteristic strength of 25 MPa. The degree of deterioration varies with the microclimate of the structural elements.

Another report on the investigation of marine structures in Singapore was the joint effort of a Japan–Asean Co-operation Programme on Materials Science and Technology, JICA[30]. In addition to the macroclimate, there is the microclimate difference between tropical and temperate regions. The five structures selected by the report were constructed under different programmes over the 7 to 22 years of the time of the investigation. Some were of reinforced concrete construction while others were of a composite construction using precast/prestressed elements with an *in-situ* reinforced topping. The grade of concrete used varied from Grade 20 for the older structures to Grade 55 for the prestressed elements.

Both beams and slabs that were subjected to severe splashing effects had a designed cover ranging from 50 mm to 75 mm and concrete with an estimated *in-situ* strength ranging from 20.5 MPa to 27.5 MPa. Where the presence of cracks had been caused by heavy loading, corrosion was observed. For beams subjected to less severe splashing, those of reinforced concrete had a higher degree of corrosion compared to the precast and prestressed beams. This is attributed to the better quality of the cover concrete in prestressed elements. In general, where corrosion was observed, most of the reinforced concrete beams were found to have less than the specified cover. On the other hand, all prestressed concrete beams were found to have greater than specified cover as a result of being precast under better quality controlled conditions. The depths of carbonation and chloride penetration were found to be lower in the prestressed elements.

Only a few of the piles supporting the deck structures were found to have deteriorated. This may well be due to the generally compressive stress present compared to the flexural elements in the deck structures. It was noted that areas where corrosion was observed were mainly in the tension zone of the reinforced concrete elements.

Wei et al.[34] reported on a site exposure test conducted with specimens exposed to different lengths of immersion time in the sea. Concrete prisms of 150 × 150 × 400 mm were cured in water for 7 days. The prisms were then taken to the exposure site located on the southern shore of Singapore and placed on different platform levels of a multi-tier rack. In order to vary the exposure time when the specimen is immersed in seawater, different levels of the platforms were selected to provide immersion times of 10%, 20%, 30% and 80% of a regular tidal cycle. The specimens were split after one year and the total chloride ion concentration at various depths was determined by X-ray fluorescence spectrometry. The mixes studied are shown in Table 2.5 and the chemical compositions of the OPC, GGBS and CSF used are indicated in Table 2.6.

The replacement percentages of OPC with GGBS are 30%, 55% and 65% and for CSF (fineness of 25 000 m^2/kg) at 5% or 10%. The fineness of GGBS used consisted of 300, 450, 600 and 800 m^2/kg. Detailed results were presented by Wei et al.[34]. The results showed that the improvement in performance in the submerged and tidal zones under tropical exposure conditions is generally achieved by using a mix with a lower water/binder ratio and a higher percentage of GGBS or CSF (Fig. 2.1). The use of GGBS with higher fineness at higher replacement percentages tends to improve compressive strength but contributes little to the increase in resistance to chloride ion ingress (Fig. 2.1).

The variation in exposure for the range of immersion time used, did not show any significant effect on chloride ion ingress. This is due possibly to the lower rate of evaporation during the drying cycle as a result of the high relative humidity. Based on Fick's 2nd law for diffusion, a solution may be expressed in the form:

$$x_s = k\sqrt{t} + a, \text{ and } k = 4\sqrt{D_c}$$

Where a is an empirical constant, x_s the depth of chloride penetration (mm) after the exposure time of t (weeks), D_c is the apparent chloride ion diffusion coefficient and k

Table 2.5 Mix proportions of concrete

Material	Quantity (kg/m^3)	
	Mix A	Mix B
Cement (OPC+65% GGBS)	567	450
Water	170	180
Sand	662	738
Coarse aggregate	992	978
Water/cement ratio	0.3	0.4
% fine	40	43

Table 2.6 Chemical compositions of cementitious materials

Material	OPC	GGBS(3000)	GGBS(6000)	GGBS(8000)	CSF
CaO – %	64.4	42.9	42.4	42.4	4.2
SiO_2 – %	21.1	32.5	32.3	32.7	93.0
Al_2O_3 – %	5.2	13.8	14.0	13.8	0.2
Fe_2O_3 – %	3.1	0.2	0.6	0.2	0.0
MgO – %	1.1	5.8	6.0	5.9	0.5
SO_3 – %	2.5	–	–	–	0.05
Na_2O – %	0.2	–	–	–	0.2
K_2O – %	0.6	–	–	–	0.2
S.G.	3.15	2.90	2.90	2.90	2.00
Fineness – cm^2/g	3170	3220	6010	7900	250000*

*Fineness by Blaine method except for CSF (N_2 absorption method).

is the chloride penetration constant. Table 2.7 shows the relative value of k after a period of 1.5 years of exposure in seawater for the mixes studied.

Results in Table 2.7 and Fig. 2.1 show the effectiveness of including 10% CSF compared with GGBS replacement. However, this is achieved at a much higher cost for CSF. For cost effectiveness, 65% replacement with GGBS is considered as optimum for marine structures. Wee et al.[35] reporting on comparative testing at 30°C compared to 20°C concluded that even though at the higher temperature, a higher early strength and a denser microstructure were obtained, the rate of chloride penetration under the higher temperature is faster.

Based on the findings that partial replacement with GGBS at a high percentage (65%) increases the resistance of concrete against chloride ion ingress, the Housing and Development Board of Singapore had already used GGBS in two of its completed projects, namely Marina Bay Reclamation (1991–1994) and Tuas Reclamation (1993–1995). In addition, the blended cement often provides better workability for the same mix proportions. To extend the service life concept further, for a new project the North Eastern Coast Reclamation Phase 4 (1997–2001) adopted an innovative concept of protection using ferrocement panel facings for seawalls for both submerged and tidal zones. The tidal water levels ranges from +0.1 m CD during low tide to +3.3 m CD during high-tide conditions. For this reason, the top level of the

Table 2.7 Relative chloride penetration coefficient after 1.5 years of exposure (Relative to mix with OPC at water/binder ratio = 0.4)

Mix	Percentage replacement	Water/binder ratio	Relative chloride penetration coefficient
OPC	–	0.3	0.59
GGBS(4500)	65	0.3	0.50
GGBS(6000)	65	0.3	0.40
OPC	–	0.4	1.00
GGBS(3000)	30	0.4	0.82
GGBS(3000)	55	0.4	0.58
GGBS(3000)	65	0.4	0.51
GGBS(4500)	30	0.4	0.80
GGBS(4500)	55	0.4	0.53
GGBS(4500)	65	0.4	0.49
GGBS(6000)	30	0.4	0.70
GGBS(6000)	55	0.4	0.62
GGBS(6000)	65	0.4	0.51
GGBS(8000)	30	0.4	0.66
GGBS(8000)	55	0.4	0.50
GGBS(8000)	65	0.4	0.40
CSF	5	0.4	0.74
CSF	10	0.4	0.48

retaining seawall structures without a coping wall ranges from +1.53 m CD to +2.1 m CD. The maximum tidal velocities during the spring tide and neap tide are of the order of 0.3 to 0.75 m/s and 0.25 to 0.50 m/s, respectively. Grade 60 PBFC was specified for the ferrocement panel facings on the vertical retaining buttress walls to give better durability characteristics against chloride penetration and sulphate attack. Grade 40 PBFC has been specified for all other marine concrete works that are exposed to the marine environment. The provision of adequate concrete cover is a good practice for marine structures. Hence, the following criteria were specified:

seaside and bottom: 70 mm minimum
landside and top: 50 mm minimum

For a well-graded high quality, angular, crushed granite coarse aggregate with a maximum size of 20 mm, and well-graded, low silt natural sand with low organic impurities, a typical mix design is shown in Table 2.8. This is a Grade 40 mix, low permeability concrete with an expected compressive strength of at least 20 MPa at 24 hours for early demoulding of the precast elements (Fig. 2.2). The achievement of such high early strength was critical to meet the seawall installation schedule. Similarly, prior to the installation of the seawall, the concrete strength must reach 43 MPa after ten days of curing.

There has been widespread use of ferrocement in the construction of silos, tanks, roofs, boats and other non-structural elements but this adoption of a ferrocement panel facing is the first of its kind in marine structural works.

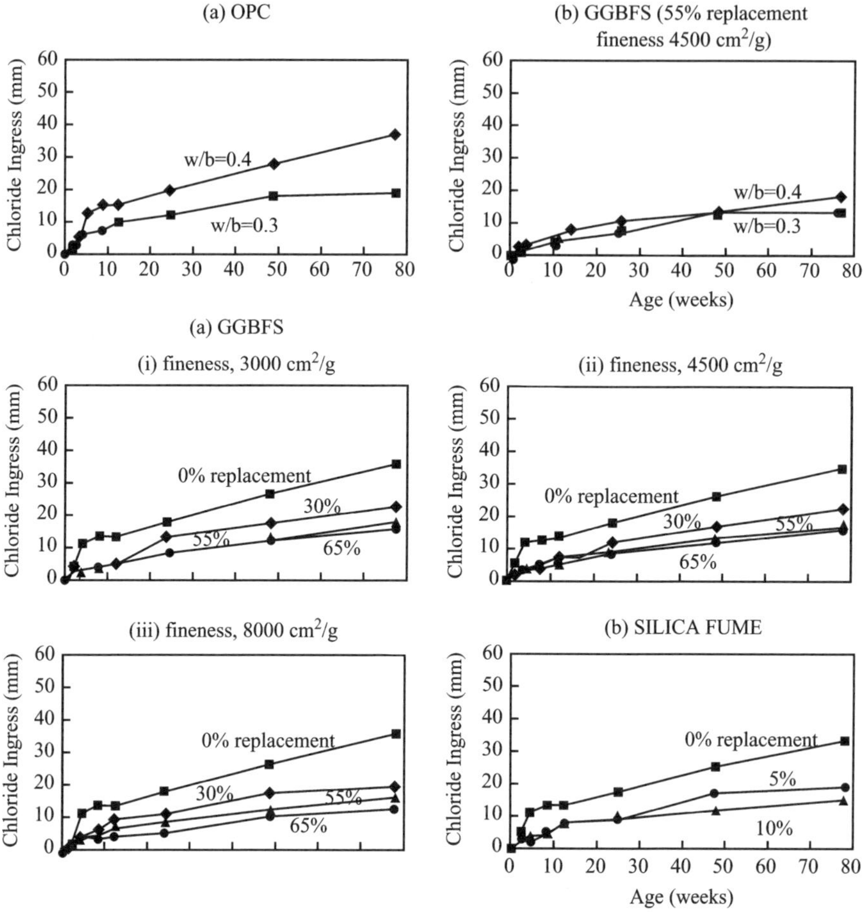

Figure 2.1 *Effect of water/binder ratio and fineness of GGBS on chloride ingress.*

The incorporation of ferrocement panel facings and their method of construction in the design of the seawall is attributed to the following factors:

1. As a separate panel facing located at the wave splashing zone of the L-Block (depending on Block type, from –0.5 m CD to 2.1 m CD), the panel functions as a sacrificial layer.
2. The splash zone lying within the tidal range is fully submerged at high tide and exposed to drying at low tides.
3. Concrete is most vulnerable when subjected to repeated drying and wetting cycles.

Ferrocement panel facing is a form of reinforced concrete but differs from the conventional type in that the reinforcement consists of closely spaced, layers of mesh

Table 2.8 Typical mix design

Materials	**Quantity**
Grade 40 – low permeability concrete	
Cement [OPC with 65% GGBS] (kg/m^3)	380
Water (kg/m^3)	145
20 mm aggregate (kg/m^3)	1050
Sand (kg/m^3)	800
Retarding plasticiser (ml/m^3)	740
High range water reducing admixture (ml/m^3)	3880
Target slump (mm)	125±25
Water/binder ratio	0.38
Grade 60 – low permeability mortar	
Cement [OPC with 65% GGBS] (kg/m^3)	830
Water (kg/m^3)	220
Sand (kg/m^3)	1245
Retarding plasticiser (ml/m^3)	2790
High range water reducing admixture (ml/m^3)	5530
Target slump (mm)	150±50
Water/binder ratio	0.27

completely surrounded by cement mortar. The advantages of incorporating the ferrocement panel facing are:

1. The high tensile strength-to-mass ratio and superior resistance to cracking compared with reinforced concrete enables the adoption of only a thin, light and watertight layer.
2. It functions as a strong sacrificial cover layer for the precast reinforced concrete seawall behind it.
3. It facilitates repair as the panel is a separate entity that can be removed or replaced without affecting the structural integrity of the L-Block.
4. When a ferrocement panel facing is combined with reinforced concrete to act as its sacrificial cover, the higher cost of material and of fabrication is cost effective.

The sequence of construction and the technical specification for the ferrocement panel facings are illustrated in Appendix A. A low permeability mortar mix of proportions in accordance with the recommendation of ACI Committee 549[36], i.e. sand/cement ratio of 1.5 and water/cement ratio not exceeding 0.35 are as shown in Table 2.8.

The durability characteristics of ferrocement in the marine environment has yet to be established. Based on the concepts of minimum micro-cracking and low permeability, ferrocement meets these durability requirements. Together with the provision for ease of repair or replacement, in terms of the life cycle cost for its protection for reinforced concrete elements, ferrocement facings offers a cost-effective alternative for durable concrete structures in the marine environment.

Until such time as explicit design methods are developed for the durability design life, qualitative requirements remain, in practice, the available approach. The factors affecting the durability of tropical concrete structures are similar to those in a temperate climate but due to the constant higher ambient temperature, deteriorating chemical processes tend to take place at a faster rate. In particular, the constant high relative humidity provides the moisture needed for chemical reactions to occur. Thus better resistance must be provided by concrete of low permeability for moisture and gases. For the marine environment, concrete should have a low diffusion coefficient for chloride ions. In addition to the higher resistance by the concrete, sacrificial protective cover which can be replaced (Appendix A) may well be a cost-effective strategy to extend the durability service life of concrete structures.

References

[1] Rahman, A.U., 'Rainfall characteristics and their applications – in Singapore', *Seminar on Wind and Rainwater Penetration in Buildings*, Singapore: Department of Building Science, National University of Singapore, 28 September 1984.

[2] Klieger, P., 'Effect of mixing and curing temperature on concrete strength', *Journal, American Concrete Institute*, Detroit, MI., Vol. 54, No. 12, 1958, pp. 1063–1081.

[3] Tam, C.T., 'Concrete construction in the tropics', *Proceedings, First Conference of ASEAN Federation of Engineering Organisations*, Jakarta, Indonesia, August 1982, pp. 23–35.

[4] Yamamoto, Y. and Kobayashi, S., 'Effect of temperature on the properties of superplasticized concrete', *Journal, American Concrete Institute*, Detroit, MI., Vol. 83, No. 1, 1986, pp. 80–87.

[5] British Standards Institution, BS 5075:1982, *Part 1: Specification for accelerating admixtures, retarding admixtures and water-reducing admixtures*, London: BSI.

[6] American Society for Testing and Materials, 'ASTM C 403: Time of setting of concrete mixtures by penetration resistance', *Annual Book of ASTM Standards*, Vol. 04.02, West Conshokocken, PA: AMSTM, 1996.

[7] Neville, A.M., *Properties of Concrete*, Fourth edition, London: Longman, 1995.

[8] Verbeck, G.J. and Helmuth, R.H., 'Structure and physical properties of cement pastes', *Proceedings, Fifth International Symposium on the Chemistry of Cement, Tokyo*, Vol. III, 1968, pp. 1–32.

[9] Powers, T.C., 'The physical structure and engineering properties of concrete', Portland Cement Association Research Department Bulletin 90, Chicago, 1958.

[10] Price, W.H., 'Factors influencing concrete strength', *Journal, American Concrete Institute*, Detroit, MI., Vol. 47, No. 2, 1951, pp. 417–432.

[11] Ackroyd, L.W. and Rodes, F.G., 'An investigation of the crushing strengths of concrete made with three different cements in Nigeria', *Journal, Institution of Civil Engineers*, London, February 1964.

[12] Quao, H.N.O., 'The age–strength relationship of concrete under tropical conditions', RILEM Bulletin No. 24, September 1964.

[13] Tam, C.T., 'The relationship between strength and maturity of concrete in tropical conditions', *Journal, Department of Engineering, University of Malaya*, Kuala Lumpur, Vol. 7, 1968, pp. 60–74.

[14] Mani, A.C., Tam, C.T. and Lee, S.L., 'Influence of high early temperatures on properties of PFA concrete', *Cement and Concrete Composites*, Vol. 12, No. 2, 1990, pp. 109–115.

[15] Bamforth, P.B., 'An investigation into the influence of partial Portland cement replacement using either fly ash or ground granulated blast-furnace slag on the performance of mass concrete', *Proceedings, Institution of Civil Engineers*, London, 1982, pp. 777–800.

[16] ACI Committee 305, 'Hot weather concreting', *ACI Manual of Concrete Practice*, Part 2, 1994, Detroit, MI.
[17] Uno, P.J., 'Plastic shrinkage cracking and evaporation formulas', *Journal, American Concrete Institute*, Farmington Hills, Vol. 95, No. 4, July–August 1998, pp. 365–375.
[18] Tam, C.T., Swaddiwuddhipong, S., Ho, D.W.S., and Seow, S.S. 'Strategy for casting of raft foundation in tropical climate', *Journal of The Institution of Engineers*, Singapore, Vol. 42, No. 6, 2002, pp. 6–11.
[19] Kropp, J. and Hilsdorf, H.K. (Eds) 'Performance criteria for concrete durability', RILEM Report No. 12, London: E. & F.N. Spon, 1995.
[20] Concrete Society, 'Discussion document on developments in durability design and performance-based specification of concrete', Concrete Society Special Publication CS 109, London: Concrete Society, 1996.
[21] British Standards Institution, BS 5328:1997, *Part 1: Concrete – Guide to specifying concrete*, London: BSI.
[22] Thaulow, N., Johansen, V. and Jakobsen, U.H., 'What causes delayed ettringite formation?', Part Four, *Mechanisms of Chemical Degradation of Cement-based Systems*, K.L. Scrivener & J.F. Young, (Eds) London: E. & F.N. Spon, 1997, pp. 211–218.
[23] Lewis, M.C. and Scrivener, K.L., 'Microchemical effects of elevated temperature curing and delayed ettringite formation', Part Four, *Mechanisms of Chemical Degradation of Cement-based Systems*, K.L. Scrivener & J.F. Young, (Eds) London: E. & F.N. Spon, 1997, pp. 243–252.
[24] Lawrence, C.D., 'Physical and mechanical properties of Portland cements' Chapter 8, *Lea's Chemistry of Cement and Concrete*, P.C. Hewlett, (Ed.) London: Arnold, 1998, pp. 343–420.
[25] St. John, D.A., Poole, A.W. and Sims, I., *Concrete Petrography, a Handbook of Investigative Techniques*, London: Arnold, 1998.
[26] Chin, M.S., 'Carbonation of concrete', Master of Engineering Thesis, National University of Singapore, Singapore, 1991.
[27] Roy, S.K., Poh, K.P. and Northwood, D.O., 'The carbonation of concrete structures in the tropical environment of Singapore and a comparison with published data for temperate climate', *Magazine of Concrete Research*, Vol. 48, No. 177, December 1996, pp. 293–300.
[28] Papadakis, V.G., Fardis, M.N. and Vayenas, C.G., 'Fundamental concrete carbonation model and application to durability of reinforced concrete', *Proceedings, Fifth International Conference on Durability of Building Materials and Components, 7–9 November 1990, Brighton, UK*, pp. 27–38.
[29] Roy, S.K., Private communication.
[30] JICA, Japan–Asean Co-operation Programme on Materials Science and Technology, *Singapore Project on Prevention of Corrosion in Structures*, Vol. III, C.T. Tam, (Ed.) Singapore, National Science and Technology Board, September 1990.
[31] British Standards Institution, BS 8110:1997, *Part 1: Structural Use of Concrete*, London: BSI.
[32] Norzan M.Y. and Clark, L.A., 'An overview of environmental loading in a hot–humid (Malaysian) marine environment', *Bulletin, Institution of Engineers*, Malaysia, March 1995, pp. 5–14.
[33] Norzan, M.Y. and Clark, L.A., 'Preliminary investigation of a structure for in-service environmental load monitoring in Malaysia', *Journal, Institution of Engineers*, Malaysia, Vol. 55, 1994, pp. 95–117.
[34] Wei, J., Ho, S.K., Khoo, T.K., Lourdesamy, I., Wee, T.H., and Lim, H.B., 'High performance concrete for marine structures', *Proceedings, Twenty-first Conference on Our World in Concrete and Structures, August 1996, Singapore*, (Ed.) C.T. Tam, Singapore, CI–Premier, 1996, pp. 263–272.

[35] Wee, T.H., Yong, K.Y., Wong, S.F. and Lim, H.B., 'Strength development and chloride ingress in concrete under tropical and temperate conditions', *Proceedings, Twenty-first Conference on Our World in Concrete and Structures, August 1996, Singapore*, (Ed.) C.T. Tam, Singapore, CI–Premier, 1996, pp. 251–261.

[36] ACI Committee 549–93, 'State-of-the-art report on ferrocement', *ACI Manual of Concrete Practice*, Part 5, 1994, Detroit, MI.

Appendix A
Innovative Application of Ferrocement for Durable Marine Structures

(North Eastern Coast Reclamation Phase 4
Area at Punggol and Coney Island,
Housing Development Board, Singapore)

2.A1 Introduction

The North Eastern Coast Reclamation Phase 4 is a four-year project (1997–2001) with a cost of S$256 million (US$160 million). The project consists of the main reclamation works and the construction of ten various types of shore protection which include retaining seawall structures, stone revetments and sand embankments which stretch over a total length of 9 km. The construction of vertical retaining seawall structures alone covers 7 km.

Prior to the commencement of the project, Housing Development Board (HDB) engineers were confronted with the question of how to construct, at a reasonable cost, a high-performance concrete structure. This concrete combines high-workability, high-strength and high-durability characteristics that require low maintenance and is suitable for the submerged, tidal and splash zone subject to tropical conditions.

An interesting and innovative concept was eventually derived to protect and prolong the service life of the seawalls using ferrocement panel facings. The adoption of ferrocement panel facings on a large-scale basis is the first of its kind in marine structural works, although it has had widespread use in the construction of silos, tanks, roofs, boats and other non-structural elements. The main objective of the ferrocement panel facing in the design of the seawall is for it to function as a sacrificial protective cover layer within the wave-splashing zone. As a separate element it will facilitate maintenance and future repair since it can be removed and replaced without affecting the structural integrity of the main seawall.

On the specification of the concrete mix, findings showed that partial replacement with ground granulated blastfurnace slag (GGBS) at high percentages, increases the resistance of concrete to chloride ion ingress. Furthermore, Portland blastfurnace slag cement (PBFC) is found to increase workability in the concrete compared with ordinary Portland cement for the same mix proportions due to the lower water demand of the GGBS particles. Based on these findings, a high percentage slag content (65%) partial replacement Grade 60 PBFC mortar has been specified for the ferrocement panel facings on the vertical retaining sea walls. This will give better durability characteristics against chloride penetration and sulphate attack. At the same time, Grade 40 PBFC concrete has been specified for all other concrete works that are exposed to the marine environment. Since the exposed face of the seawall is provided with a high performance panel facing of only 30 mm thickness (minimum cement content of 800 kg/m^3, effectively, the seawall (300 mm thick) benefits from this higher level of protection at a significant cost saving.

2.A2 Construction techniques/concreting details

The fabrication of the precast vertical retaining seawall is divided into two main stages:

Stage 1: Casting of the ferrocement panel facing (Fig. 2.A2).
Stage 2: Casting of the L-shaped, vertical seawall together with the completed ferrocement panel facing in a steel mould.

2.A2.1 Casting of ferrocement panel facing

The ferrocement panel facing is cast 'face-down' as shown in Fig. 2.A3(a). The dimension of 2030 mm or 2600 mm is for different types of panels. Its construction can be summarised in 4 steps:

Step 1: Fabricating the reinforcement mesh to form a skeletal framing system.
Step 2: Fastening of the mesh and the bars to the framing system.
Step 3: First stage casting with a Grade 60 mortar mix (with one vertical side for the second stage).
Step 4: Curing (10 days specified).

The position of the ferrocement panel in the vertical seawall is shown in Fig. 2.A3(b).

Fig. 2.A3(c) shows the bar and mesh arrangement together with the lifting hooks. The square mesh pattern was used since it offers equal strength in both directions, parallel to the wires, as shown in Fig. 2.A4. The typical properties of the mesh are listed in Table 2.Al. A minimum cover of 10 mm was maintained on the exposed (mould) face. The design of the ferrocement cross-section subjected to either bending or combined bending and axial load is similar to the design of a reinforced concrete beam or column. The allowable cracking strength for flexure and direct tension is 7.1 MPa and 3.1 MPa, respectively.

As the cement component is normally higher in ferrocement than in reinforced concrete, mineral admixtures, in this case GGBS, is used to maintain a high volume fraction of fine filler material. At the same time, the high strength mortar (Grade 60) with the addition of a specified mineral admixture aims to improve flowability, long-term strength gain, lower mortar permeability increase and resistance to sulphate attack and to chloride penetration. The quality of the mortar and its application to the mesh are the most critical, unless plasticity is maintained. Shrinkage is not an issue in ferrocement because of the high reinforcement content and its uniform distribution within the thickness. In view of this, a low permeability mortar mix of proportions in accordance with the recommendations of ACI Committee 549-93[A1], with sand/cement ratio of 1.5 and water/cement ratio not exceeding 0.35, is adopted.

Moist or wet curing is a requirement specified for the ferrocement panels. Continuous wetting of the surface with covers over the mortar surface holds in the moisture and assists with the curing process. This is necessary due to the low water/cement ratio and high cement content requiring a large quantity of free water for the hydration process. Typically, concrete cube strength of 60 MPa was reached by 7 days and not less than 75 MPa at 28 days.

2.A2.2 Casting of the L-shaped vertical seawall units

The L-shaped vertical seawall unit (L-Block) is cast with the buttress wall placed parallel to the horizontal base mould (Fig. 2.A3(b)). The partially completed ferrocement panel facing (Fig. 2.A5) is lifted and secured to the L-Block steel mould (Fig. 2.A6) for the second stage of the casting operation. The balance of the ferrocement panel facing is cast using mortar of the same mix proportions as for the first stage. The minimum lapping length of the welded reinforcement mesh and fine wire mesh, as shown in Figs 2.A5 and 2.A6, is 200 mm.

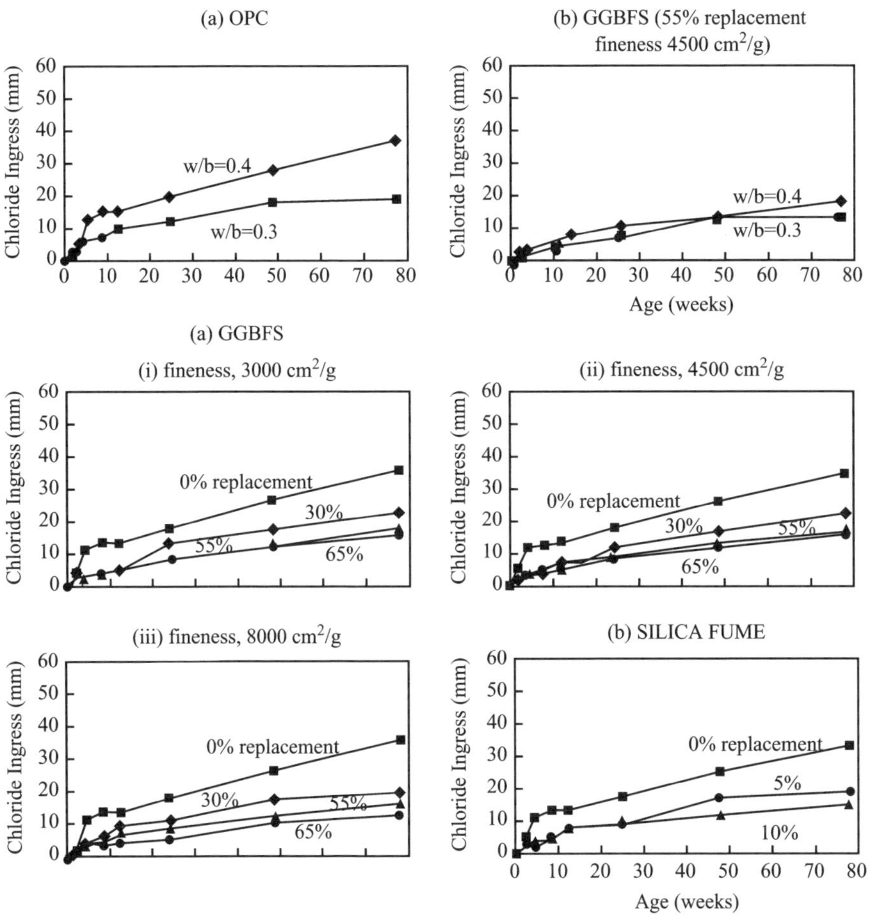

Figure 2.A1 *Effect of water/binder ratio and fineness of GGBS on chloride ingress.*

A typical completed L-Block is shown in Fig. 2.A7 with the dimensions being given in Fig. 2.A8. In view of the severe exposure conditions of the marine environment, adequate concrete cover has been specified as 70 mm for the sea-side/bottom face and 50 mm for the land-side/top face of the seawall unit. A low permeability concrete mix has been specified with water/cement ratio of 0.38 in combination with PBFC to achieve low porosity and low permeability. A superplasticiser and a retarding plasticiser were used to satisfy slump and casting requirements. The properties of the PBFC comply with the requirements of BS 4246: 1997[A2].

The coarse aggregate consists of well-graded, high-quality angular crushed granite with a maximum size of 20 mm. The fine aggregate is a well-graded, high-quality low-silt natural sand with low organic impurities.

A temporary batching plant was set up next to the casting yard to ensure that the concrete was conveyed from the mixer to its point of placing as rapidly as possible to avoid segregation and loss of ingredients. The elapsed time between the adding of water to the mix and when the concrete is finally consolidated in position is kept at less than 30 minutes. The concrete is

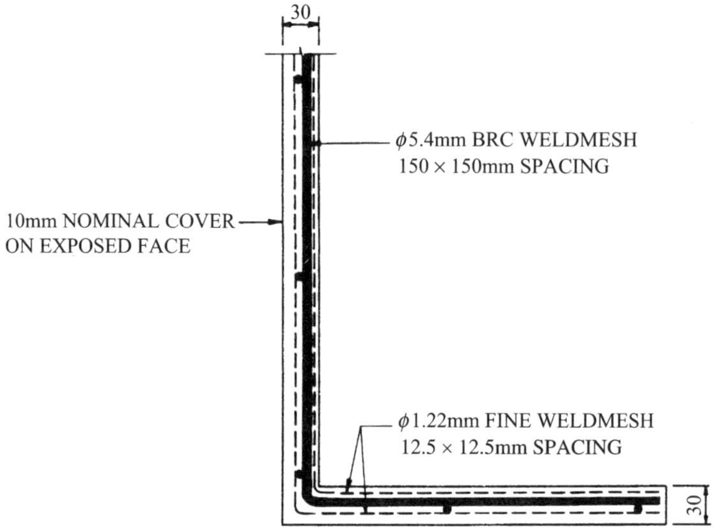

Figure 2.A2 *Detail of precast ferrocement panel facing.*

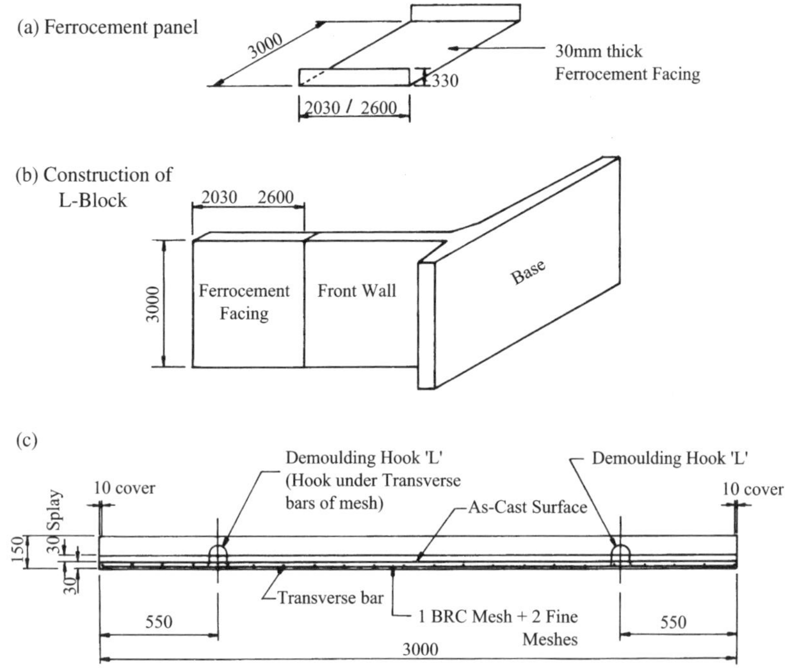

Figure 2.A3(c) *Bar and mesh arrangement of ferrocement.*

Figure 2.A4 *Square mesh pattern of ferrocement panel.*

Table 2.A1 Typical properties of mesh

Properties of mesh	Fine wire mesh (Galvanised)	Skeletal steel (Welded wire mesh)
Wire diameter (mm)	1.22	5.4
Grid size (mm)	12.5 ×12.5	100 × 100
Yield strength (MPa)	350–400	500–550
Modulus of elasticity (GPa)	200	200

Table 2.A2 Typical concrete strength

Age (days)	Specified (MPa)	Actual (MPa)
3	25	37
7	32	46
28	43	65

placed in a single operation to the full thickness of the members with concreting carried out continuously for each unit until completion.

A compressive strength exceeding 20 MPa at 24 hours after casting has been specified for demoulding. The achievement of this strength within the specified time is critical to meet the seawall installation schedule.

Adequate and sufficient curing time is ensured, especially during the initial hardening stage of the concrete. After demoulding, the top and side surfaces of the concrete are kept moist to enable curing to take place for a period of ten days. During the process of setting and the early period of hardening, care is taken to protect the concrete from direct sunlight and from vibration.

The age of the concrete, to achieve adequate strength, has always been emphasised especially for marine structures. Thus, prior to the installation of the seawall units, it is a requirement that

Figure 2.A5 *Completed ferrocement panel facing showing splicing for casting into L-Block.*

Figure 2.A6 *Portion of ferrocement panel facing in position for second stage casting.*

Figure 2.A7 *Precast L-shaped vertical seawall (L-Block).*

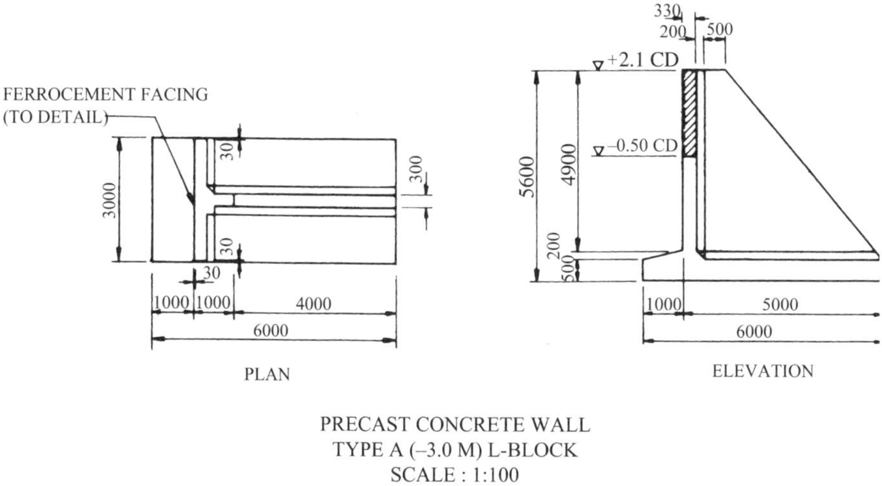

Figure 2.A8 *Plan and elevation view of vertical seawall.*

the concrete cube strength must exceed 43 MPa, after 10 days of curing. Typical strength results are shown in Table 2.A2. The results show that the strength achieved is higher than specified. Thus a final product of high performance has been achieved successfully. Fig. 2.A9 gives an overview of the installed seawall.

Figure 2.A9 *Overview of seawall installed.*

Appendix references

[Al] ACI Committee 549 (1993), 'State-of-the-art report on ferrocement', *ACI Manual of Concrete Practice*, Part 5, 1994, Detroit, MI, 1994.

[A2] British Standards Institution, BS 4246:1997, *Low Heat Portland Blastfurnace Cement*, London: BSI.

3 The Design of Concrete Structures to Increase Durability

Dr Steen Rostam

3.1 Structural design versus durability design

The generally accepted aim of a design is 'to achieve an acceptable probability that the structure will perform satisfactorily during its intended life'.

When designing a structure the designer first defines the loads to be resisted. Since these loads are changeable, the designer applies some safety factors. These loads as factored must then be resisted by the structure through selecting a combination of structural systems, element geometry, material types and material strengths. As it is difficult to predict the precise properties of the materials, the geometry and the material qualities which will be achieved in the structure, other safety factors are applied that effectively limit the maximum allowable stresses in the structure. Mathematical equations are then used to verify that the probability that a load will exceed a resistance is maintained at a statistically acceptable low level.

In durability design, the situation is quite different. When verifying that the intended service life can be achieved with an acceptable level of reliability, it seems to be acceptable to use a grossly over-simplistic approach. The codes provide only qualitative definitions of exposure and fail to define the design life in relation to durability (Bamforth 1998). In particular, the codes fail to quantify the durability limit states that must be exceeded for the design life to end.

The traditional design follows a deem-to-satisfy approach. This means that in order to specify requirements the designer must substitute parameters for such things as the cement type and quantity, concrete mix, maximum water/cement ratio, minimum air content, concrete cover, type of curing, control of early-age cracking, and the limitation of crack widths. The values chosen depend upon the assumed aggressivity of the environment but are selected using engineering judgement without any factual calculation to verify the consequences.

Previous approaches fail to recognise that, in relation to durability, it is not the properties of the materials or components alone that define performance, but the condition of the structure in its environment as a whole, and its need for maintenance. Thus performance can be defined by functional requirements such as fitness-for-purpose, which includes issues such as deflections, cracks, spalling, vibrations, aesthetics and structural integrity.

When it comes to residual service life assessment of existing structures the same situation of using over-simplistic procedures prevails as for the service life design of new structures. This is in spite of the fact that once the structure is a reality a large part of the inherent uncertainties associated with new designs can be eliminated, or limited to required levels, through appropriately planned inspections, measurements and testing. The longer the structure has served, and the longer it has interacted with the environment, the more reliable the residual service life forecast becomes. This is the condition which must be kept in mind when designing new structures as well as when performing assessment and residual service life evaluation of existing structures.

3.2 Durability design and service life design

The operational way of designing for durability is to define durability as a service life requirement. In this way the non-factual and rather subjective concept of durability is transformed into a factual requirement of the number of years during which the structure shall perform satisfactorily (Fagerlund 1979).

Designing for a specified service life requires knowledge of the parameters determining the ageing and deterioration of concrete structures. Hence, the need (as a pre-condition) for scientifically sound mathematical modelling of the:

- environmental loadings;
- materials and structural resistances.

This is the only rational way of performing a quantified service life design for new concrete structures – and a residual service life design for existing structures. The materials and structural resistances include transport mechanisms for substances into and within the concrete, and the deterioration mechanisms of both concrete and reinforcement.

When deciding upon a specific service life the main issue is to clarify the event which will identify the end of the service life. The requirement for a specific service life performance of a structure is therefore closely associated with the short- and long-term costs of fulfilling this requirement. The owner must therefore acknowledge that decisions must be taken on both the service life and on the performance requirements, and the owner must accept both the short- and the long-term costs associated with these decisions.

For everyday buildings, national codes and regulations have defined society's service life requirements – often not explicitly but implicitly through the standards and codified design requirements. A 50 year service life seems to be the general objective of most codes.

The 'technical service life' is the time in service until a defined unacceptable state of deterioration has been reached by the structure (Fig. 1.7). The real service life may also reach its end when the structure becomes functionally obsolete, meaning it no longer satisfies the initial or possible new functional requirements. Similarly, if the costs of keeping the structure in service become excessive compared with the cost of replacing it, the economic service life is exhausted. In the following the main focus will be on how to design for a specific technical service life, this being the 'target service life' for the design.

The 'functional and economic service life' reflects the more common reality that structures are replaced or upgraded when they are considered obsolete for other than purely technical reasons. Usually no prior engineering efforts can forecast such situations. The engineering input in design must concentrate on providing a technically sound and durable structure that is able, technically, to outlive, within reason, the expected functional and economic service life. The latter are based on educated guesswork. To assist this guesswork, benefit can be gained from past experience, e.g. the rate at which axle loads on roads and bridges have been increased; design loads in dwellings have been raised; clearances of bridges have been increased; accidental design loads have been changed; etc.

The process of creating a design basis for service life is similar to the process of creating a design basis for the structural design of new structures. When preparing a design basis for the assessment and re-design of existing structures the factual information to be collected needs to target the requirements of the service life updating (see Section 3.9; Rostam 2005b).

The end of the target service life for principal structures and their main structural components is usually defined as the end of the initiation period and onset of the identified critical deterioration mechanism(s). However, for some types of structure a part of the propagation period may have already been taken into account at the design stage. Examples could be the robust parts of harbours and marine structures risking reinforcement corrosion when safety and performance is not jeopardised by some initial corrosion, cracking and spalling.

For non-structural elements the end of their service life indicates only the time until renewal or replacement is necessary before their reduced performance or malfunction creates secondary damage to the principal structure. This refers to, for example, the re-laying of coatings, replacement of drainage systems, bearings, joints or wearing courses.

Typical examples of expected – or specified – service lives for structures and main structural elements in temperate climates have been:

- 35 years: offshore structures;
- 50 years: structures designed according to international and national codes;
- 100 years: bridges, tunnels, harbours (UK 120 years);
- 200 years: storm surge barriers;
- 300 years: Messina Strait Bridge.

The service life depends equally on the behaviour of structural and non-structural elements. Both must be considered during design, construction and use of the structure. Typical examples of accepted service lives of non-structural components in temperate climates are:

- 10 years: surface coatings;
- 25 years: drainage, bearings, expansion joints, wearing courses;
- 50 years: membranes, linings.

In hot, humid and saline environments service lives may be, in the worst cases, only one-fifth or one-tenth of those indicated above if special precautions are not taken.

3.2.1 Durability design according to the CEB-FIP Model Code 1990

The CEB-FIP Model Code 1990 (MC 90) is the first code of practice, which states specific service life requirements for new designs, and indicates how this shall be achieved in the design process (CEB-FIP 1992). The importance of the level of workmanship during execution, the integration of anticipated future maintenance procedures, and the need for quality assurance schemes throughout the processes of designing, constructing and operating structures are integral elements of a modern service life design.

MC 90 expresses the basic requirements for service life design as follows:

> *Concrete structures shall be designed, constructed and operated in such a way that, under the expected environmental influences, they maintain their safety, serviceability and acceptable appearance during an explicit or implicit period of time without requiring unforeseen high costs for maintenance and repair.*

The Model Code introduces three new features in the design process, compared to current national codes of practice. These are:

- the influence of the period of use on the service life achieved;
- the need to maintain an acceptable appearance, meaning that structures should 'grow old gracefully', thus enhancing the reputation of concrete in the eyes of the public;
- the costs for maintaining the structure in satisfactory use – which lead to the design of accessible, inspectable, maintainable, repairable, and replaceable structures.

Measures are necessary to ensure the required service life is chosen according to the environmental conditions and the significance of the structure. The service life depends equally upon the behaviour of structural and non-structural elements. Both must be considered during design, construction and use of the structure.

Non-structural elements such as drainage, joints, bearings, installations, etc. may require specialist attention other than that of structural engineers. Certain structural components (such as anchorages, couplers and deviators for prestressing tendons) and their location in the structure may require particular attention.

The Model Code clauses on design, execution and maintenance will generally lead to a service life in excess of 50 years for structures – based on the knowledge and experience behind the clauses. However, some structures will require a substantially longer service life, say 100 years or more, and other structures may need only a considerably shorter service life, say less than 25 years.

3.2.2 Deterioration mechanisms

In order to understand, and cope with, the true and expected future durability performance of concrete structures, the deterioration mechanisms must be known. In fact, knowing and understanding the transport mechanisms of liquid and gaseous substances into and within concrete structures is the most important element in ensuring sufficiently durable concrete structures. This is also the fundamental basis for quantifying durability in the form of service life performance (CEB 1992).

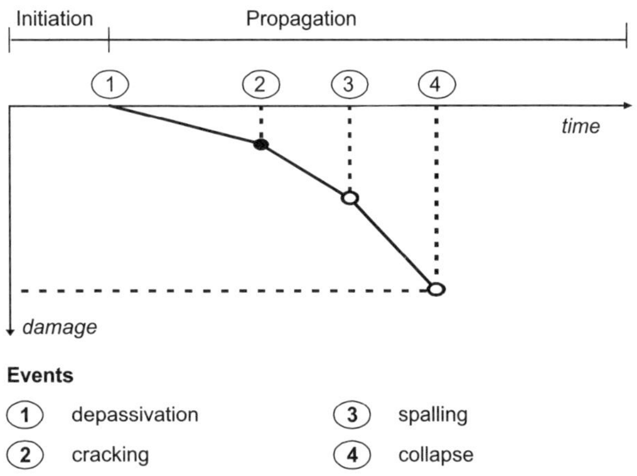

Figure 3.1 *Corrosion dependent durability-related events within the technical service life of structures.*

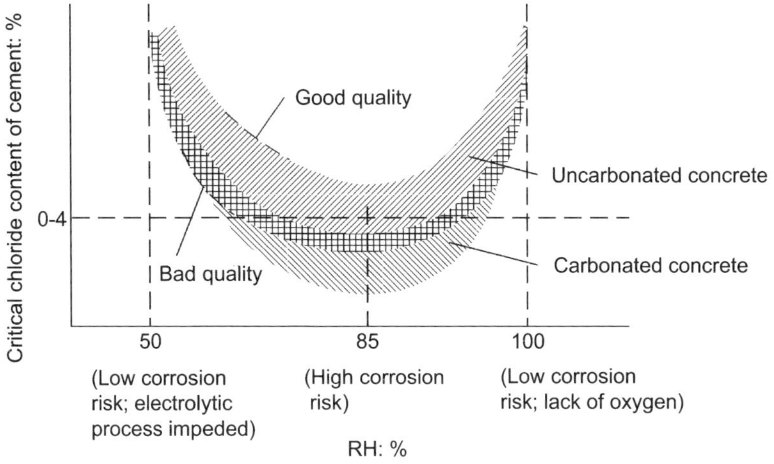

Figure 3.2 *Apparent critical chloride concentration for corrosion initiation depending on moisture level in the concrete (CEB 1992).*

The transport mechanisms which are decisive for durability are discussed in Chapter 1. A detailed description and modelling is presented in FIB (1999). In this chapter additional information focusing specifically on the design aspects of durable structures has been included.

Chloride-induced corrosion is probably the most serious and widespread type of deterioration of concrete structures, as highlighted in Chapter 1.

Fig. 3.1 is an extension of Fig. 1.7 and shows in principle the performance of a concrete structure with respect to reinforcement corrosion and related events. Points 1 and 2 represent events related to the serviceability of the structure, point 3 is related to both serviceability and ultimate limit states, and point 4 represents collapse.

The interrelations just described have the following consequences. Corrosion will not occur, neither in dry concrete (where the electrolytic process is impeded – or the electric resistivity is too high for the current to flow), nor in water-saturated concrete (where oxygen cannot penetrate), even if the passive layer at the surface of the reinforcement has been destroyed (Fig. 3.2). The highest corrosion rate will occur in concrete surface layers subjected to rapidly changing wetting (which ensures a low resistivity) and drying (allows diffusion of oxygen) conditions.

Both CO_2 and chlorides may penetrate to the steel surface some order of magnitudes faster through cracked than through uncracked concrete. This may influence the initiation period but with respect to carbonation it has been shown convincingly that with well distributed cracks and with relatively large cover, corrosion of the reinforcement at the crack usually will self-heal. For chloride penetration at a crack the initiation period may be reduced but increased cover will considerably enhance durability.

3.2.3 FIB-Model Code for Service Life Design (MC-SLD)

Currently FIB Commission 5 (Structural Service Life Aspects) has two active task groups developing guidance documents on service life design. Task Group 5.6 'Model Code for Service Life Design' has already produced a code-type document (FIB 2006). Task Group 5.7 'Service Life Design Guide' is in the process of developing a designers guide to design for durability and service life respecting the requirements of the MC-SLD.

The MC-SLD is divided into five chapters (Fig. 3.3):

1. Introduction/general
2. Basis of design
3. Verification of service life design
4. Execution and quality control
5. Maintenance and condition control

The flow chart illustrates the flow of decisions and the design activities needed in a rational service life design process with a chosen level of reliability. The MC-SLD has identified four different levels of sophistication in the performance-based design:

1. A *full probabilistic design*, corresponding to design strategy B (see Section 3.3.1).
2. A *partial factor design* (semi-probabilistic), corresponding also to design strategy B, but with factors calibrated with level 1 above.
3. A *deem-to-satisfy design* corresponding to methods in current codes, but also corresponding to design strategy B and with requirements calibrated, to the extent possible, with level 1 above.
4. *Avoidance of deterioration*, corresponding to design strategy A.

The design guide will provide background descriptions and guidance to assist the designer select the appropriate design strategy in individual cases. This will include detailed descriptions on how the owner or client should be part of the design process. In particular the owner should take decisions defining the design service life

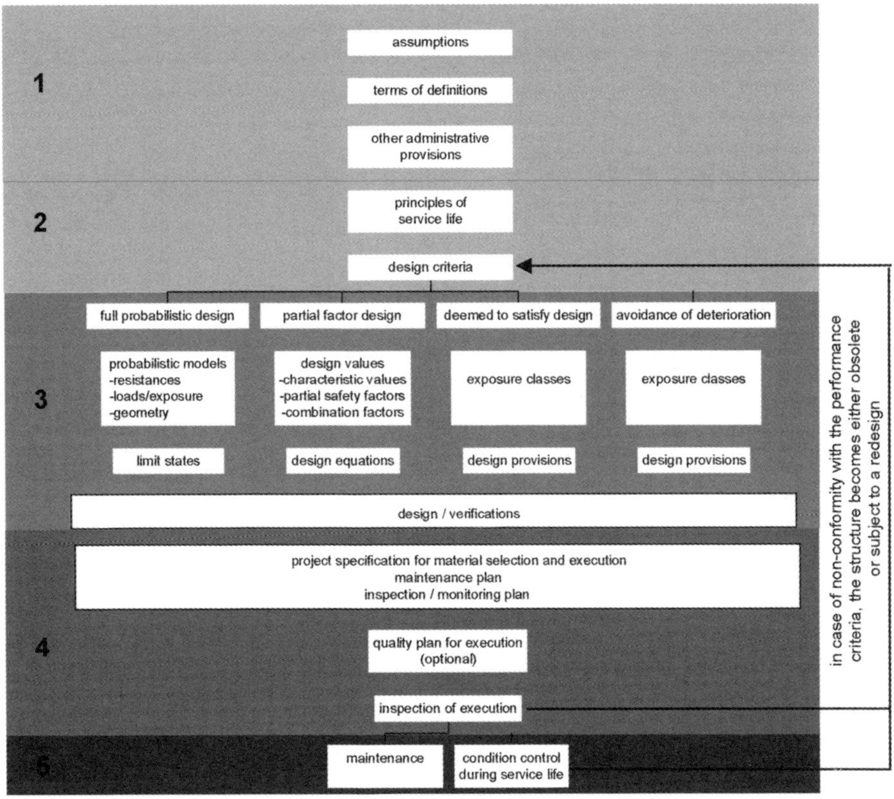

Figure 3.3 *Flow chart of MC-SLD and the related chapters 1–5. Chapter 3 identifies the four different levels of sophistication adopted in the MC-SLD.*

verification method and fixing the acceptance criteria so this becomes operational in the design process.

Currently FIB is working on a full revision of the CEB/FIP Model Code 1990. The draft is expected to be presented at the FIB congress in Naples in June 2006. The 1990 Model Code deals mainly with structural design of concrete structures. The present work aims to include aspects of execution, materials and conservation in a consistent manner, as well as service life design as summarized in this chapter.

The MC-SLD prepared by Task Group 5.6 and the Design Guide prepared by Task Group 5.7 are both to be published as a separate FIB bulletin and in part serve as input to the full FIB new model code concerning service life aspects, and as a supportive guidance document.

Examples: SLD following Strategy A (see Section 3.3.1)

The avoidance of deterioration is the most used design to preserve the integrity of the concrete itself. This is exemplified through the following:

1. Air entrainment is adopted to avoid freeze-thaw deterioration, both bulk freezing and surface scaling, the latter typically occurring on structures exposed

to de-icing salts. This is a very well established and reliable mitigating measure, provided the air bubble distribution intended is also achieved in the hardened concrete and has not been destroyed through excess pumping and vibration.

2. Non-reactive aggregates are selected to avoid harmful alkali-aggregate reactions. This is also a well established and reliable mitigating measure. If some uncertainty remains with a particular aggregate, the use of low-alkali cement and possibly also the addition of a pozzolanic additive, such as fly-ash or a limited amount of silica fume, may provide further enhancement of the avoidance of deleterious reactions. Alternatively slag cement may be used.
3. Low C_3A cements are selected to avoid harmful sulphate attack. Possibly an addition of a pozzolanic additive, such as fly-ash or a limited amount of silica fume, may provide further enhancement of the avoidance of deleterious sulphate reactions. Alternatively slag cement may also be used in this case.
4. Surface coatings may be adopted to avoid disintegration of the concrete due to acid attack. However, such coatings need maintenance and full re-coating at regular intervals, an issue needing detailed consideration during the initial design, in particular in connection with the LCC evaluation. Blended cements or pozzolanic additions may also contribute to an increased resistance against acid attacks.
5. Corrosion resistant reinforcement is the most recent measure to eliminate the risk of reinforcement corrosion, also in the most corrosive chloride-containing environments. The adoption of corrosion resistant steel reinforcement (CRSR) is the simplest solution as this not only solves the corrosion problem in a foolproof manner, but also leaves the site activities nearly unchanged. This is particularly advantageous as serious changes to the site operations always introduce difficulties and uncertainties. Using non-metallic reinforcement may solve the corrosion problem, but most such reinforcements cannot be adjusted on site, some may be brittle (e.g. carbon fibre bars) and sensitive to impacts from a vibrator. Walking on the reinforcement should usually be avoided, and most disturbing from a practical point of view is that all these types of reinforcements are light and require anchoring to the formwork to avoid them floating.

3.3 Service life design principles

The requirements of national codes and standards will usually be satisfied through the traditional structural design procedure. This must satisfy the Ultimate Limit State requirements as well as the Serviceability Limit State requirements.

If the codes or standards do not specify the target service life, then this needs to be specified by the owner or fixed in some way in order to form the basis of the service life design.

When the required target service life has been fixed, there are an additional number of special design issues, which should be considered, such as:

- structural layout and geometric form;
- structural system;
- location of expansion joints and construction joints;
- construction methods;

- environmental aggressivity;
- decisive deterioration mechanisms;
- selection of materials and concrete mix;
- concrete cover;
- maximum design crack widths;
- special protective measures;
- placing of installations and equipment.

The details of several of these design issues lie outside the scope of this chapter, but they are equally important within the overall design and construction of durable concrete structures (CEB 1992; Markeset & Rostam 1999).

The basis for the service life design will differ substantially whether a new structure is to be designed or whether the residual service life of an existing structure is to be determined and possibly prolonged through rehabilitation or upgrading (Rostam 1991, 2005b). Nevertheless, the actual methodology and the calculations are the same.

Through the service life design process the designer must consider the main elements of design indicated above by taking the following actions:

- Identify the specific physical actions and aggressive chemical substances, which may be expected from the environment during the foreseen service life.
- Design the geometrical form of the exposed parts of the structure to be robust in resisting the ingress of the aggressive substance from the environment.
- Select the cement type and concrete quality which would be able to resist the possible deleterious actions identified, provided the design, construction and assumed maintenance are adapted accordingly.
- Select the type of reinforcement considered optimal in combination with the chosen concrete, the detailing of the structure, and the reinforcement layout.
- Design the concrete cover and consider crack widths and crack orientations relative to the reinforcement.

3.3.1 Basic design strategies

The first and foremost task is to identify and quantify (to the extent possible) the type and aggressivity of the environment in which the structure is to perform during its service life. This will then clarify which type of deterioration could be a threat to the future performance and service life of the structure. The design strategy will then be to create a structure which has sufficient resistance towards the identified environment and its resulting threats of deterioration.

In principle there are two basic design strategies which can be followed (Rostam & Schiessl 1994):

A: Eliminate the risk of the degradation reactions under consideration.
B: Select optimal material compositions and structural detailing to resist the degradation reaction under consideration.

Strategy A can be subdivided into three alternative means of eliminating the risk of deterioration:

A1: Change the environment on the exposed concrete surfaces, e.g. by tanking, membranes, coatings, etc.

A2: Select non-reactive materials, e.g. stainless steel reinforcement, coated steel reinforcement, non-reactive aggregates, sulphate-resistant cements, low-alkali cements, etc.

A3: Inhibit the reaction, e.g. cathodic protection or cathodic prevention. The avoidance of frost attack by air entrainment is also classified in this category.

Strategy B represents different types of design provision. For example, corrosion protection could be achieved by selecting an appropriate cover and a suitable dense concrete mix. In addition, a structure can be made more resistant against aggressive environments of different sorts by appropriate detailing – such as minimising the exposed concrete surface, by rounded corners, and by adequate drainage.

Modelling of deterioration processes is only applicable for strategy B. An outline of a procedure for strategy B could be the following:

- The ideal procedure would start with the definition of the performance criteria related to the environmental conditions to be expected.
- The next important element is the realistic modelling of the actions (environment) and the material resistance against these actions.
- Based upon the performance criteria, performance tests are indispensable for quality control purposes. The performance tests must be suitable both to check the potential quality of the material under laboratory conditions and, even more importantly, the *in situ* quality.

Based upon all these elements the design procedure can be established.

3.3.2 Identification of environmental aggressivity

Depending on the type and use of the structure, information on the physical actions for which the structure is to be designed would usually be readily available, and the strength and abrasive resistance needed to resist these actions can be determined straightforwardly.

With respect to durability one of the most important decisions to be taken by the designer is the determination of the exposure conditions for which each member of a structure must be designed. Due to the dominating influence of the micro-climate on the type and severity of the environment, different parts of a structure may be categorised in different exposure classes. Obvious examples are the submerged, the tidal, the splash, and the atmospheric zones of a marine structure, but different geographic orientations (north/south/east/west, or seaward/landward) also may warrant different exposure class categories. Even very local differences can be taken into account such as vertical faces, horizontal surfaces facing upward (risk of ponding) or downward (protected against wetting with risk of increased rate of carbonation).

The structure itself has a decisive influence on the future micro-climate to be expected. This emphasises the need for a conscious evaluation by the designer of the structure–environment interaction depending upon the chosen structural form and geometric detailing of the different exposed members.

Clearly, it is not easy to decide upon the conditions of exposure of particular elements in particular environments. However, the factors listed below are known to have a dominating influence on the aggressivity of a particular environment.

Water or moisture

Aggressive substances, which may be present in the moisture:

- carbon dioxide – necessary for carbonation;
- oxygen – necessary for corrosion;
- chlorides – promote corrosion;
- acids – dissolve cement;
- sulphates – produce expansive reaction with cement and some aggregates;
- alkalis – produce expansive reaction with some aggregates.

Atmosphere

Aggressive substances, which may be present in the atmosphere:

- carbon dioxide – necessary for carbonation;
- sulphur – promotes neutralisation of the alkalinity of concrete.

Temperature

Aggressive effects of temperature may include the following:

- without frost – a main deterioration rate determining factor;
- with frost – temperature variations determine type of frost damage.

One important observation is needed when defining exposure classes. That is, to relate the exposure to the type and severity of deterioration that may result from the exposure. Hence, detailed knowledge of the governing deterioration mechanisms are needed to define exposure classes. In this respect a differentiation is needed between mechanisms deteriorating the concrete and mechanisms leading to reinforcement corrosion.

The designer may find valuable help in defining the design basis for the exposure of different elements of a structure from the classifications presented in different national or international codes or design guidelines such as the European Standard (CEN 1999).

3.3.3 Multi-stage protection approach

A design strategy A and a following strategy B, should consider possible measures to protect the structure against premature deterioration. A set of appropriate measures (one or more) can be combined to ensure that the required service life is obtained with a sufficiently high probability.

The protective measures available may influence either the duration of the initiation phase or the rate of propagation of active deterioration. The strategy of the service life design is to select, intelligently, the appropriate number and types of co-operating measures to ensure the required service life, considering the environment in question.

This design strategy is considered a 'multi-stage protection strategy' or a 'multiple barrier design' which leaves the selection of individual protective measures to the designer (Rostam 1993).

Protective measures may be established by:

- the selection of the structural form;
- the concrete composition, including special additions or admixtures;
- the reinforcement detailing including concrete cover;
- a special skin concrete quality, including skin reinforcement;
- limiting or avoiding crack development and crack widths, e.g. by prestressing;
- additional protective measures such as tanking, membranes or coatings, including coating of reinforcement;
- specified inspection and maintenance procedures during in-service operation of the structure, including monitoring procedures;
- special active protective measures such as cathodic protection or monitoring by way of sensors.

A different level of reliability is associated with the protective ability of each type of measure. This level of reliability depends much upon the quality assurance scheme associated with the establishment and possible maintenance of each protective measure.

The different measures may act simultaneously in contributing to the protection, or one measure may be backed up by the next.

3.3.4 Geometrical form of exposed parts of the structure

An architectural design that is well thought of from the point of view of required long service life may lead to excellent performance with respect to durability and appearance. Complexity in structural form will usually increase the sensitivity of the structure to an aggressive environment, shorten the service life, or require increased

Figure 3.4 *Architectural design of a building facade with a high durability risk factor.*

Figure 3.5 *Architectural design of a building facade with a low durability risk factor.*

efforts in maintenance. Also, ageing may increase with increased geometric complexity.

The robustness of an exposed structure or structural component is related partly to the ratio of the exposed surface area to the volume of the concrete member, or to the projected area of the facade. The larger this ratio the greater the risk of some deleterious substance penetrating locally into the concrete in sufficient quantity to initiate deterioration.

An architectural design based on permanently exposed precast concrete elements is illustrated in Fig. 3.4. This solution creates large column-free areas within the building but the result leads to a large ratio of exposed surface to projected facade area. Furthermore, the large number of joints and connections collect dust and dirt which holds water for long periods of time. This structure therefore has a high durability risk factor.

Another architectural facade layout is shown in Fig. 3.5. A very smooth surface with no windowsills may cause uncontrolled rain run-off patterns. However, the drainage of rain from the window areas is very efficient and allows the concrete to dry quickly after rain. The durability risk factor of this facade is low when judged from the geometric form alone.

Near outgoing edges and corners, aggressive substances can penetrate into the concrete from more than one side thus leading to local concentrations. If the concrete or the reinforcement is prone to deterioration, this could lead to the early development of damage at the outgoing corners and along the edges. A so-called 'corner effect' develops.

Cyclic wetting and drying effects will strongly accelerate the rate at which dissolved agents enter the concrete and concentrate near the surface of evaporation. In such cases the selection of rounded corners and edges will reduce the concentration effects and enhance the durability of the structure.

3.3.5 Cement type and concrete quality

Cement is the main component of concrete, and requires special attention when the concrete mix is composed. The main differences to consider are whether a pure Portland cement based mix or a slag cement based mix will be chosen, or whether special mineral admixtures such as microsilica and/or flyash should be introduced as a blending agent to the Portland cement.

The water/binder ratio is in practice the most important single parameter to be controlled. It is the dominating factor controlling permeability and diffusivity of the concrete. The permeability and diffusivity of concrete reduces, as the water/binder ratio (w/b ratio) lowers, with specified maximum values as low as 0.35. Such low values reduce or complicate the workability of the concrete increasing the risk of plastic shrinkage cracking and autogenous shrinkage, which increases the tendency of thermal cracking. In highly aggressive environments the w/c ratio (w/b ratio) should be low, preferably with a maximum value of between 0.40 and 0.45.

Industry codes and standards usually indicate minimum values which cover the large majority of cases encountered in practice. This does impose some evaluations that must be made by the designer when faced with particularly aggressive environments, complex or sensitive structures, or requirements for long service lives beyond the service life assumed in the codes and standards.

In order to ensure a high-quality concrete with good long-term resistance against the ingress of deleterious substances and with slow rates of degradation once the initiation period has passed, the following can be recommended for all concrete components (apart from the cement):

- The aggregates must be non-reactive when tested with approved methods and, in addition, low alkali cements as well as blended cements will improve further the resistance against alkali–aggregate reaction (AAR).
- Fine aggregates dredged from the sea must be washed so that chloride contamination of the concrete is avoided.
- No chloride-containing additives (e.g. as accelerators) should be used.
- Seawater must never be used as mixing water or for curing.

3.3.6 Reinforcement

Normal reinforcement (black steel or carbon steel) as well as prestressed reinforcement will, under all usual conditions, be ensured sufficient protection against premature corrosion, provided the design and the construction fulfil the detailed requirements for durability.

Nevertheless, practice has shown that apparently very well designed and built structures sometimes exhibit locally serious premature corrosion damage. Investigations have shown that very local defects such as honeycombing, leaking through-cracks, and local unintentionally small cover on single bars or links may lead to local deterioration in an otherwise very well performing structure. This is part of the nature of concrete structures which are designed and constructed following current practice.

Possibilities to avoid or reduce the risks of such premature local corrosion defects would be to increase the quality control of the concrete cover, to ensure uniformly good compaction of concrete, e.g. using self-compacting concrete (see Section 1.8.6) or non-corroding reinforcement (see Section 1.7.1).

3.3.7 Concrete cover

The susceptibility of reinforcement to corrosion, together with the thickness of the concrete cover protecting the reinforcement and the quality (i.e. the permeability and alkalinity) of the cover, interact with the environment in a way which determines whether or not the environment is aggressive for the reinforcement. The values in the codes usually refer only to corrosion protection of reinforced and prestressed concrete structures. However, trading-off cover (reducing cover) against reduced w/c ratio or intensified curing should be avoided or limited to avoid error-sensitive solutions. This is well illustrated in Fig. 3.6, which shows the consequences for the service life if the concrete cover is not achieved locally (see also Figs. 3.8 and 3.9).

The modelling of the penetration of carbonation or chlorides into concrete, based on present day knowledge, can illustrate the considerable importance of the concrete cover and its quality (Fig. 3.7; Schiessl 1976). The surface layer of concrete, the outer 5 to 10 mm, is especially susceptible to increased permeability caused by the increased cement paste content and locally increased w/c ratio at surfaces cast against

Figure 3.6 *Corrosion damage of mainly aesthetic nature due to lack of concrete cover.*

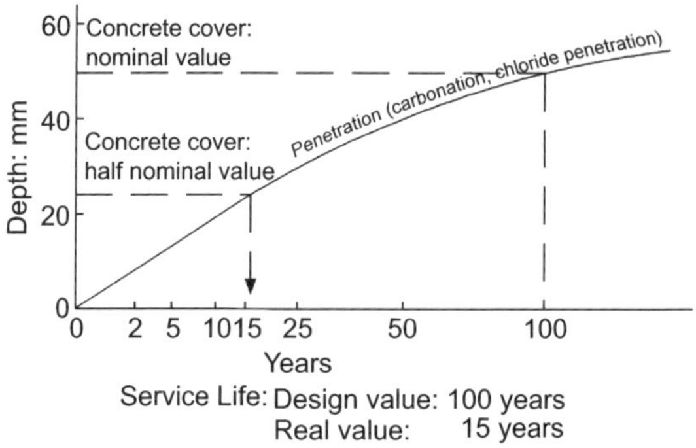

Figure 3.7 *Time to corrosion initiation due to carbonation or chloride penetration depending on the real concrete cover as achieved in the finished structure. The example corresponds to high-quality uncracked concrete (Schiessl 1976).*

impermeable formwork. Similarly, trowelled or floated horizontal surfaces may be more porous than the bulk of the concrete due to bleeding and settlement of the coarse aggregates. In such cases, any locally reduced concrete cover will considerably reduce the durability of the structure.

The permeability of concrete cover is influenced by:

- the w/c ratio – the lower the w/c ratio the lower the permeability;
- the type of cement and mineral admixtures – blended cements can lead to concrete with much improved permeability compared to similar concretes based only on Portland cement;
- compaction – honeycombing will increase the permeability of concrete to the extent that protection of reinforcement becomes non-existent;
- the quality of curing – curing determines the quality of the hardening of the surface layer of the concrete and controls the occurrence of thermal cracking plus plastic shrinkage cracking;
- the thickness of cover achieved in practice (Figs 3.6 to 3.9).

The values specified in codes are usually the absolute minimum values with no downward tolerances allowed and no upward tolerance being specified. This is valid for MC 90 as follows:

The nominal values, c_{nom}, are equal to the minimum values plus tolerance according to the following rule:

$$c_{nom} = c_{min} + \text{tolerance}$$

Tolerance should be taken as 10 mm unless, in the individual case, it can be demonstrated that a lower value is reliably obtainable in practice. The tolerance should not be less than 5 mm. Requirements for spacers are treated in Section 1.8.7.

Spacers must provide the required thickness of protective cover to the reinforcement. Spacers are preferably made of concrete of at least the same quality as the structural concrete and the geometry and fixing of the spacers must be reliable.

Figure 3.8 *Small concrete spacer with inadequate fixing (North American project).*

Figure 3.9 *Inadequate size and fixing of too small concrete spaces (European project).*

Fig. 3.8 illustrates inadequate form and fixing of a spacer for a large North American construction project. Fig. 3.9 illustrates the same problem for a large contemporary European Construction project. A two-point fixing of such spacers is essential.

By selecting non-corrodible reinforcement the non-durability related factors listed below may be accepted to govern the selection of concrete cover. Reasons other than durability may warrant different – and sometimes larger – covers. For example:

- ensuring bond strength;
- ensuring fire protection;
- using larger aggregate sizes.

3.3.8 Cracking

Cracking is inevitable in reinforced concrete structures subjected to bending, shear, torsion or tension. Cracks do not indicate undue lack of serviceability or durability provided the crack width is limited and the locations of probable crack occurrence are controlled.

The fact that the calculated tensile stresses, or the calculated crack widths, do not reach the specified values does not mean that cracks will not be produced, or that those which are produced will have widths smaller than the specified limits. The checking of the limit states for cracking should, therefore, be considered solely as a conventional means of obtaining a graduated system for measures of control against cracking (CEB-FIP 1992).

Four types of cracks may be considered:

1. Load-induced cracks, which are considered and controlled through the design process.
2. Early-age cracking, i.e. thermal cracking, and plastic shrinkage cracking, which can be minimised or avoided through an appropriate pre-planned curing technology.
3. Micro-cracking, i.e. cracking occurring in the cement paste during hydration and continued hardening of the concrete. Micro-cracks do not necessarily reduce the durability, if small enough and randomly distributed. They are

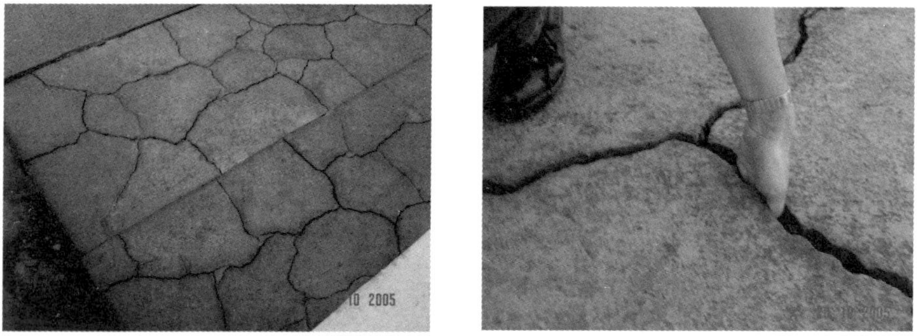

Figure 3.10 *Unreinforced concrete blocks in a seawall heavily cracked after 3–4 years of exposure due to DEF (left: overview; right: close-up).*

controlled through selecting an appropriate cement type, concrete mix and curing procedure.

4. Delayed technological cracks, i.e. cracks developing after a longer period of exposure to a wet environment. One such special type is due to delayed ettringite formation (DEF). A recent example of DEF is shown on Fig. 3.10. It is an expansive reaction that takes place if the concrete becomes very hot during early hardening and curing (temperature > 65–70°C), resulting in dissolution of ettringite, i.e. sulphate in larger than normal contents is dissolved in the paste. The cracking shown was only observed after 3–4 years of exposure. The concrete blocks were unreinforced.

In order to control load-induced crack widths, and possibly to limit the number of joints in a structure, prestressing is an option to be considered. Normally the required 'decompression' is sufficient to ensure that cracks due to e.g. temperature gradients within the sections not covered by direct calculations, will be limited to the surface areas of the concrete, and will not cross the prestressing steel. However, this may not be ensured in more complicated cross-sections such as box girders. Further temperature gradients may arise e.g. from sunlight.

Specifying small allowable crack widths can have very adverse effects on durability. The design formulae used to calculate crack widths will inadvertently lead the designer to think along the lines of:

- using as small a cover as allowed by the codes and specifications;
- using many small diameter bars with small bar spacing.

In addition, the designer would usually be forced to place an excessive amount of reinforcement in sections with locally high bending moments, such as at the corners of frames and above intermediate supports.

All these measures are clearly counterproductive with respect to obtaining a long service life with low maintenance costs. Increased concrete cover using good quality concrete, and fewer but larger diameter bars – both leading to increased crack widths at the concrete surface – will result in structures being more resistant and robust

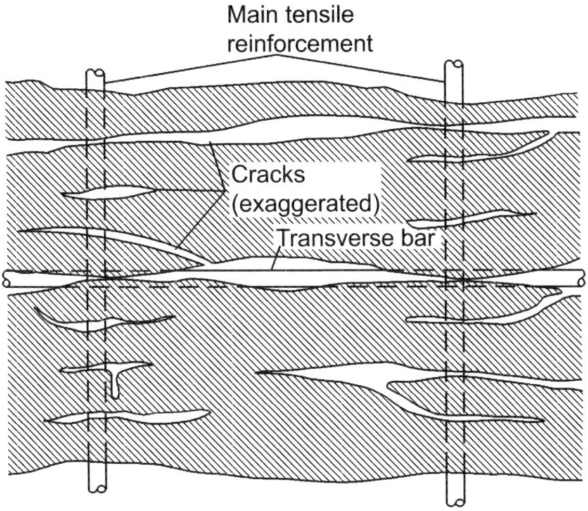

Figure 3.11 *Alignment of cracks relative to reinforcement (redrawn from CEB 1992).*

against deterioration than limiting the crack widths through the traditionally used measures given above.

Due to tensile cracks, the concrete structure will become water permeable to a degree, depending upon the crack width, the crack length, the hydraulic gradient, etc.

Practical experience has demonstrated that cracks have the ability to heal themselves, e.g. water flow will be reduced with time. In extreme cases cracks can seal completely. The autogenous healing of cracks in concrete seems to be a complicated chemical/physical process. Although the term 'autogenous healing of cracks' has become part of the civil engineering slang, the information about the process itself and the influencing parameters is limited. A quantification of the possible flow reduction due to self-healing has been investigated on a large scale through both theoretical studies and verifying experiments in Edvardsen (1996), and summarised in Edvardsen (1999). It has also been shown that autogenous healing occurs in active cracks. The healing is due mainly to calcium hydroxide from the hydrated cement being deposited as crystals on the side faces of the cracks. Hence, it may be expected that some healing also will occur in fatigue cracks, although full sealing is not to be expected.

Early-age cracking such as plastic shrinkage cracking and thermal cracking, and drying shrinkage cracking can be controlled, and often avoided, by using steel fibre reinforcement. However, other measures could be more cost effective, e.g. appropriate choice of cement, concrete mix, curing, cooling of the hardening concrete, membranes, and specific coatings.

Plastic shrinkage cracking should not be accepted as this reflects insufficient curing which also has other adverse effects. A small amount of polypropylene fibres, say 0.5 kg/m^3 will reduce thc risks of plastic shrinkage cracking.

The importance of cracking related to the durability of a structure may be critically influenced by whether the cracks follow the line main or the transverse reinforcing bars (Fig. 3.11).

3.4 Realism in service life design

3.4.1 Responsibilities

The service life of a structure represents its durability and is determined by specialised input from all the parties involved:

- the owner (client) by defining the requirements, demands and wishes;
- the designer by preparing the design, specifications and conditions;
- the contractor, who will follow the requirements of the owner and designer in the construction work;
- the user (who is often the owner) who is responsible for timely maintaining the structure to ensure that the intended service life is obtained without unforeseen high costs.

Any of these four parties may – by their actions or lack of attention – contribute to an unsatisfactory state of durability of the structure. Also, interactions between two parties, the so-called interface problems, may cause faults, which can have an adverse effect on durability. Thus, each of these parties has well defined responsibilities within the process of creating, operating and maintaining concrete structures (Jessen 1983).

But, first and foremost the owner must ask for quality, and check the quality received. The owner must maintain the structure, and pay the building costs as well as the costs for maintaining the structure in a satisfactory condition (Rostam & Schiessl 1994; Rostam 1996).

3.4.2 Codes and standards as design basis

The minimum requirements to be fulfilled are stated in the national codes and standards. However, it is recognised that traditionally these documents do not yet include service life aspects in sufficient detail to provide what would be considered the most up-to-date durability related quality of structures exposed to the aggressive environments. Historically and for reasons of tradition, codes and standards differ considerably from country to country.

Only in very few cases can systematic differences in national requirements explain noticeable differences in the performance and durability of concrete structures in the different countries. Examples could be the acceptance or non-acceptance of using calcium chloride as an accelerator, allowing high water/cement ratios, adopting very small concrete cover for exposed concrete structures. All of these are counterproductive to a long service life.

However standards cannot, by themselves, ensure quality as they define only the minimum requirements. Standards can serve only as helpful tools in support of more specific recommendations and project specifications developed to ensure the fulfilment of the individual owner's requirements.

Change in design paradigm needed

Due to the severe drawbacks of existing codes and standards in reliably providing durable and long-lived concrete structures, and due to the nature of structural concrete, a change in design paradigm is urgently needed (Rostam 2005a). Recent designs have shown that ultimate limit state design (ULS) is not decisive, serviceability limit state design (SLS) governs more and more, and now the durability and service life aspects are rapidly governing the designs of important buildings and infrastructure constructions (Rostam 2005b).

3.4.3 Deem-to-satisfy design

The deterministic design for durability will govern for some time yet, but with regular updating of the characterisations of the environment and improvements in the modelling of transport and deterioration mechanisms. Hence, the currently used means of ensuring durability and service life in national codes as well as according to MC 90 (CEB-FIP 1992), will to a large extent maintain the well known deem-to-satisfy design approach.

The deem-to-satisfy approach specifies requirements to parameters such as the cement type and quantity, maximum w/c ratio, minimum air content, concrete cover, type of curing, control of early-age cracking, limitation of crack widths, etc. The values chosen depend on the assumed aggressivity of the environment.

The main parts of the service life design procedures described in this chapter are based on the deem-to-satisfy approach but applied within a rationale where the time dependent features of the actions (environment) and resistances (layout, detailing and concrete quality) are taken into account in a deterministic way.

3.4.4 Probabilistic performance-based service life design

Due to the nature of concrete and the very individual ways concrete and concrete structures are constructed, it is unavoidable that variations in properties occur. The variation in properties do not occur only between different structures, different types of concrete compositions, and different geographic locations, but also within different parts of the structure itself.

In the past it has been difficult to quantify these variations. It has also been difficult to develop design procedures able to treat the variations in a rational and logic way aimed at service life design similar to that of the well-known structural designs. Therefore, more simplistic methods and experience-related requirements have until now governed the design for durability and service life.

Service life design is currently in an interesting transitional period. The modelling of the environment and deterioration mechanisms is being developed on a probabilistic basis allowing reliability based service life designs. Service life design methods similar to the load-and-resistance-factor design procedure used for structural design are now being perfected and practical applications have already been made.

In a few years such probabilistic performance-based durability design of concrete structures is expected to be fully operational and may thus drastically change the design procedures for concrete structures (see Section 3.11 and Rostam 2000).

3.4.5 Other corrosion prevention measures

When considering the adoption of SSR to eliminate the corrosion risk problems within a stipulated long service life, the additional costs for the SSR obviously become a key issue and alternative solutions are frequently tabled. The alternatives usually considered are described below with their 'pros and cons'. The level of reliability of alternative solutions becomes an important factor, though difficult to quantify, and only a profound knowledge of the governing deterioration mechanisms can assist in such evaluations, together with long-term practical experience, as also illustrated.

Due to the cost implications and the level of reliability of the alternative solutions, these issues should be addressed at a very early stage of the design process in order for the owner and client to select the final solution being the optimal for him.

3.4.6 General

A number of corrosion prevention measures, apart from concrete quality, cover and SSR, may be considered when ensuring long service life of concrete structures. Some of these methods are described below.

- Surface treatment of concrete
- Corrosion inhibitors
- Cathodic prevention
- Non-metallic reinforcement
- Corrosion resistant steel (not stainless) reinforcement

Coatings for concrete

Surface treatment and coatings for concrete are available in many types and qualities, from invisible hydrophobic silane impregnations to thick, pigmented acrylic, polyurethane or epoxy based coatings. The experience with silane impregnations varies from country to country. In some countries such treatment is compulsory for bridges exposed to de-icing salts. In other countries where the hydrophobic effect cannot be verified, silanes are not generally used.

Coatings may have durability enhancing effects by slowing down the rate of carbonation or chloride ingress. However, practical verification of the beneficial effects is very difficult. It has been demonstrated that pinholes in a thick carbonation resistant coating reduces the protective ability to just fractions of the effect of a pinhole-free coating – and it is not realistic to make pinhole-free coatings on concrete – even after having applied a surface wash using a smooth polymer-modified cementitious material. In addition, surface treatment to concrete is exposed to degrading effects from the environment and from UV light. Hence such coatings shall be maintained or re-applied after a limited number of years, depending on the type of material and exposure.

Corrosion inhibitors

The newest addition to the concrete mix is the introduction of corrosion inhibitors. They can be anodic, cathodic or mixed inhibitors. Corrosion inhibitor technology is well-known within the chemical industries but adding inhibitors to concrete has resulted in diverging reports on their efficiency. It seems that adding too little nitrite corrosion inhibitor compared to the future concentration of chlorides may even cause increased corrosion rates. The most usual inhibitors also act as concrete accelerators, which is not always an advantage, particularly in hot environments where high temperatures often cause casting difficulties due to rapid loss of workability and early hardening.

The risk of the inhibitor being washed out of concrete exposed to splashing water is also an issue under discussion as the inhibitors seems more mobile than chlorides. The effect of washing out has not been verified in laboratory tests, and in situ verification after long-term exposure has yet to be produced. In this connection it should be noted that inhibitors are dormant in the concrete until inhibitor and chlorides are present at the same time at the surface of the steel reinforcement. The inhibitor must, for several reasons, be added to all the concrete in the exposed components although it is only needed in the vicinity of the reinforcement. This has adverse cost implications for the construction.

With these uncertainties the level of long-term reliability of the use of inhibitors is questionable.

Cathodic prevention

Cathodic prevention has become an interesting option to protect parts of structures exposed to high chloride concentrations. The cathodic protection system is installed within the new structure ready to be used to protect the reinforcement against chloride corrosion. This option seems most promising in marine structures as a sacrificial anode system can be installed simply by placing anodes in the water near or on the structure and linking them to the reinforcement. Similarly, impressed current systems may be used, but they require the owner or operator to permanently monitor the current and protection level. The impressed current system can comprise inert anode meshes or bands cast into the new concrete structure, or they can be added later as needed. However, installing cathodic protection at a later stage will only stop further corrosion; there will be no restoring of lost steel. Therefore such a solution must be implemented promptly when the need has been identified.

The unavoidable life-long monitoring of the installations is also an issue needing careful consideration. With required design lives in excess of 50 to 100 years the reliability of such a solution seems to diminish with time.

Non-metallic reinforcement

Non-metallic-fibre reinforcing bars, based on glass, aramid or carbon fibres, have been developed and provide corrosion resistance. However, the different mechanical characteristics of fibre and steel reinforcement, together with the different conditions for practical use on site, seems to indicate only rather limited or special applications in concrete structures. Fibre reinforcing bars have very limited strength when loaded transversely to the direction of the fibres. This is an issue for the connection between main bars and stirrups. Some bars are also very sensitive to hard impacts. Carbon fibre bars may fracture if hit by a falling hammer or a dropped vibrator. During casting of concrete the workers cannot walk on the reinforcement and temporary work bridges must be used.

All such fibre reinforcing bars with very low mass must also be anchored to the form to prevent floating due to their buoyancy. Currently the precast industry seems to present the largest potential use of fibre reinforcing bars.

3.4.7 Corrosion resistant steel reinforcement

Epoxy coating of reinforcement

Epoxy coating of reinforcement was a protective measure against chloride-induced corrosion introduced in North America in the mid 1970s. Fusion bonded epoxy coating has since then been used extensively in North America, and subsequently in other countries. However, from the early 1990s, reports emerged indicating that an undercutting chloride-induced corrosion may develop from local pinholes and damage in the coating without necessarily causing cracking and spalling of concrete cover, just slowly disintegrating the reinforcement. The first examples were from the bridges on the Florida Keys (Fig. 3.12 upper). At that time the technology was slowly spreading to Europe, the Middle East, and to some parts of the Far East.

The first major application in Europe was for the precast concrete segments for the twin bored East Tunnel of the Great Belt Link in Denmark. The large number of pinholes allowed by the original specification for epoxy coating was of concern, as was the need for patch repairs of all cut ends and cracks formed at sharply bent coated bars. Using epoxy-coated single bars would be to 'put all the eggs in one basket', hence after several full scale tests it was decided to develop a fusion-bonded epoxy coating technique using (for the first time recorded) the fluidized bed dipping technique of 3D full size welded reinforcement cages. A special cleaning, heating, coating and testing plant were incorporated in the indoor concrete segment production plant producing 62,000 concrete segments.

Thus, before the problems with epoxy-coated reinforcement became public in North America, a viable and reliable technology for precasting plants was developed in Europe to eliminate cutting, bending and patch repairing of coated bars, and at the same time maintain the possibility of introducing cathodic protection if corrosion was initiated.

The North American experience of the traditional technology of epoxy coating together with additional testing and site investigations, has not gained a foothold in Europe, and the technology is now slowly being phased out (also in the Middle East and Gulf states). Fig. 3.12 (lower) shows examples from a marine structure in the Gulf reinforced with epoxy coated reinforcement.

Hot-dip galvanised reinforcement (HDG)

Little practical experience is available for this technique regarding enhanced performance in very aggressive environment. However, it has been shown in tests that HDG steels may have a higher threshold value for chloride-induced corrosion (initiation) compared to black steel. The protective ability depends on the reactions at the iron-zinc interface. An evaluation of the available data and practical experience indicate that galvanising of appropriate reinforcing steel alloy with stringent quality and thickness control, will provide:

- reliable corrosion protection of reinforcing steel in concrete exposed to carbonation but without chlorides and sulphates;

Figure 3.12 *upper: Deleterious effects in the Florida Keys bridges. lower: Adverse corrosion protection performance of epoxy-coated reinforcement in the vicinity of the saline waters of the Arabic or Persian Gulf under hot and humid environments.*

- lightly increased threshold level for chloride-induced corrosion compared to non-coated steel. However, once corrosion starts the rate of pitting corrosion may be higher, and at times much higher, than for uncoated black steel. For immersed conditions this effect may be further aggravated.

As stated previously stainless steel reinforcement can be mixed with ordinary black steel (carbon steel) reinforcement without causing galvanic corrosion. However, zinc coated steel reinforcement (galvanised steel, gray galvanising) cannot be mixed with carbon steel reinforcement as this will cause galvanic corrosion of the black steel. Therefore, a simple replacement of stainless steel reinforcement with galvanised steel reinforcement is not a viable solution. Realistically all reinforcement should then be galvanised. Nevertheless any post-coating mechanical treatment like cutting and bending will reduce the reliability of the galvanising to provide adequate corrosion protection.

Stainless steel clad black steel reinforcement

Recently a hybrid solution has emerged, stainless steel clad carbon steel reinforcement, (Nuovinox, produced by Stelax Industries Ltd). The stainless layer provides it with the corrosion resistance and the inner carbon steel core gives the necessary physical and mechanical properties. The outer stainless layer can be made from different grades of stainless steel.

The composite nature can be seen in the cross-section of the rebar as the outer stainless steel appears brighter than the carbon steel core. The cladding itself is of uniform thickness varying between 0.5–1.0mm around the perimeter and along the length for a 16mm rebar. Core composition and finishing rolling temperatures can be adjusted to achieve the tensile properties required. The stress strain curve for Nuovinox is the same as for a carbon steel bar.

Four main drawbacks have been identified:

- the reinforcement can only be produced in short lengths, usually 6m and by special order 11.6m;
- the reinforcement cannot be produced in coils, which would have simplified on site automatic production of prescribed lengths and bending of stirrups and links;
- the cut ends expose the carbon steel core and would be exposed to local chloride ingress. Therefore it is recommended by the producer that such bar ends are covered with a stainless steel cap (Fig. 3.13). The bars can then be cut by means that do not deform the ends. Usual scissor type cutting cannot be applied. This could be considered a major practical drawback of this technology;
- currently available maximum dimension is 20mm.

Microcomposite multistructural formable steel

Microcomposite multistructural formable (MMFX) steel is manufactured without the use of coating technologies as a result of a patented chemical composition and proprietary steel microstructure that is formed during the production associated with controlled rolling and cooling of the steels. This physical feature should minimize the formation of micro galvanic cells in the steel structure, hence minimizing corrosion

Figure 3.13 *Stainless steel clad carbon steel reinforcement provided with a stainless steel cap at the cut ends.*

initiation. MMFX steel has a low carbon content (<0.15%), and contains around 8–10% chrome.

For the MMFX steel reinforcement to be acceptable to the concrete community, a number of design and detailing characteristics must be determined. The mechanical characteristics of the reinforcement alone, and in conjunction with concrete must be independently validated. The Concrete Innovation Appraisal Service (CIAS) published a report confirming, among others, that the mechanical and physical properties of MMFX materials grade 60 exceed the requirements of the ASTM A 615 standard by about a factor 2. Prestressing strengths are also available in MMFX.

The threshold value for chloride-induced corrosion initiation is claimed to be a factor of nine greater for this steel. However, when comparing the published data with European experience this increased threshold value would only be about 2–4 times higher when adopting a critical value of 0.10–0.05% total chloride contents by weight of dry concrete.

'Top 12' - steel grade 1.4003

TOP12 steel reinforcement corresponds to the well-known stainless steel X2CrNi12 with the material number 1.4003. The fine-grained ferritic-martensitic structure of the steel is a result of the specific manufacturing process and leads to an advantageous combination of properties. The martensitic phase varies depending on the rolling parameters (e.g. finished rolling temperature, cooling rates) and the chemical composition. Optimal mechanical properties can be achieved by controlling the volume fraction and the morphology of the martensitic phase.

As a result of the optimised chemical composition as well as the above-mentioned manufacturing process the mechanical properties are equal to or better than ordinary carbon steel reinforcement.

3.5 Interaction between durability design and execution

At the design stage possible methods of construction must be considered and fixed, as this will influence the durability design.

It is important that this interaction between the foreseen execution and the provision to provide durable structures is identified at an early stage of the design to optimise the design and prepare the structure for easy inspection and maintenance.

In short, the structure must be designed so it is:

- constructable (buildable);
- accessible;
- inspectable;
- maintainable;
- repairable.

In view of the increasing intensified focus on all aspects of environmental impact, the list should be completed by adding:

- replaceable;
- able to be demolished;
- recycleable.

The above issues constitute different elements of energy saving and sustainability, issues now gaining major international importance. The production of cement is the single largest source of CO_2 generation within the construction sector. Considerable effort is now being focused on addressing these issues, which will impose considerable changes in cement and concrete compositions in future structures. This will constitute a major challenge for concrete construction where durability and life cycle cost optimisation are expected to gain increased influence on all elements of construction and use of concrete structures.

Recycling of concrete by crushing it to provide aggregates for new concrete is a well-known technology with well-known potential benefits. Also the separation of reinforcement from concrete in demolished structures is a well-known technology. The precast industry considers the possible re-use of concrete elements as an asset. However, it must be recalled that one of the greatest advantages of concrete is its adaptability to any form and shape. This is where concrete is highly advantageous compared to other construction materials. Recycling complete concrete elements from an old structure to a new structure would unreasonably restrict the beneficial exploitation of concrete.

3.6 Robustness in design and construction

One of the main obligations of the conscientious designer is to adapt the design to the conditions, under which the structure is to be constructed, operated and maintained. To avoid structures being sensitive to variations in the assumed qualities, a degree of robustness in the design is very advantageous – with respect to the future performance and durability of the structure. The following sections indicate the means of increasing the robustness of concrete structures.

3.6.1 Adapt requirements to local conditions

The conditions under which a structure will be constructed and used differ from case to case. It is essential therefore to adapt the requirements for the concrete mix, casting, compaction and curing to what can realistically be achieved at the individual location. The available concrete components, the local workmanship and the prevailing climatic conditions must be considered. This is particularly important for remotely located structures and in geographic regions with little or no alternatives to the local cement, aggregates and water, and only local, maybe unskilled, labour.

Such situations could occur in hot, arid and saline environments with unavoidable high chloride content in the available fine aggregates and in the mixing water. Difficult curing conditions and the unavoidable high initial chloride content in the concrete could then warrant the application of stainless steel reinforcement selectively in the most critically exposed parts of the structure. Such a solution would with 'one stroke' eliminate the reinforcement corrosion problem but with an initial cost increase, which would be marginal compared to the overall costs of construction and equipping the whole building. In the long term this solution could become a remarkable cost saving decision, which could prolong the service life of the structure many fold compared to current experience in such environments (see Section 1.7.1). The service life would in such cases be determined by the durability of the concrete, or by the structure becoming functionally obsolete.

3.6.2 Consequences of designing slender or minimal structures

The concept of minimal structures has developed into a challenging engineering and architectural discipline. By selecting combinations of structural form and materials, which minimise the total consumption of materials, very light structures can be achieved.

This full utilisation of the material strength throughout the structure assumes *á priori* that no deterioration takes place while the structure performs its long-term load carrying duty under the influence of the prevailing environment. If local weakening or deterioration occur in a fully optimised structure there are no built-in alternative paths for the loads to follow while maintaining the overall safety level. Such structures, without reserves or redundancies, have no robustness. Therefore they impose very high demands – and costs – on the inspection and maintenance operations (Rostam 1998).

The engineering attractiveness of optimising concrete structures is gaining momentum through the continuous development of still stronger high strength/high performance concretes. This development is valuable and attractive as long as the overall time-dependent performance of the structure in its foreseen environment is taken fully into account at the design stage.

However, the more engineers learn to optimise the structural exploitation of concrete and reinforcement strengths, the greater becomes the demand to master ageing effects and control the deterioration mechanisms that threaten structures.

3.6.3 Adequate dimensions

The designer can ease the construction phase, and thus increase the probability of achieving good quality concrete, by using dimensions of individual members which enhance the placing of reinforcement and the placing and compacting of the concrete. Tight dimensions, small sized members and difficult accessibility during casting, often lead to poor quality concrete in the final structure – despite the use of a good quality initial concrete mix.

Abrupt changes in geometry should be avoided. Locally, concrete quality along edges and at corners usually becomes somewhat inferior compared to the bulk of the concrete and to that along smooth faces. In addition, the concrete along edges and at outgoing corners is exposed to the aggressive environment from two or three sides simultaneously. This is often referred to as the so-called 'corner effect', where cracks due to deterioration first develop along the exposed edges. Hence, edge zones and corners are more prone to premature deterioration than the remaining parts of the exposed faces. Where feasible, rounded edges and corners can improve the robustness of concrete structures (see also Section 3.3.4).

3.6.4 Congested reinforcement

The reinforcement (non-prestressed as well as prestressed) should be sufficiently distributed within the concrete to ensure a good and reliable casting and compaction of the concrete, especially in the outer concrete layer constituting the skin of the concrete structure.

Congestion of the reinforcement may lead to difficult casting conditions as highlighted in Section 1.8.2. Where possible the reinforcement layout should constitute a closed three-dimensional reinforcement cage. When using links or stirrups, these should have a geometric form which complies with a three-dimensional closed cage action. In case of some unforeseen deterioration, which would crack or split the concrete, e.g. bulk frost damage or alkali aggregate reactions, then the three-dimensional reinforcement cages would tend to hold the pieces together. There are cases where concrete members have been cracked into pieces by alkali–aggregate reactions so that they look like a three-dimensional 'puzzle', but have maintained full load-carrying capacity due to the valuable confining effect of the reinforcement cage.

3.6.5 Fibre reinforcement

The beneficial effect of adding a small amount of polypropylene fibre to the concrete mix to reduce the risk of plastic shrinkage cracking has been mentioned. By adding steel fibres both the risk of plastic shrinkage cracking and risks of later drying shrinkage cracking may be reduced. By adding steel fibres to concrete floors, relatively large areas can remain crack-free without expansion joints.

While polypropylene fibres in concrete influence only the performance of the very young (green) concrete (due to their low strength), steel fibres and other high-strength fibres, such as carbon fibres, can have a major structural effect on the finished structure. In cases where concrete without reinforcing bars is considered because of corrosion risks, e.g. for some foundations, retaining walls or seawalls, then steel fibres can be both a durability and a structural enhancement. It should be noted, in particular, that while steel fibres in exposed concrete may corrode, practice has shown that only the outer exposed fibres corrode. Due to the size of the single fibre, the expansive forces of the rust are insufficient to damage good quality concrete. The only disadvantage seems to be a rusty discolouring of the surface but in some circumstances this may be acceptable e.g. in marine structures.

3.7 Aesthetics

The visual appearance of a structure may deteriorate to a level where the structure becomes so unsightly that the users lose confidence in the structure, or for other reasons deem the structure prematurely unfit for use.

The design should consider detailing which increases the self-protection and the robustness of the structure against an aggressive environment. This includes provisions to ensure satisfactory weathering and ageing of exposed surfaces so allowing buildings to grow old gracefully without expensive maintenance. An appropriate selection of structural form should be ensured at an early, conceptual stage of the project.

Unsightly or dirty structures reflect apparent poor quality, low maintenance, reduced safety and serviceability, and low social standard. Such structures contribute to a bad reputation for concrete structures and reflect a similar opinion on the concrete and on the people creating such structures. This has lead to the standing of concrete being much lower than this versatile material deserves.

It is the responsibility of all involved: architects, engineers, contractors and owners, to contribute to an improved reputation of concrete as our most important construction material. A pleasant visual appearance and attractive ageing properties may contribute substantially to an improved reputation (Fig. 1.1).

3.8 Inserts and fixtures

Inserts and fixtures constitute elements which are especially prone to early deterioration. Deterioration often starts from such elements with corrosion followed by cracking and local spalling, then develops into more widespread damage to the structure. The interaction between inserts and the cast-in reinforcement can enhance rapid corrosion, and special precautions must be taken to counteract such deleterious reactions.

When selecting the type of inserts and fixtures, the material composition and the way of embedding these elements in the concrete and e.g. fixing them to the reinforcement, the following shall be considered:

Exposure conditions

- chlorides – no chlorides
- submerged
- splash zone
- atmospheric zone

Contact to the reinforcement

- electrochemical potentials

Material composition

- black steel
- galvanised steel
- stainless steel

There is a major risk of macro-cell corrosion which is inherent with inserts that are in electric contact with the reinforcement and exposed to a corrosive environment at the concrete surface. In such cases a reliable long-term solution would be to use stainless steel for the cast-in part of the insert and then either hot-dip galvanised steel for the member fixed to the insert, or use stainless steel for this part, if no maintenance is foreseen.

3.9 Updating of service life

It has become clear from the previous descriptions that the use of specified concrete properties, or properties determined from separate test specimens, cannot be used for realistic service life predictions. Only the testing of the finished structure and its materials will provide reliable data for such predictions due to the dominant influence of the construction process on the *in situ* quality of the concrete. This will determine if an acceptable service life can be realistically achieved, or whether additional durability-enhancing measures will be needed in the future.

Currently, the codes of practice requirements for durability are given in a prescriptive form. Thus, requirements aimed at the concrete mix (cement type and content, w/c ratio, etc.); at the design parameters (cover, detailing, etc.); and at the construction (compaction, curing, etc.); while being essential to determine the accept/reject criteria for the product, will not serve as a basis for realistic service life calculations.

However, calculations predicting future performance will be valid and useful when based on an assessment of existing structures, provided adequate test methods exist to define the actual material properties in the structure. More work needs to be done on the development of *in situ* test methods, but it is already possible to make very useful and valid service life predictions.

For example, if the carbonation depths are measured in a structure that is 5 years old, the time when carbonation will reach the reinforcement can be predicted with reasonable reliability and any necessary protective procedures can be planned. Similarly, the time to the initiation of chloride-induced corrosion can be realistically estimated using the newest chloride transport models, if the calculation is based on real chloride profiles determined after the structure has been in service for some years (Fig. 3.14).

As a result of these considerations, a design procedure according to strategy B (Section 3.3.1), should contain the following:

a) Models describing the deterioration processes with quantification as far as possible.

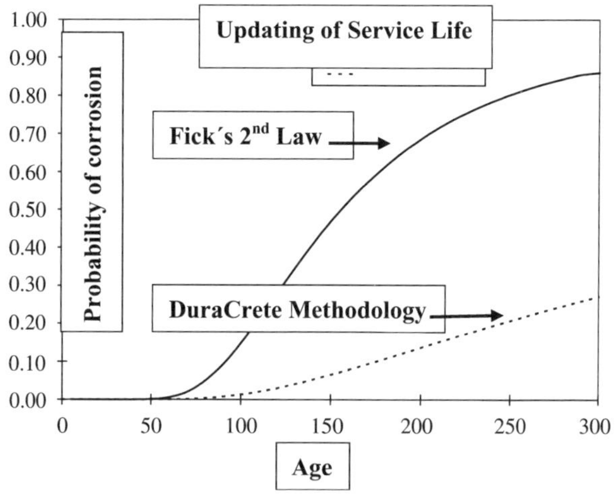

Figure 3.14 *Updating of the service life estimation based on experience and in situ testing of the ageing effect, or continued maturing, of concrete with respect to chloride penetration. The DuraCrete Model is based on an ageing factor of 0.3 (DuraCrete 1999).*

b) Prescriptive requirements to ensure durability, such as maximum w/c ratio, minimum concrete cover (including tolerance), curing procedures (moisture and temperature control), etc.
c) Quality control procedures covering all steps of the material production and the construction processes. Specifically important are interfaces where the responsibility for the quality of the product changes (e.g. handing over the ready-mixed concrete at the site).
d) Definition of a 'reference durability period', e.g. 5 years. After that period relevant durability parameters such as porosity, permeability, diffusivity, carbonation, chloride penetration, etc. can be determined to verify the achieved quality and resistance of the structure under the given environment. Already on completion of construction, surveys of some of the durability parameters such as cover, porosity, permeability and diffusivity could be carried out and entered into the 'birth certificate' (see Section 1.9.2). Procedures equivalent to this are already in normal use in some countries.
e) Using (a) – models, and (d) – the durability parameters measured, the future performance of the structure can be estimated realistically and an update of the achievable service life obtained.

Improved specifications, which include long-term qualities, are needed. However, they will not by themselves improve *in situ* quality without a changed attitude towards on-site production and towards the interaction between design and production. The competence and experience of the workforce is crucial. As stated in Section 3.4.1, the owner must ask for quality and must check the relevant quality of the structure as built. If this were to happen, quality would improve considerably just through the quality checks on durability made by the owner.

3.10 Durability monitoring

Designing for long service life needs some form of verification during the use of the structure. This becomes a natural part of the operation and maintenance activities. The service life design is based on some assumptions regarding deterioration mechanisms related to the aggressiveness of the environment. Due to the uncertainties between the prior assumptions made at the design stage and the reality of the service conditions of the finished structure, a need arises with time to verify – or adjust – the initial assumptions and revise the service life forecast accordingly.

This need for adjustment, and the benefits of obtaining more and more reliable forecasts of the service life to be expected, leads naturally to a need to monitor non-destructively the durability performance of the structure in service (see Section 1.10.4).

3.11 Recent advances in service life design

In recent years, extensive developments in modelling deterioration mechanisms and in introducing reliability theories to handle the unavoidable uncertainties associated with the problems of durability have brought the possibilities of performing service life designs in practice that are close to what happens in practice. It is estimated that within five to ten years, reliability-based overall service life designs may well have taken over from today's general design procedures for concrete structures (CEB 1997b; DuraCrete 1999; Rostam 2000b).

3.11.1 Probabilistic performance-based service life design

The theories of probability and reliability in structural design have been developed and have matured remarkably during the past five years. These theories have been transformed from the level of research and development to being directly applicable and operational in practical engineering design. The methodology has been recognised internationally and used for many decades as the basis for structural safety design through the well-known load-and-resistance-factor design. However, the factors governing the durability and performance of structures throughout their service life have only recently been developed in similar ways (Siemes & Rostam 1996; DuraCrete 1999; Siemes et al. 1998; Høj & Rostam 1999).

This has allowed the treatment of transport and deterioration mechanisms to be modelled on a probabilistic level and introduced into the general design of structures. Thus, design for safety and for durability can be performed using similar procedures. This raises the awareness of owners who are now forced to take decisions which include the long-term performance and consequences regarding maintenance and costs.

3.11.2 Modelling

Throughout this chapter the need to model environmental aggressivity, the penetration of aggressive substance into concrete, and the corresponding deterioration mechanisms has been emphasised. The indication of specific models is beyond the scope of the text. The reader is therefore referred to Helland et al. (1995), HETEK 6, 7 and 8 (1997), Engelund & Sørensen (1998), CEB (1997) and DuraCrete (1998).

The importance of such modelling follows from the fact, that no prediction of future developments can be made without a model. For practical purposes such modelling must be a rational mathematical model to be effective. In this respect it must also be noted, that any such model, no matter how crude, is better than pure guesswork. With a lack of actual data, educated guesswork, in the form of experience-based engineering judgement, is valuable but should be presented in a mathematical format to be effective.

Such a probabilistic reliability-based treatment of the models and parameters governing the durability performance and service life of concrete structures exposed to different types of environmental aggressivity has been developed in a European research project called DuraCrete. The results are presented in a series of reports, with DuraCrete (1999) presenting the general guidelines for a probabilistic performance-based durability design of concrete structures.

3.11.3 Deterministic versus probabilistic service life design

The merits of the probabilistic approach to durability design are illustrated by the following example of a marine structure. Two different environments are considered, representing yearly average temperatures of 10°C (Northern Europe) and 30°C (Middle East), respectively. The design requirement is for a 50 year service life. For illustrative purposes the service life is defined as the length of the initiation period,

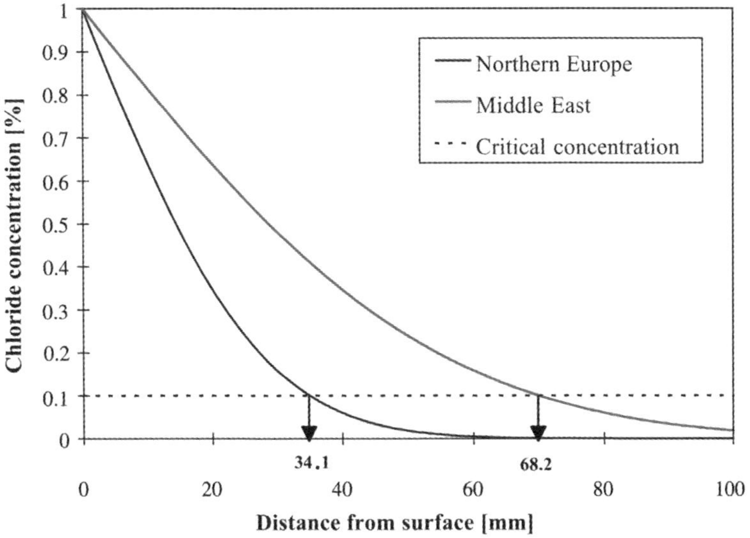

Figure 3.15 ***Deterministic approach.*** *Required concrete cover to ensure 50 years service life and assuming a chloride threshold value of 0.1% by weight of concrete.*

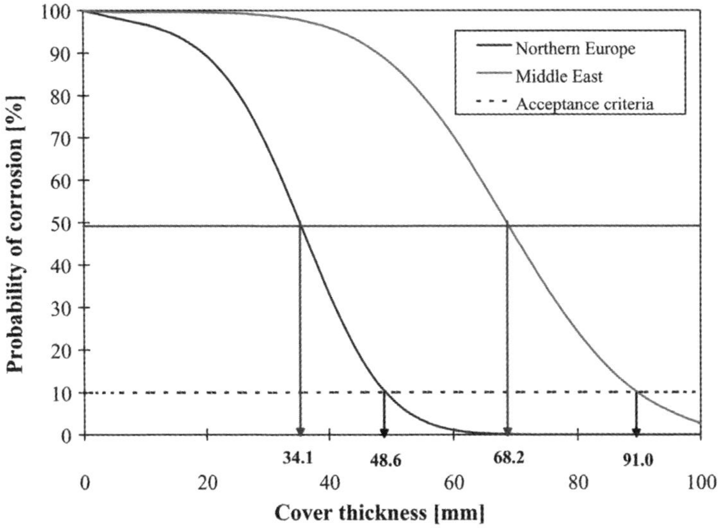

Figure 3.16 ***Probabilistic approach***. *The deterministic approach provides only 50% probability of avoiding corrosion at the age of 50 years. Accepting 10% probability of having corrosion initiated after 50 years results in considerably larger covers.*

i.e. the time until corrosion initiation due to chloride ingress (Rostam 1999; Engelund & Sørensen 1998).

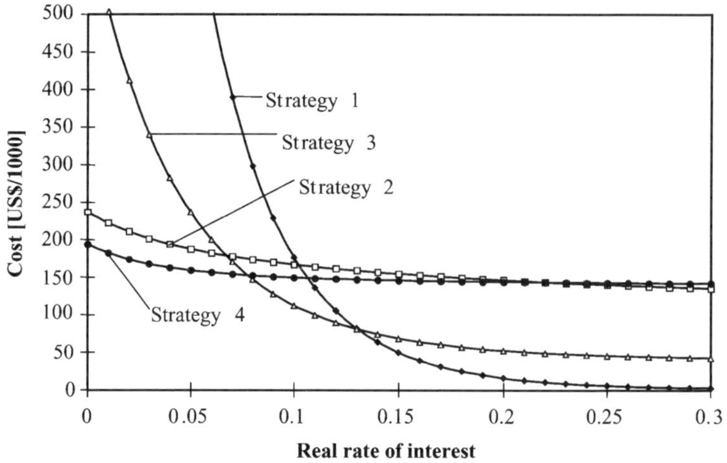

Figure 3.17 *Life cycle costs for the four different design strategies.*

Fig. 3.15 depicts the required concrete covers in each of the two environments, based on a traditional deterministic approach. Fig. 3.16 highlights the fact that the deterministic approach only provides a 50% probability of achieving the required 50 years corrosion-free service life. This fact is often overlooked in the usual design for durability. If, say, only a 10% risk of having corrosion initiated before 50 years is considered acceptable, then a much larger cover is required (Fig. 3.16).

3.11.4 Life cycle cost optimisation

The deterministic approach described earlier is based on the mean values of the governing parameters. In the probabilistic approach, the mean values and their known or assumed standard deviations are used together with the relevant distribution functions. This latter approach makes it possible not only to relate cover thickness to the probability of corrosion but also to quantify the consequences of choosing different levels of risks of corrosion. These consequences relate not only to concrete quality and cover thickness, but more importantly to the economic consequences.

This can be illustrated by the following simple demonstration of the life cycle costing of constructing a 300 m × 10 m concrete pier on concrete piles along the North Atlantic coast.

The following four different design strategies to ensure 50 years service life, i.e. time to corrosion initiation, are considered:

Strategy 1: A traditional design assuming ordinary concrete, usual construction methods and repairs only when extensive corrosion damage leading to cracking and spalling has been observed through regular inspections.

Strategy 2: A design using very high-performance concrete, large concrete cover, optimal execution and curing procedures, intensive quality assurance procedures and quality control on site. Regular inspections are made but only minor local repairs are foreseen.

Strategy 3: A traditional structural design and using ordinary concrete but providing preventive cathodic protection as sacrificial anodes in the water to protect the submerged and tidal zones, and prepared for cathodic protection (CP) by ensuring electric continuity of the reinforcement and providing cast-in anode bands in the parts above water. The external components of the impressed current system are only provided and energised when local corrosion initiation is observed.

Strategy 4: A traditional design assuming ordinary concrete and usual construction methods. 50% of the reinforcement is replaced with stainless steel reinforcement. Regular inspections are performed.

Realistic costs have been chosen for all elements of the above strategies. A probabilistic calculation of the service life costs has been made for each strategy, relating these costs to the real rate of interest and using the well-known discounting method. To illustrate the influence of the real rate of interest, values from zero up to 30% have been analysed. The results are presented in Fig. 3.17 which illustrates the two major consequences of choosing between the different design strategies.

The first observation is, that the most usual approach, Strategy 1 will incur very high accumulated life cycle costs when using relatively low rates of interest. With high interest rates it can be seen how it can become economically advantageous to postpone costs as long as possible. Thus Strategy 1 leads to high future maintenance costs. The same observations are seen to govern Strategy 3.

3.11.5 Reliability updating

A very valuable consequence of adopting the probabilistic method is that the uncertainties to be expected are taken into account at the initial design stage, where the concrete and the construction and maintenance conditions are not known. When these important, but initially assumed values are known, or can be determined through testing, the tools are already available to update the service life expectations, and in due time take necessary precautions. Hence, the service life and the economic consequences of any decisions taken (savings or additional costs) can be readily updated when inspections are made and additional information becomes available.

3.12 Examples from practice

3.12.1 Great Belt Link, Denmark

The Great Belt Link comprises:

- a twin-bored 7.9 km long railway tunnel (the East Tunnel) (Storebælt 1997);
- a 6.6 km long, low level combined road and railway concrete bridge (the West Bridge) (Storebælt 1998–1);
- a 6.8 km long, high-level road bridge (the East Bridge) with concrete pylons rising 254 m above sea level carrying a suspension span of 1624 m (Storebælt 1998–2).

Figure 3.18 *The Great Belt East Bridge. Main free span of 1624 m. 100 year service life design (Søren Madsen).*

The West and the East Bridges are Europe's two longest bridges and the main span of the East Bridge is the second longest span in the world (Fig. 3.18).

The Great Belt Link has been designed with the specific service life requirement of 100 years. This meant that the concrete structures had to be designed, constructed and operated under rather severe environmental conditions without requiring unforeseen high costs for maintenance and repair.

A multistage protection approach providing successive barriers against incoming chlorides and sulphates was set up to ensure the 100 year service life requirements in the most aggressive environments (Rostam 1993). These are the splash zones of the pier shafts, the tunnel lining segments and the edge beams of the road girders. The most important factor in this multi-stage approach was the use of high-quality concrete. A three-powder mix with Portland cement, flyash and microsilica was used (Fig. 3.19; Storebælt 1999).

The main provisions to avoid deterioration were the use of non-reactive aggregates (very low reactivity) and pozzolanic additions (microsilica and fly ash) to reduce the risk of alkali–silica reactions to negligible levels. Air entrainment eliminates the risks of frost damage. Highly dense and impermeable concrete is achieved by a special mix chosen using the three-powder mix and a very low water/powder ratio of a maximum of 0.35. This would give a very slow ingress of chlorides. But if – or when – chlorides in sufficient quantities to depassivate the reinforcement reach the level of the steel, the high electric resistivity of the concrete will only allow a very limited and slow rate of corrosion to develop, if any at all. Finally, special efforts were made to ensure a high degree of electric continuity of the reinforcement thus facilitating the possibility of installing cathodic protection some time in the future, if the need arises.

Figure 3.19 *Proportions of a high-performance concrete mix. The water/powder ratio is 0.33. In addition to course and fine aggregates the mix contains three powders (Portland cement, flyash and microsilica), plasticiser, superplasticiser, air entraining agent – and a little water (Photo: Idorn Consult).*

The East Tunnel is the most critical structure from a durability point of view. The reason being that if serious deterioration develops, it is extremely difficult to make repairs. This is due not only to restricted access in the tunnel, but more so because the segments of the tunnel lining are critical elements which cannot be replaced without very substantial sealing operations. This would have to be in the form of grouting or freezing of the soil surrounding the damaged zone, or establishing a temporary artificial island and shaft down to the tunnel from above sea level.

The multi-stage approach to protect the tunnel lining against incoming chlorides and sulphates is provided by the following consecutive barriers:

1. Annular grouting of the void between the lining and the soil, formulated so it has a high chemical and physical binding capacity towards chloride and sulphate ions.
2. Dense, low permeability and low diffusion concrete as described above.
3. Fully welded three-dimensional reinforcement cages cleaned, heated and dipped in fluidised epoxy powder providing a very reliable epoxy coating without having to cut and bend bars after coating, as is usually done with epoxy coating.
4. The possibility of installing cathodic protection some time in the future, if the other provisions prove insufficient. The fluidised-bed dipping technique for epoxy coating allows cathodic protection, if needed. This is not the case with the traditional coating of single bars being electrically isolated from each other when tied together and cast into the concrete.

As a final measure, a number of rings through the tunnels are provided with corrosion sensors (Fig. 1.23), in order to provide an early warning of chloride ingress, knowing that the chlorides also have to overcome the epoxy coating before corrosion may be initiated.

The concrete structures of the West and East Bridges are of a more conventional design and are more easily accessible for maintenance and repair. Thus simpler protective measures were selected.

The concrete cover for the permanently submerged caissons is 50 mm. The cover in the splash zones is 75 mm. For the superstructure of the West Bridge the outer concrete cover is 50 mm and inside the box girders the cover is 35 mm. All these values are absolute minimum values and due to the high level of quality control performed by the contractor the addition for tolerance was throughout +5 mm.

In addition efforts were made during binding the reinforcement to ensure electrical continuity of the reinforcement. This was to allow the easy installation of cathodic protection if necessary at some time in the future.

A special protective measure was introduced for the caissons of the West Bridge. Slipforming of the caissons revealed extensive micro-cracking of the concrete. Investigations revealed this to be systematic for most slipformed structures but to be a particularly potential risk when using high-performance concrete. To compensate for this slightly less impermeable concrete and to assist in early passivation of the reinforcement it was decided to install cathodic protection on all caissons using simple sacrificial aluminium anodes placed on the bottom slab of the caissons (Fig. 1.20).

Finally, a possible alternative protective measure considered during the design process and part of the tender design is mentioned. The importance of the concrete cover to ensure long-term durability in an aggressive environment such as a marine structure has been clearly recognised. As an alternative it was considered including a very large concrete cover for the piers of the Great Belt Link, of approximately 100 mm. However, this could result in some single cracks with excessive large crack widths in the large unreinforced cover. Therefore, a stainless steel reinforcement mesh with a cover of 35 mm was specified.

Due to the novelty of such an approach and the contractors foreseeing difficulties in casting and compacting the concrete with so much reinforcement, the costs indicated at the tender made this solution prohibitively expensive.

3.12.2 Western Scheldt Tunnel, the Netherlands

The first practical application of the DuraCrete design approach is the Western Scheldt Tunnel in the Netherlands. The Ministry of Transportation, the owner, required a service life for the bored reinforced concrete tunnel of at least 100 years. The tunnel consists of two tubes with an external diameter of 11 m and a length of 6.5 km. The tunnel will cross the Western Scheldt in the south-west part of the Netherlands.

The bored tunnel has to be designed against variously induced rebar corrosion. The bored tunnel is located in chloride-contaminated soil and therefore is under chloride attack. The walls inside of the tunnel are not only ventilated, and therefore under carbonation influence, but are also exposed to a road traffic induced chloride contaminated salt fog and splash environment. Eventually, leaking joints will also lead to a chloride attack within the joints as well as partly, especially in deep points of the tunnel, on the inside surfaces.

The decision about the thickness of concrete cover within the joints on the Western Scheldt Tunnel project provides a useful case study example of the application of an effective durability design method. Due to structural design constraints the concrete cover within the joints must be minimised. Therefore the joint with its restricted cover was identified as the most critical detail with regard to durability. In the case of permanent leakage, the concrete is subjected to chlorides coming from the chloride-contaminated soil. Constantly humid conditions are assumed. Carbonation can be neglected. The identified deterioration mechanism is in this case the chloride-induced corrosion. This meant that it had to be established which material's resistance against chloride penetration in combination with the level of concrete cover would be necessary to fulfil the owner's requirements.

The different steps of the durability design to determine the necessary concrete cover for a serviceability limit state defined as the onset of corrosion: $T_{\text{Service Life}} = 100$ years with a minimum reliability index of $\beta_{0,SLS} \geq 1.50 - 1.80$, were as follows [Gehlen and Schiessl (1999)]:

Selection of the deterioration model. In the case of chloride-induced reinforcement corrosion, deterioration models are required, describing the initiation period as well as the propagation period. For this example, only the initiation period has been taken into consideration.

Input data for the probabilistic analysis. The type of data needed to perform such a probabilistic service life design of these tunnel segments are listed below:

- cover;
- chloride migration coefficient;
- critical chloride content;
- age factor;
- environment factor;
- execution factor;
- surface chloride concentration factor;
- reference time.

With the results of the probabilistic durability design it was then possible to calculate whether the stochastic variable representing the cover applied in the design was sufficient to prove a serviceability limit state related minimum service life of $T = 100$ years. The result of the calculations proved that, with the finally chosen concrete mix and the planned concrete cover, a durable structure could be obtained with a reliability index of $\beta_{SLS} = 1.50$ (probability of failure = 0.068) as far as chloride induced steel corrosion was concerned. This was in agreement with the required design criteria, a minimum reliability index of $\beta_{0,SLS} = 1.50$.

3.12.3 Experts' recommendations for durability design

In 1998 an international specialist workshop on 'Durable Marine Structures' was organised in Norway. The participation was limited to just 12 invited European specialists, six of them from Norway. Many of these specialists were also very well acquainted with the problems encountered in hot, humid and saline environments around the world and were thus well known and acknowledged experts not only in

Europe but also in these regions. The outcome of the workshop was reported in Markeset & Rostam (1999).

At the end of this workshop each participant was requested to summarise in one single statement their main message towards achieving the objectives of the workshop. These statements have been grouped into the categories targeted by the individual statement. Although these statements focus specifically on marine structures, their messages are of a much more general nature. Therefore, these statements provide a valuable summary of how to design concrete structures for durability.

The 'Twelve Commandments' of durability

Targeting the owners:

1. Convince the management that design and construction of long-term durable structures is the most economical way to act (Finn Fluge, Norway).
2. The unprofessional owner must be helped to set up simple requirements or guidelines identifying the service life assumptions for the structure. There is also a need for revision of the traditional design requirements and specifications to take into account the life expectancy of the structure (Rikard Karlstrøm, Norway).
3. The owner cannot expect to obtain a durable structure when the lowest bid is selected. As the owner is not an expert in these matters he has to be advised or educated into making the right decisions (Gro Markeset, Norway).
4. The true intention to design and construct durable concrete structures in marine environments is to be initiated by the owners and be formulated through agreed contracts between the owner, the consultant and the contractor. These requirements, and the means to fulfil them, must be specified in a manner that leaves some creativeness to the contractor and provides a fair commercial approach. This would be a very good incentive (John Regtop, The Netherlands).
5. Durability-related quality needs must be enforced by the owner. The owner needs to **define** the required quality (i.e. service life), to **check** the quality received, to **pay** for the quality requested (i.e. not go for the lowest bid), and monitor the future performance of the structure. Standards cannot ensure quality – not even good standards. They can only serve as helpful tools. The contractor needs incentives to produce quality. The most effective incentives are reflected through the owner's clearly formulated requirements and the corresponding remuneration (Peter Schiessl, Germany).

Targeting the designers and consultants

6. Designers must be provided with the tools, procedures and data required to enable them to consider all of the options for achieving a specified design life and to select the most cost effective option in relation to the owner's needs (Phil Bamforth, UK).
7. Make concrete more tolerant regarding chloride-initiated corrosion. Apply chloride-resistant cementitious binders with proven on-site performance. Provide cover of at least 50 mm, preferably more (Jan Bijen, The Netherlands).

8. When designing concrete marine structures the consultants should, in co-operation with the owner, develop an operation and maintenance program for the specific structure. The system must be able to detect damage, chloride penetration and ongoing corrosion at an early stage, so that maintenance and repair can be carried out at a time that gives the owner the lowest annual maintenance cost. Annual maintenance costs for a marine structure should be calculated on the basis of the required service life of the structure, the assumed maintenance costs and the real rate of interest for the required service life (Trygve Isaksen, Norway).

Targeting the researchers

9. A consistent model for the ingress of chlorides in concrete should be agreed upon. This should include test methods and interpretation of the results. Such a system is a precondition for the later development of performance-based requirements and service life predictions (Steinar Helland, Norway).
10. Take into account practical measures for reducing or eliminating the corrosion of reinforcing steel in marine concrete structures and in other very aggressive environments by e.g. hot-dip galvanising and epoxy coating, or by austenitic stainless steel reinforcement, respectively (Audun Hofsøy, Norway).
11. When designing for a required service life the scientific basis for understanding the deterioration mechanisms must be kept in mind. In this respect it should be noted that submerged parts of marine structures are not fully saturated (Lars-Olof Nilsson, Sweden).

Targeting the contractors

12. Concrete is the only important construction material where the real quality and performance in the finished structure cannot be determined at the design stage but is established during construction. Hence, contractors should be encouraged to study and understand the interaction between their structures and the environment, and understand the deterioration mechanisms so that they can concentrate on providing high quality work at the right places and at the right time during the construction process (Steen Rostam, Denmark).

References

Bamforth, P. (1998) 'Double Standards in Design', *Concrete*, December 1998.

CEB–RILEM (1984) 'Durability of Concrete Structures' *Workshop Report*, 18–20 May 1983, Copenhagen, (Ed.) Steen Rostam.

CEB (1997a) *Design of fastenings in concrete; Design Guide – Parts 1 to 3*, CEB Bulletin 233, January 1997

CEB (1997b) *New Approach to Durability Design. An example for carbonation induced corrosion*, Bulletin 138, May 1997.

CEB (1994) *Fastenings to concrete and masonry structures*, CEB Bulletin d'Information No. 216, July 1994.

CEB (1992) *Durable Concrete Structures. Design Guide*, London: Thomas Telford Services Ltd. (First draft published as CEB Bulletin d'Information No. 166, 1985.)

CEB–FIP (1992) *Model Code 1990*, London: Thomas Telford Services Ltd.

CEN (1999) Draft prEN 206–1/24. *Concrete – Part 1: Specification, performance, production and conformity*. April 1999.

DuraCrete (1999) General Guidelines for Durability Design and Redesign, The European Union – Brite EuRam III, Project No. BE95–1347, Probabilistic Performance based Durability Design of Concrete Structures, Report No. T7–01–l, 1999.

DuraCrete (1998) Modelling of Degradation, The European Union – Brite EuRam III, Project No. BE95–1347, Probabilistic Performance based Durability Design of Concrete Structures, Report No. 4–5, 1998.

Edvardsen, C.K. (1996) Wasserdurchlässigkeit und Selbstheilung von Trennrissen in Beton. *Deutscher Ausschss für Stahlbeton Heft 455*. Beuth Verlag GmbH, Berlin, 1996.

Edvardsen, C.K. (1999) 'Water permeability and autogenous healing of cracks in concrete', *ACI Materials Journal*, 1999.

Engelund, S. & Sørensen, J.D. (1998) 'A probabilistic model for chloride-ingress and initiation of corrosion in reinforced concrete structures', *Structural Safety*, Elsevier 20, 1998.

Fagerlund, G. (1979) 'Service life of structures', General Report, Session 2.3, *Proceedings, RILEM Symposium on Quality Control of Structures, June 1979, Stockholm, Sweden.*

FIB (1999) *Structural Concrete. Textbook on Behaviour, Design and Performance*. Updated knowledge of the CEB/FIP Model Code 1990. Manual – Textbook Volume 3. Steen Rostam: Chapter 5: 'Durability', and Chapter 8: 'Assessment, Maintenance and Repair'. fédéretion internationale du béton, 1999.

Gehlen, Ch. & Schiessl, P. (1999) 'Probability-based durability design for the Western Scheldt Tunnel', *Structural Concrete, Journal of the **fib***, Vol. P1, No. 2, June 1999.

Helland, S. et al. (1995) 'Service life prediction of marine structures', *Proceedings, ACI Fall Convention*, 07.11.95.

HETEK 6 (1997) 'The effect of the w/c ratio on chloride transport into concrete, Immersion, migration and resistivity tests', *Report No. 54, 1997, Road Directorate Denmark, Ministry of Transport.*

HETEK 7 (1997) 'Chloride penetration into concrete, Relevant test methods', *Report No. 94, 1997, Road Directorate Denmark, Ministry of Transport.*

HETEK 8 (1997) 'Chloride penetration into concrete, Manual', *Report No. 123, 1997, Road Directorate Denmark, Ministry of Transport.*

Høj, N.P. & Rostam, S. (1999) 'Optimisation of concrete for durability and accident resistance', *Proceedings, IABSE Symposium "Structures for the Future – the Search for Quality", 25–27 August 1999, Rio de Janeiro, Brazil.*

Jessen, J.J. (1983) 'Durability of concrete structures – The building process and the period of use', CEB–RILEM Workshop "Durability of concrete structures", 18–20 May 1983, Copenhagen, pp. 137–143.

Markeset, G. & Rostam, S. (1999) International Specialist Workshop on Durable Marine Structures, *Workshop Report*, Norwegian Defence Construction Service.

Markeset, G., Rostam, S. & Klinghoffer, O. (2006) 'Guide for the use of stainless steel reinforcement in concrete structures', *Scandinavian Project sponsored by the Danish, Swedish and Norwegian Road Authorities, the Norwegian Defence Estates Agency, and by Nordic Innovation* (in print).

Rostam, S. & Schiessl, P. (1994) 'Service life design in practice – today and tomorrow', *Proceedings of the International Conference 1994: Concrete across borders, Odense, Denmark.*

Rostam, S. (2005a) 'Concrete structures are facing a shift in design paradigm', *Keynote Paper at the New Zealand Concrete Industry Conference* 22–24 *September 2005, Auckland.*

Rostam, S. (2005b) 'Durability and the Impact of the Execution Process on the Useful Service Life of Concrete Structures', *Proceedings of the New Zealand Concrete Industry Conference* 22–24 *September 2005, Auckland.*

Rostam, S. (2005c) 'Design and construction of segmental concrete bridges for service life of 100 to 150 years', *Keynote Paper at the American Segmental Bridge Institute (ASBI) November 2005 Convention, Washington, DC.*

Rostam, S. (2004) 'Service life of concrete structures – Challenging the steel reinforcement industry. Shall concrete remain the dominating means of corrosion prevention?' ANIFER, *International Calavera Award paper 2004.*

Rostam, S. (2003) 'Quantitative service life design methods for new and repaired concrete structures', *Proceedings 7th International Conference on Concrete in Hot and Aggressive Environments,* 13–15 October 2003, *Bahrain,* Vol. 1, Keynote lecture, pp. 59–89.

Rostam, S. (2000a) 'Service life design of concrete structures – an experience-based discipline becoming scientific', *Papers in Structural Engineering and Materials – A Centenary Celebration, Department of Structural Engineering and Materials, Technical University of Denmark*, 2000.

Rostam, S. (2000b) 'Service life performance of concrete structures – reliable design and maintenance for the future', *Proceedings, Sixth International Conference on "Deterioration and Repair of Reinforced Concrete in the Arabian Gulf"*, 20–22 *November 2000, Bahrain,* pp. 9–32.

Rostam, S. (1999) 'Performance-based design of structures for the future" *Proceedings, IABSE Symposium "Structures for the Future – The Search for Quality"*, Rio de Janeiro, Brazil, 1999.

Rostam, S. (1998) *Aesthetics and Service Life Performance of Concrete Bridges*. Bouwkroniek, special edition, Belgian Concrete Day – 22 October 1998.

Rostam, S. (1996) 'High performance concrete cover – why it is needed, and how to achieve it in practice', Reprinted from *Construction and Building Materials*, Vol. 10, No. 5. Elsevier Science Ltd.

Rostam, S. (1996) 'Service life design for the next century', *Proceedings of the International Workshop "Rational Design of Concrete Structures under Severe Conditions"*, Ed. K. Sakai London: E. & F.N. Spon.

Rostam, S. (1993) 'Service life design – the European approach', *ACI Concrete International*, Vol.15, No.7, July 1993, pp. 24–32.

Rostam, S. (1991) 'Philosophy of assessment and repair of concrete structures, and the feedback into new designs', *Regional Conference on Damage Assessment, Repair Techniques and Strategies for Reinforced Concrete held in Bahrain 7–9 December*, 1991.

Schiessl, P. (1976) Zur Frage der zulassigen Rissbreite und der erforderlichen Betondeckung im Stahlbetonbau unter besondere Berücksichtigung der Karbonatisierung des betons. Deutcher Ausschuss für Stahlbeton Heft No. 255, 1976.

Siemes, A.J.M. & Rostam, S. (1996) 'Durability, safety and serviceability. A performance based design', IABSE Colloquium "Basis of Design and Actions on Structures", 27–29 March 1996, Delft, The Netherlands, TNO–report 96–BT–R0437–001.

Siemes, A.J.M. & Vrouwenvelder, A.C.W.M. (1998) 'Durability of buildings; a reliability analysis', *HERON*, Vol. 30. No. 3, Delft.

Siemes, T., Polder, R. & de Vreis, H. (1998) 'Design of concrete structures for durability – an example', *HERON*, Vol. 43, 1998, No. 4. TNO, the Netherlands.

Storebælt (1999) *Concrete, The Storebælt Publications*. Published by A/S Storebæltsforbindelsen.

Storebælt (1998–2) *East Bridge, the Storebælt Publications*. Published by A/S Storebæltsforbindelsen.

Storebælt (1998–1) *West Bridge, The Storebælt Publications*. Published by A/S Storebæltsforbindelsen.

Storebælt (1997) *East Tunnel, The Storebælt Publications*. Published by A/S Storebæltsforbindelsen.

Tuutti, K (1982) *Corrosion of Steel in Concrete*, Cement och Betong Institutet, Stockholm.

4 The Durability of Concrete Structures in the Marine Environment

Odd E Gjørv

4.1 Introduction

Since John Smeaton (1791) published his experiences from the building of the Eddystone Lighthouse in 1791, all the published literature on durability of concrete structures in the marine environment makes a comprehensive and fascinating chapter in the long history of concrete technology. During the last 150 years, a number of professionals, committees and national authorities have been engaged on the problem. Numerous papers have been presented to international conferences such as the International Association for Testing Materials in Copenhagen (1909), New York (1912) and Amsterdam (1927); the Permanent International Association of Navigation Congresses (PIANC) in London (1923), Cairo (1926), Venice (1931) and Lisbon (1949); the International Union of Testing and Research Laboratories for Materials and Structures (RILEM) in Prague in 1961 and 1969; the RILEM–PIANC in Palermo in 1965; and the Féderation Internationale de la Précontrainte (FIP) in Tibilisi in 1972. By 1923, Atwood and Johnson (1924) had assembled a list of approximately three thousand references, and today, the durability of concrete structures in the marine environment continues to be the subject of continuing research and international conferences (Malhotra 1980, 1988, 1996; Metha 1996; Sakai, Banthia, Gjørv 1995, 1998, 2004).

Although deteriorating processes such as chemical seawater attack, freezing and thawing, and expansive alkali reactions also present some problems, it is not the disintegration of the concrete itself, but rather electrochemical corrosion of the embedded steel which poses the most critical and greatest threat to the durability and safety of concrete structures in the marine environment (Gjørv 1975, 1989). Already in 1917, this was pointed out by Wig and Ferguson (1917) after a comprehensive survey of concrete structures in US waters.

In addition to conventional structures such as bridges and harbour structures, reinforced and prestressed concrete has been applied to an increasing number of very important ocean structures and vessels. This development was foreseen as early as 1972, when a technological forecast on the use of concrete up to the year 2000 was published by an ACI Committee (American Concrete Institute 1972). This report also pointed out the great potential for the utilisation of concrete in installations related to offshore oil and gas exploration.

In Norway where most of the offshore concrete construction has taken place, long traditions exist in the utilisation of concrete in the marine environment. In 1910, the two Norwegian engineers Gundersen and Hoff obtained their patent for underwater placing of concrete through a submerged pipe and called it the 'tremie method'. This led to a new generation of piers and harbour structures along the rocky shore of the Norwegian coastline (Gjørv 1968). A large number of open structures were constructed, typically consisting of a reinforced concrete deck on top of slender, reinforced concrete pillars cast under water (Fig. 4.1). This is still the most common type of construction work in Norwegian harbours.

Due to the broken coastline, many fjords and numerous inhabited islands, Norway also has a long tradition of using concrete in coastal bridges. The latest trend even uses

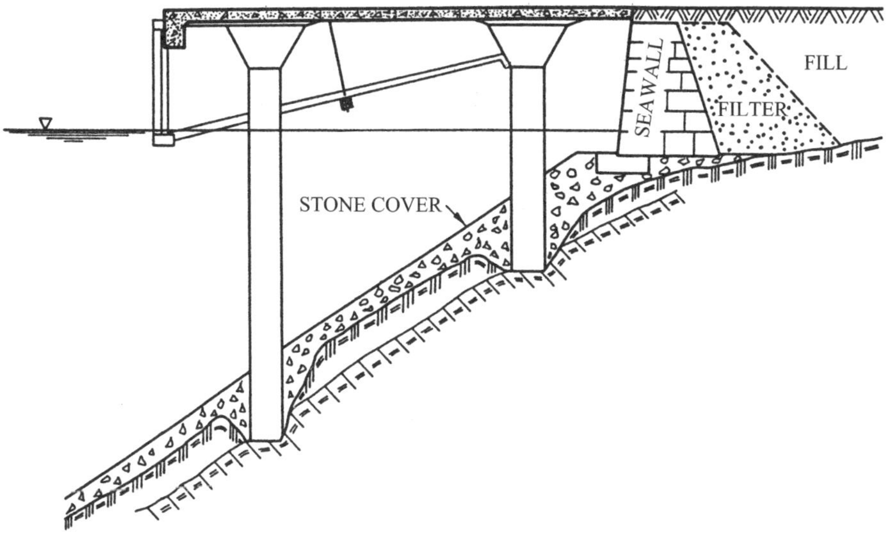

Figure 4.1 *A section through an open concrete wharf with a reinforced concrete deck on slender, reinforced concrete pillars cast under water (Gjørv 1968).*

Figure 4.2 *A floating concrete bridge in a Norwegian fjord.*

floating bridges for strait crossings, while submerged tunnels are also being considered (Fig. 4.2) ('Strait Crossings' 1986, 1990, 1994).

The rapid development which took place on utilisation of concrete in offshore installations is well known. From 1973 to 1995, 28 major concrete structures containing more than 2.5 million cubic metres of concrete were installed in the North Sea, most of which were produced in Norway (Fig. 4.3). The largest of them, the Troll

Figure 4.3 *The Ekofisk Tank (1973) on its way from the construction site in Stavanger.*

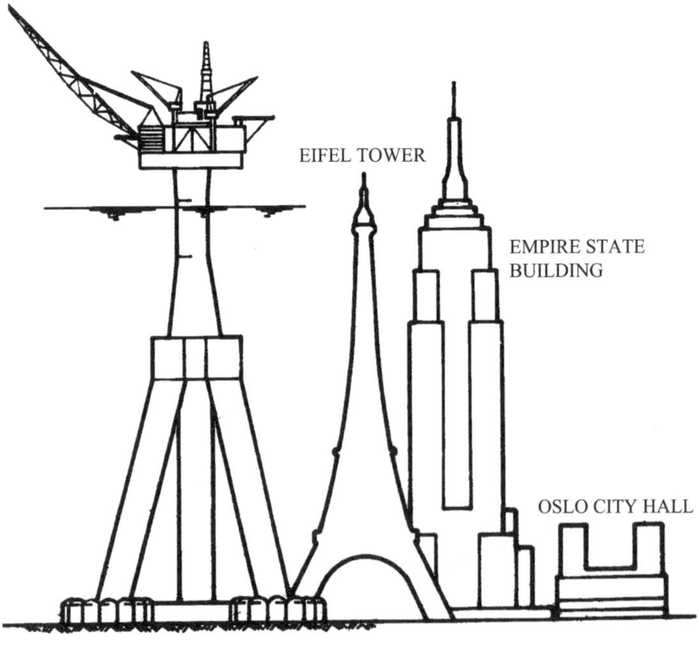

Figure 4.4 *An indication of the size of the Troll Platform in the North Sea (1995).*

Platform completed in 1995 and installed at a water depth of more than 300 m, is taller than the Empire State Building (Fig. 4.4).

The remainder of this chapter discusses the current experience on the long-term performance and durability of concrete structures in the marine environment, mostly in Norwegian waters.

4.2 Field performance

4.2.1 Harbour structures

From 1962 to 1968, a comprehensive research program on the long-term performance of concrete harbour structures along the Norwegian coastline was carried out (Gjørv 1968, 1970). The majority of the 219 structures investigated were of the open type with a concrete deck on top of tremie-cast, slender concrete pillars (Fig. 4.5). This field investigation included more than 190 000 m^2 of decks on more than 5000 pillars with a total length of approximately 53 km, of which more than 20 km was under water.

The research program revealed that the overall condition of the structures was very good. Even after a service period of 50 to 60 years, the majority of the structures showed a high ability to withstand the combined effect of the most severe marine exposure and heavy structural loads (Figs 4.6 and 4.7). The observations showed no particular trend for development of damage under water. Within the tidal zone, only 35–40 year old structures exhibited some pillars with cross-sectional reductions of more than 20%. Also, only 35–40 year old structures (1925–30) had severely weakened deck beams due to steel corrosion, while on the whole, both the deck slabs and the seawalls behaved much better.

However, almost all of the concrete structures investigated had steel corrosion to a varying extent or had been repaired due to steel corrosion. Also, the first visible sign of steel corrosion typically occurred after a service period of 5–10 years. It was primarily those parts of the structures that were the most exposed to intermittent

Figure 4.5 *An industrial wharf in a Norwegian harbour.*

Figure 4.6 *Environmental exposure of a Norwegian harbour structure. (Courtesy of B. Skarbøvik).*

Figure 4.7 *A 50-year-old concrete harbour structure carrying aluminium bars representing a deck load of up to six times the design load (Gjørv 1968).*

wetting and drying such as deck beams and girders, which were the most vulnerable to steel corrosion (Fig. 4.8).

For the oldest structures, the requirement for compressive strength in the concrete deck ranged from 25 to 30 MPa, but successively, the specified concrete strength was increased to 35 MPa. The required concrete cover for the steel in deck slabs and deck

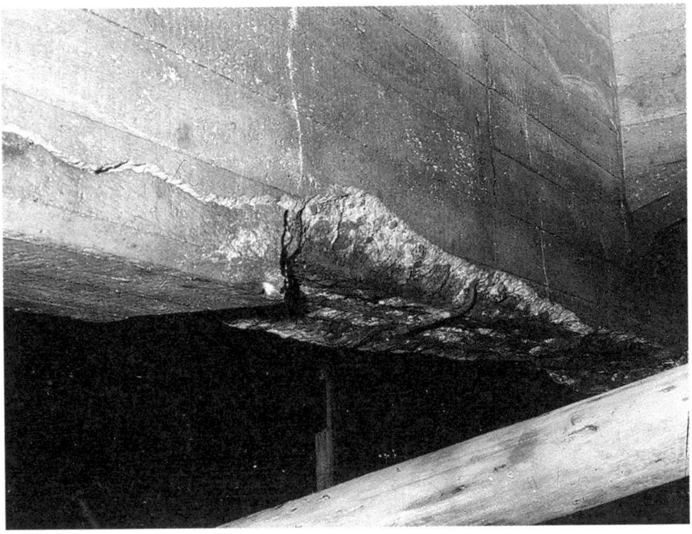

Figure 4.8 *Steel corrosion in a deck beam.*

beams was 25 and 40 mm, respectively. For the pillars, a somewhat higher concrete strength was applied, as the tremie concrete required a minimum cement content of 400 kg/m^3. The cover thickness in the pillars was typically 70 mm.

Before 1930, experience indicated that the deck slabs performed much better than the deck beams. This was assumed to be due to better casting and compaction of the concrete in the deck slabs compared to that in the deep and narrow beams and girders. The practical consequence of this was drawn in 1932, when the first flat type of deck was introduced in Norwegian harbour construction. From then on, a number of structures with flat decks were constructed. Since such a design was often more expensive, the slab and beam type of deck was gradually introduced again. It was assumed that if only the beams were made shallower and wider, it would be equally easy to cast and compact these deck elements. After some time, however, even the shallower and wider beams showed early corrosion, while the flat type of deck still performed much better.

What was not known in the earlier days was that a concrete structure exposed to a chloride-containing environment would develop a complex system of galvanic cell activities. In such a system, the more exposed parts of the structure such as deck beams would absorb chlorides much more easily and develop anodic areas, while the less exposed parts such as the slab sections in between would act as catchment areas for oxygen and hence form cathodic areas. As a consequence, beams and girders would always be more vulnerable to steel corrosion than the rest of the deck.

The extensive repair work carried out due to steel corrosion in the deck beams also showed a very short service life, mostly less than 10 years. Usually, the locally spalled areas were first cleaned and then patched with new concrete. Also what was not known in the earlier days, was that such local changes in the electrolytic conditions of the concrete deck would create increased corrosion rates in the adjacent areas.

In spite of the extensive corrosion taking place in almost all of the concrete structures investigated, the effect of the corrosion on the load-bearing capacity was rather moderate and slow. The overall condition of the deck beams in the structures is demonstrated in Fig. 4.9. In Fig. 4.9 the beams in each structure were rated on a scale from 1 to 7, where 1 represents distinct damage, while 7 represents such a severe impairment that the intended function could not be fulfilled. At the same time, the extent of damage is represented by the number of beams within a given structure with damage observed. As can be seen from Fig. 4.9, only 35–40 year old structures (1925–30) had beams with a rating of more than 2 and an extent of damage of more than 50%.

In 1982–83, a more detailed investigation of one of the harbour structures was carried out (Gjørv, Kashino 1986). This was a 60-year-old concrete pier that was going to be demolished in order to create space for new construction work. During demolition, a unique opportunity occurred for investigating the overall condition of the embedded steel both in the concrete deck and in the tremie-cast concrete pillars.

An overall plan of the pier is shown in Fig. 4.10, where the deck area of 12 500 m^2 was supported on approximately 300 tremie-cast concrete pillars. In 1919 and 1922 when the pier was constructed, the required concrete compressive strength in the deck was 25–30 MPa, but at the time of demolition, an *in situ* compressive strength of 40–45 MPa was observed. Although the original concrete composition was not known, all concrete structures in this early period were produced with a very coarse-grained Portland cement giving a very high, long-term strength development.

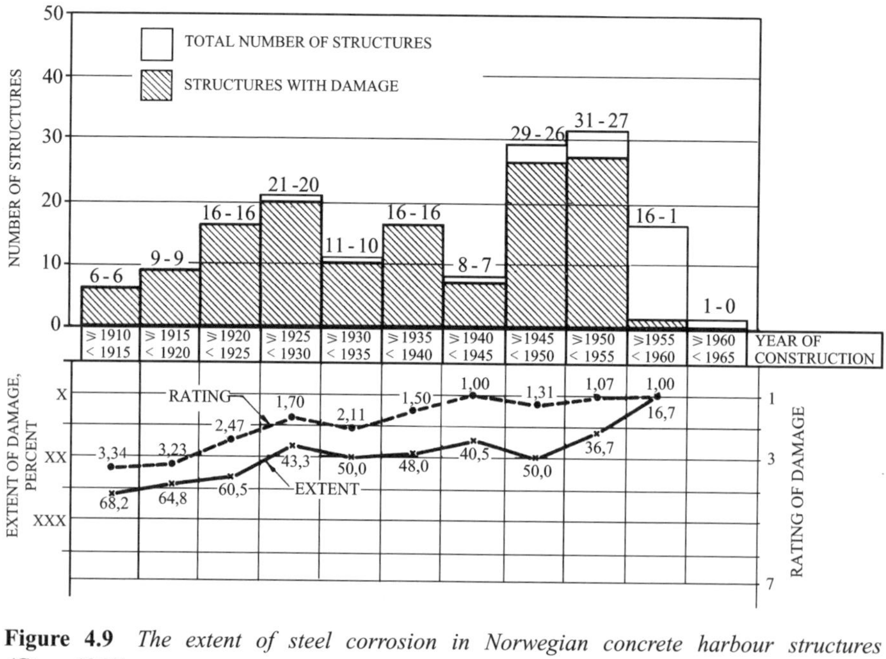

Figure 4.9 *The extent of steel corrosion in Norwegian concrete harbour structures (Gjørv 1968).*

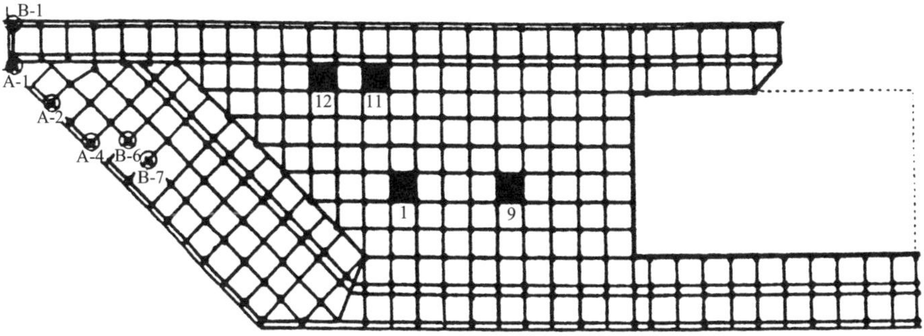

Figure 4.10 *An overall plan view of a reinforced concrete pier in Oslo Harbour built between 1919–22, which was investigated during demolition after 60 years of service (Gjørv, Kashino 1986).*

Although steel corrosion had been developing in all the deck beams throughout most of the service life, the overall condition of this particular structure was relatively good after more than 60 years of exposure. In spite of the deep chloride penetration, more than 75% of the reinforcing bar system in the deck was observed to be in good condition during demolition; practically without any corrosion. For the rest of the steel, which was located mostly in the lower part of the deck beams, the observed corrosion was very unevenly distributed. However, the cross-section of the reinforcing bars was seldom reduced by more than 30%, while most of the corroded bars had a reduced cross-section of less than 10%. These observations demonstrated clearly how efficiently the corroding steel in the lower parts of the deck beams had functioned as sacrificial anodes and thus cathodically protected the rest of the reinforcing bar system in the deck.

The embedded steel in the tremie-cast pillars was observed to be in very good condition mainly due to lack of oxygen. In the tidal zone, the deepest pitting on the steel surface was never more than 1 mm, while below the low-water level, the pitting was mostly less than 0.2 mm and only occasionally as much as 0.5 mm. For the structure as a whole, it was not possible to find any relationship between half-cell surface potentials and the condition of the steel. The depth of carbonation which was generally very small, varied according to the prevailing moisture conditions. In the upper part of the deck, a carbonation depth of 2–8 mm was observed, while for the pillars within the tidal zone and below, the carbonation depth varied from 1–7 mm. A greater depth of carbonation of 24–33 mm and 23–24 mm was observed for the lower part of the deck beams and for the top part of the pillars, respectively.

For the deck slab, the chloride content at the level of the reinforcing steel was typically 0.05–0.10% by weight of concrete. For the upper part of the pillars above the tidal zone, the chloride content ranged from 0.15–0.25%, while in the tidal zone, the chloride content was 0.20–0.25% (Fig. 4.11). For the continuously submerged part of the pillars, an even higher chloride content was observed (0.30–0.35%), and from Fig. 4.11 it is clearly demonstrated that large amounts of chlorides had penetrated the concrete far beyond the level of the embedded steel.

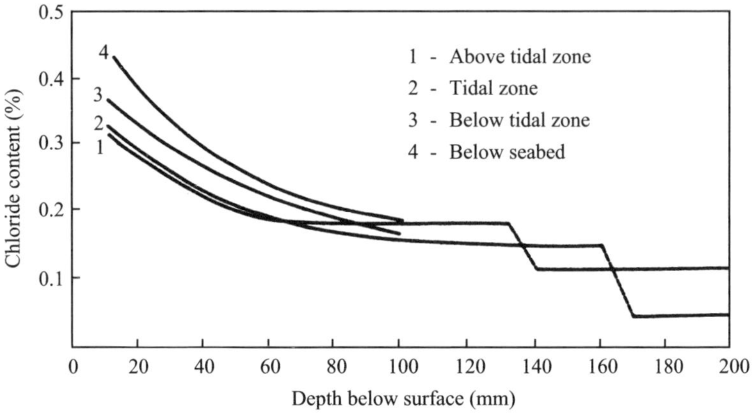

Figure 4.11 *Penetration of chlorides into tremie-cast concrete pillars after 60 years of exposure (Gjørv, Kashino 1986).*

In recent years, both the design and the execution of concrete work for harbour structures have improved. However, new and recent field investigations in Norwegian harbours have revealed that a rapid and uncontrolled chloride penetration still represents a big problem, and steel corrosion may still occur after a service period of less than 10 years (Fig. 4.12). (Lahus, Gussiås and Gjørv 1998; Lahus 1999; Gjørv 2002).

For construction work in the marine environment, recent experience has also shown that a high chloride penetration may take place during the construction period before the concrete has gained sufficient maturity and density (Fig. 4.13). At an early age, most types of concrete are very sensitive to chloride penetration, and this may represent a special problem when the construction work is carried out during cold and rough weather conditions.

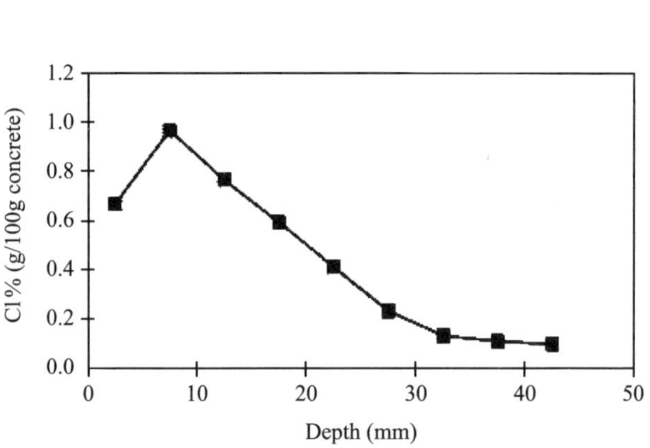

Figure 4.12 *Penetration of chlorides into a concrete harbour structure after 8 years of exposure (Gjørv 2002).*

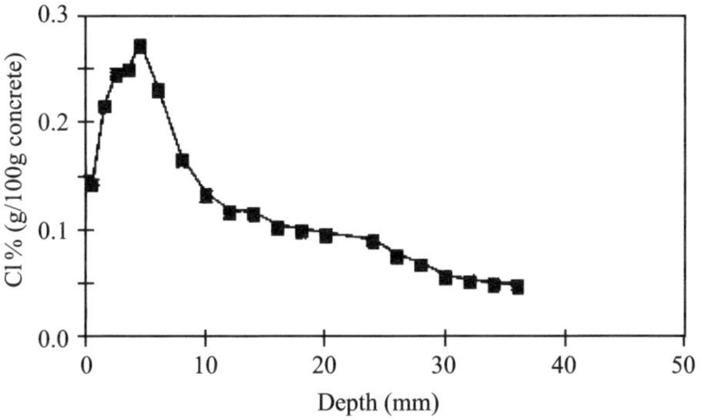

Figure 4.13 *Early-age penetration of chlorides into a concrete harbour structure (2001) during construction (Gjørv 2002).*

4.2.2 Bridges

In many countries, concrete bridges and parking garages exposed to de-icing salts have developed serious corrosion problems. (Comptroller General of the United States 1979; American Association of State Highway and Transportation Officials 1986).

The marine environment may represent even more severe conditions for chloride-induced corrosion (Malhotra 1980, 1988, 1996; Sakai, Banthia, Gjørv 1995, 1998, 2004). Recent field investigations in Norway have shown that more than 50% of the 300 concrete bridges along the Norwegian coastline either had a varying extent of steel corrosion or had been repaired due to steel corrosion (Østmoen, Liestøl, Grefstad, Sand, Farstad 1993). Most of these bridges were less than 25 years old, and one of these bridges was so heavily corroded that it was demolished at an age of 25 years (Fig. 4.14).

For most of the corroding bridges along the Norwegian coastline, large amounts of chlorides had penetrated the concrete, far beyond the level of the reinforcing steel. A typical pattern for the observed damage was that those parts of the bridges that had been the most exposed to the prevailing winds and the salt spray had the least chloride penetration (Fig. 4.15). This was assumed to be due to the rain intermittently washing off the salt from the most exposed surfaces, while on the more protected surfaces, the salt would accumulate. Another pattern appeared to be that older structures built before approximately 1970 had performed better than structures built later on.

Also in a number of other countries, extensive field investigations have revealed problems due to steel corrosion in concrete bridges exposed to the marine environment (Nilsson 1991; Stoltzner, Sørensen 1994; Beslac, Hranilovic, Maric, Sesar 1997; Wood, Crerar 1997).

Based on all the durability problems associated with concrete bridges in the marine environment, there has been an increased focus on improved durability specifications for the construction of new concrete bridges. In Norway, the Norwegian Highway Authorities introduced their own internal specifications for concrete durability that

Figure 4.14 *Ullasundet Bridge (1970) was demolished after 25 years of use due to heavy corrosion of the reinforcing steel (Hasselø 1997).*

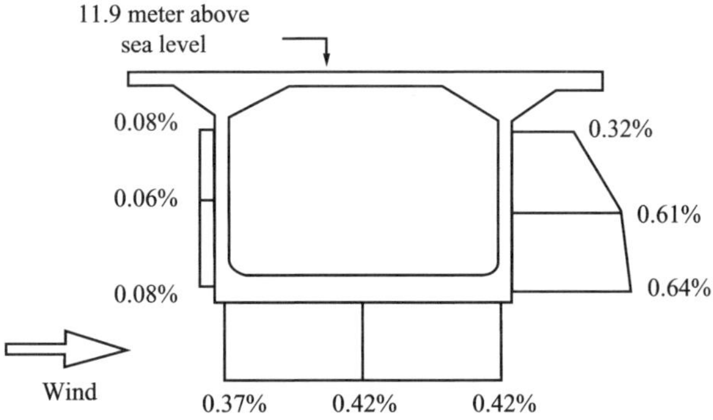

Figure 4.15 *Typical chloride penetration of those parts of a Norwegian concrete coastal bridge that are the most exposed to the prevailing winds and the salt spray compared to that of the more protected parts (Fluge 1997).*

were much stricter than the official concrete codes. In the USA, current Oregon DOT specifications call for the use of stainless steel reinforcement in all exposed parts of their new concrete coastal bridges (Cramer, Covino, Bullard, Holcomb, Russell, Nelson, Laylor and Stoltesz 2002).

4.2.3 Offshore structures

Extensive surveying programmes both above and below water are regularly carried out on all the concrete platforms in the North Sea, but in spite of the very harsh and aggressive environment, where the concrete platforms are completely splashed over by seawater several times a year, very few corrosion problems have been reported on the embedded steel so far; the most serious problems have been related to accidental loads from ships and falling items (FIP Concrete Sea Structures Commission 1994). More detailed inspections of eleven of the oldest platforms installed during 1973–78 and reported in 1982, also showed good conditions (Fjeld, Røland 1982).

Regular corrosion monitoring based on embedded steel tubes in the Frigg TCP-2 platform built in 1977 has also shown good performance without any serious steel corrosion (Elf Petroleum Norge AS). Regular inspections of relatively wide cracks in the concrete at the foundation of the Frigg CDP-1 platform built in 1975 have revealed no serious corrosion in the cracked concrete (Elf Petroleum Norge AS). Also, experimental investigations have shown that the risk of steel corrosion in cracked, submerged concrete is less than originally expected (Vennesland, Gjørv 1981; Espelid, Nilsen 1988). Recent observations from the concrete platforms on the Gullfaks Oil Field have also shown a good condition (Hølaas 1992).

However, monitoring of chloride penetration into the concrete platforms in the North Sea also demonstrates that the chlorides do penetrate this high quality concrete, but only at a slower rate. Figs 4.16 and 4.17 show the chloride penetration into the Statfjord A Platform built in 1977 and the Ekofisk Tank built in 1973 after 8 and 17 years of exposure, respectively. Figs 4.18 and 4.19 show the chloride penetration into the Brent B 1975 and Brent C 1976 Platforms, respectively, after 20 years of exposure. For the Oseberg A Platform built in 1988, where the concrete cover was partly less than prescribed, serious corrosion problems occurred at an early stage, and cathodic protection has already been installed (Østmoen 1998).

From Fig. 4.16 it can be noted that solid epoxy coatings on the concrete legs in the splash zone had very efficiently prevented the chlorides from penetrating the concrete, and even after 15 years this protection still appears to be very effective (Aarstein, Rindarøy, Liodden, Jenssen 1998). However, the much thinner surface coating applied to the Heidrun Platform (built in 1995) was not very efficient in keeping the chlorides out, even after an exposure period of only two years (see Fig. 4.20).

4.3 Codes and practice

The better performance of the offshore concrete structures compared to that of onshore structures, raises the question why offshore structures in the harsh environment of the North Sea have performed so much better than land-based concrete structures in the marine environment.

When the first concept for concrete platform construction in the North Sea was introduced in the late 1960s, the offshore technical community displayed considerable scepticism over concrete as a construction material for offshore installations. However, at the same time the results from the comprehensive field investigation of existing concrete structures along the Norwegian coastline were published (Gjørv 1968).

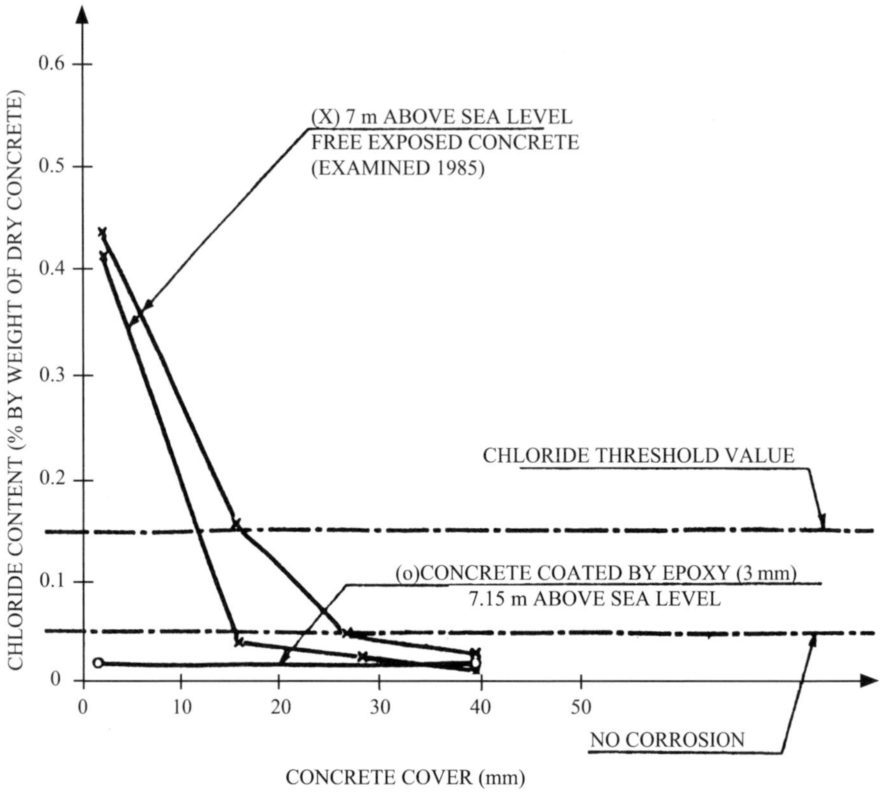

Figure 4.16 *Chloride penetration into the Statfjord A Platform (1977) after 8 years of exposure in the North Sea (Sandvik 1993).*

The overall good performance of these structures, even after a service period of 50–60 years, contributed to convincing the offshore technical community that concrete might also be a reliable construction material for offshore structures. For the first offshore installations in the North Sea, the intended service life was only 25–30 years.

After a further study of the existing experience with concrete structures in the marine environment, however, it was not possible for the international operators in the North Sea to accept a construction material that showed corrosion problems after a service period of only 5–10 years and at the same time was difficult to repair. In order to gain acceptance for the first offshore concrete platform (the Ekofisk Tank), both increased concrete quality and increased cover thickness to the reinforcing steel were required. It would also be important to introduce very strict quality assurance and quality control programs during concrete construction.

When the first FIP publication 'Recommendations for the Design and Construction of Concrete Sea Structures' was introduced in 1973 (FIP 1973), the increased durability requirements in this document were based on the experience gained from the comprehensive field investigations carried out in Norwegian waters. In 1977, when the third revised edition of the FIP Recommendations was published,

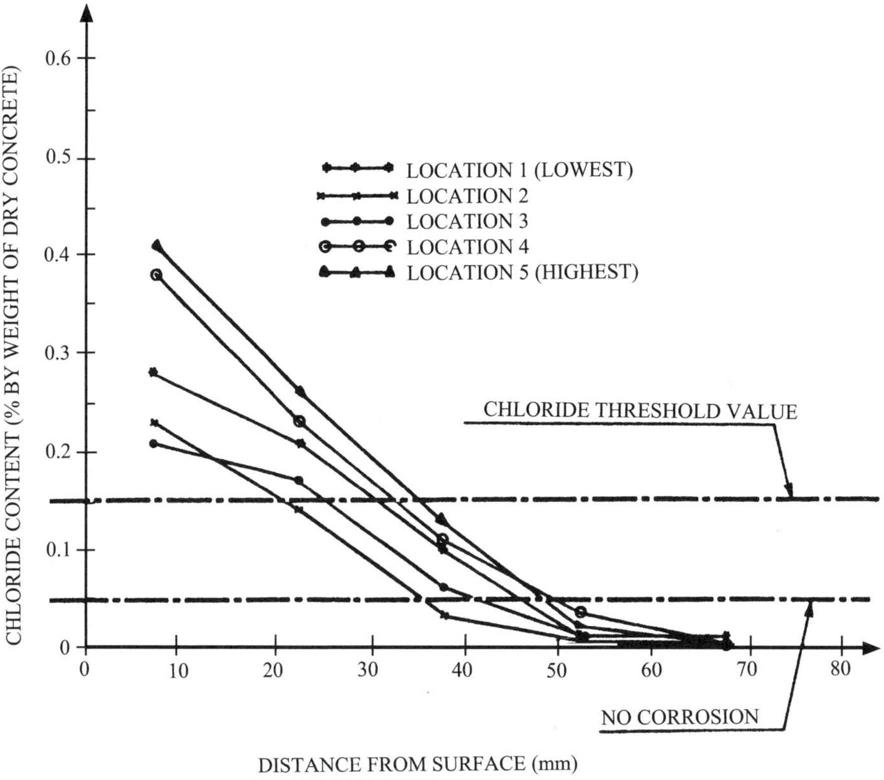

Figure 4.17 *Chloride penetration into the Ekofisk Tank (1973) after 17 years of exposure in the North Sea (Sandvik 1994).*

other new national and international guidelines and codes for offshore concrete structures had also adopted the new and stricter requirements for durability (The Norwegian Petroleum Directorate 1976; Det Norske Veritas 1976). Since the Ekofisk Tank was installed in 1973, the concrete quality requirements steadily increased. Thus, for the first concrete platforms, a water/cement ratio of 0.40–0.45, a cement content of more than 400 kg/m^3 and a concrete permeability of less than 10^{-13} kg/Pa.m.s. were specified, while later on, the water/cement ratio was successively reduced to less than 0.40. At the same time, the compressive strength was successively increased from 45 to 70 MPa (Fig. 4.21), while the minimum concrete cover was 50–75 mm. If the observed concrete cover did not meet the specified requirements during concrete construction, corrective measures in the form of surface coatings were required.

The good performance of the concrete platforms in the North Sea demonstrates that even from the early 1970s, it was possible to design and produce quite durable concrete structures even for the most aggressive and harsh marine environment. However, service life design based on a proper utilisation of existing knowledge and experience was essential.

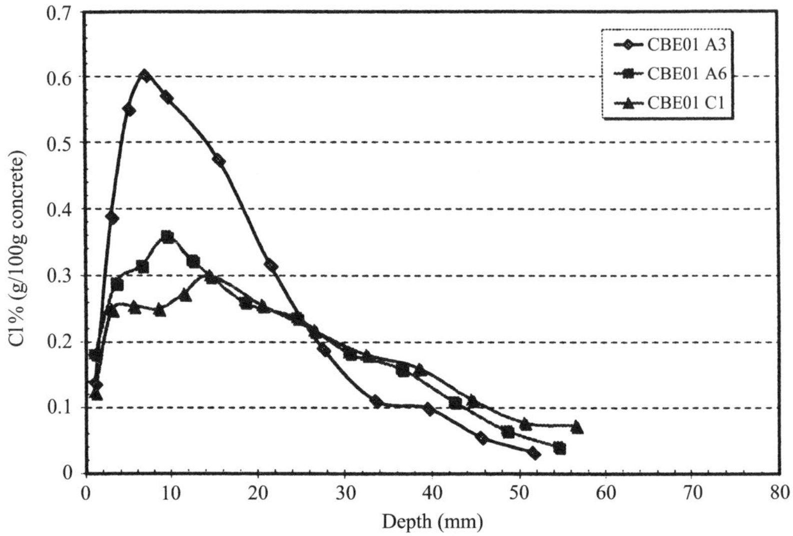

Figure 4.18 *Chloride penetration into the Brent B Platform (1975) after 20 years of exposure in the North Sea, 14 m above sea level (Gjørv 2002).*

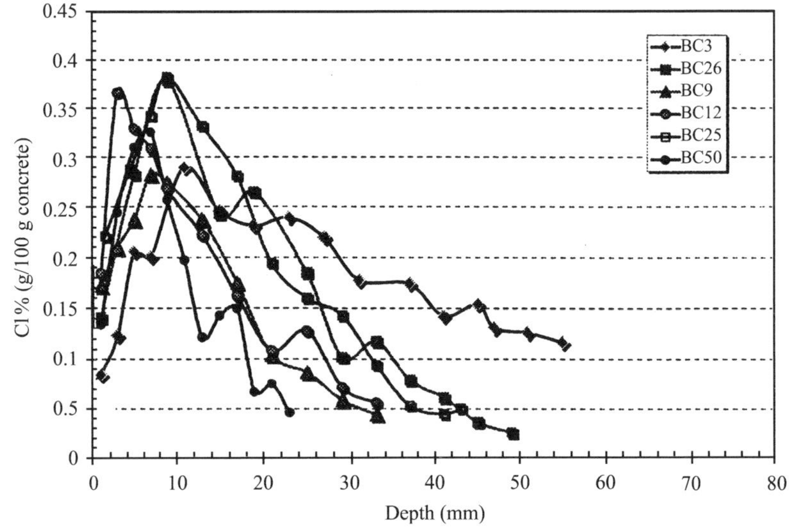

Figure 4.19 *Chloride penetration into the Brent C Platform (1976) after 20 years of exposure in the North Sea, 11 to 18 m below sea level (Gjørv 2002).*

For land-based concrete structures, however, the traditional requirement of compressive strength was almost the exclusive requirement for concrete quality until the late 1980s. During the boom in construction activity that took place in many countries during the 1970s and 1980s, both harbour structures and coastal bridges

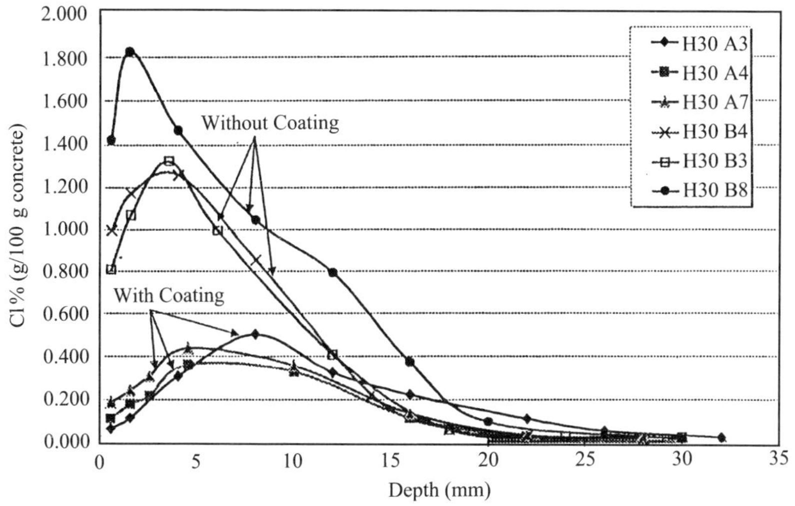

Figure 4.20 *Effect of concrete coating on the chloride penetration into the Heidrun Platform (1995) after 2 years of exposure in the North Sea (Gjørv 2002).*

were typically produced with a concrete compressive strength of 30–40 MPa and a minimum concrete cover of 25–50 mm.

The fact that older structures built before approximately 1970, performed better than structures built later on, may be related to the lack of general knowledge in the construction industry in combination with the rapid development in concrete technology that started in the early 1970s. Thus, until the late 1960s, a 30 MPa type of concrete in Norway would typically be produced with a cement content of 400–500 kg/m^3. During the 1970s and 1980s, however, new organic and mineral admixtures were introduced, and new and more efficient cements became available. At the same time, both aggregate production and concrete mixing were optimised. Therefore, it successively became possible to meet the specified requirement of compressive strength with less cement. Also, increased production schedules often resulted in poor workmanship and inadequate curing of the concrete. As a consequence, the ability of the concrete to resist chloride penetration and to protect embedded steel from corrosion became successively reduced.

Poor workmanship in combination with a lack of proper quality control on the construction site often resulted in very small concrete covers. This situation is illustrated in Fig. 4.22, where 2028 single observations of concrete cover in a Norwegian coastal bridge from 1978–81 are shown. This situation reflects a general problem in many countries as demonstrated in Fig. 4.23, where the observations from the Norwegian bridge (N) are shown together with similar observations from a Japanese bridge (J) (Ohta, Sakai, Obi, Ono 1992) and more than 100 concrete structures from the Persian Gulf (AG) (Matta 1993).

Although most codes of practice for concrete structures have been upgraded a number of times during the last 30 years, current code specifications for concrete durability are still almost exclusively based on the traditional requirements of concrete composition, construction procedures and curing conditions, and where the

C45-C50

Cement (SP30)	430	kg
Sand, 0-10 mm	900	kg
Coarse aggregate		
10-32 mm	900	kg
Betokem LP	4	ltr.
Slump	129	mm
W/C ratio	0.41	
Obtained m.s.	55	MPa

Beryl A (1973) – 75

C55-C80

Cement (SP30-4A)	380	kg
Sand, 0-5 mm	1030	kg
Coarse aggregate		
5-20 mm	8450	kg
Betokem Pa (B)	4	ltr.
Slump	220	mm
W/C ratio	0.42	
Obtained m.s.	67	MPa

Statfjord C (1981 – 83)

C65-C70

Cement		
(SP30-4A Mod)	380	kg
Sand, 0-5 mm	940	kg
Coarse aggregate		
5-20 mm	945	kg
Betokem Pa(B)	6	ltr.
Slump	240	mm
W/C ratio	0.43	
Obtained m.s.	79	MPa

Gultfaks C (1988 – 87)

C80-C90

Cement		
(SP30-4A Mod)	420	kg
Microsilica	30	kg
Sand, 0-5 mm	875	kg
Coarse aggregate		
5-20 mm	945	kg
Betokem P	2.5	ltr.
Slump	260	mm
W/C + S	0.34	
Obtained m.s.	98	MPa

Test (1983 – 84)

Figure 4.21 *Development of concrete mixtures for concrete platforms in the North Sea (Sandvik 1994).*

requirement for compressive strength is still the only performance criterion. For several reasons, this descriptive approach has shown to yield insufficient and unsatisfactory results. In recent years, therefore, much research work has been carried out in order to develop new procedures and recommendations both for durability design, improved construction quality and life cycle management. However, almost nothing from that has been incorporated in current concrete codes so far. Thus, for concrete structures in the marine environment, the durability requirements according to the latest version of the European Concrete Code EN 206-1 (CEN 2000) are still not stricter than what was specified for the first offshore concrete platforms in the North Sea more than 30 years ago (FIP 1973).

4.4 Durability design

In recent years, extensive research has been carried out in many countries in order to better understand, and to better control, the various deteriorating processes which are causing problems to the durability and long-term performance of many important

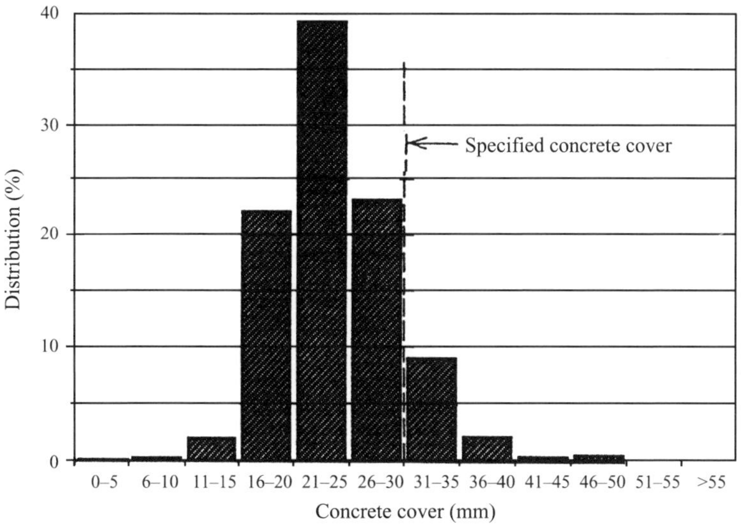

Figure 4.22 *Variation in concrete cover observed in the Gimsøystraumen Bridge (1978–1981) (Kompen 1994).*

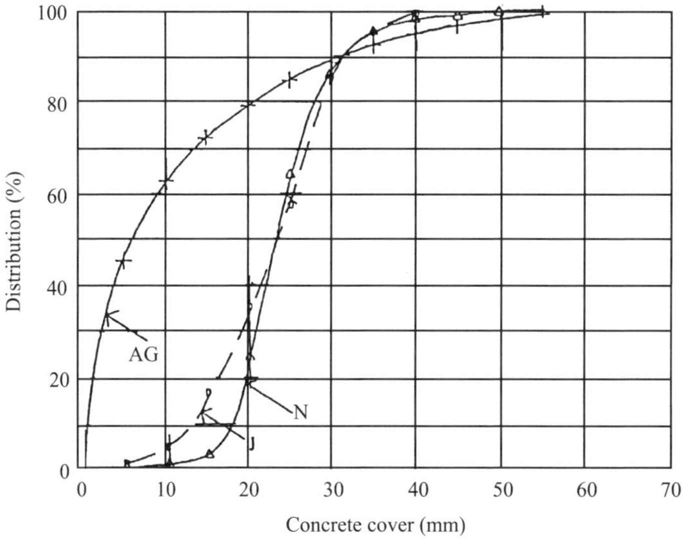

Figure 4.23 *Variation in concrete cover in the Gimsøystraumen Bridge compared to that in a Japanese bridge (J) and more than 100 concrete structures from the Persian Gulf (AG) (Kompen 1994).*

concrete structures. Especially, much research has been carried out in order to provide a better basis for durability design of concrete structures in chloride-containing environments (Gjørv 2002). A probability-based durability design has now become the basis for new recommendations and guidelines for increased durability of new concrete structures in Norwegian harbours (Norwegian Association for Harbour Engineers 2004). In principle, these design procedures include the following elements:

- probability-based durability analysis
- evaluation of alternative strategies and protective measures
- requirements for documentation of obtained construction quality and durability properties
- preparation of a service manual for regular condition assessment and monitoring of chloride penetration as well as protective measures for control of this penetration

As an overall requirement for the durability, an upper risk level of 10% for steel corrosion to occur during a certain period of service is required. This is a type of quality requirement which is based on engineering judgement of all those factors which are considered relevant for the durability, including scatter and variability of all factors involved.

On a trial basis, the new procedures for durability design have already been applied to a number of new, very important concrete structures (Gjørv 2004). For a given environment, the new procedures appear to provide a good basis for comparing the risk of steel corrosion for various combinations of concrete quality and concrete cover. If this risk appears to be too high under the given conditions, additional protective measures such as cathodic prevention or a partial use of stainless steel are considered.

As soon as the requirement for concrete quality based on chloride diffusivity is given, the relationship between chloride diffusivity and electrical resistivity for the given concrete is established. For further quality control during concrete construction, the electrical resistivity, and hence also the chloride diffusivity, is then monitored with a Wenner-electrode on all the concrete specimens as being used for the regular testing of compressive strength. In addition to this laboratory-based quality control, a control of obtained chloride diffusivity on the construction site is also carried out. Together with a regular monitoring of the obtained concrete cover, a basis for documentation of both obtained construction quality and durability is gathered during the construction period.

Even if the strictest requirements for concrete quality are specified, experience has shown that a certain rate of chloride penetration may still take place. Therefore, when the construction period is over and the service period starts, it is very important to have a service manual for regular condition assessment and monitoring of the real chloride penetration. It is such a monitoring, in combination with protective measures for control of this penetration, which provides the ultimate basis for obtaining a more controlled durability and service life.

4.5 Conclusions

An uncontrolled service life of a concrete structure together with escalating spending on repairs, rehabilitation and premature demolition does not only represent technical and economical problems. It also represents a poor utilisation of natural resources and hence, stores up ecological problems (Gjørv 2000).

It is not the disintegration of the concrete itself but rather electrochemical corrosion of the embedded steel which poses the most critical and greatest threat to the performance and service life of concrete structures. This is particularly true for concrete structures exposed to the marine or other chloride-containing environments. In such environments, it appears to be just a question of time before detrimental amounts of chlorides reach the embedded steel even through a solid cover of high-quality concrete.

For concrete structures where durability and safety are important, a proper durability design in combination with a regular monitoring of the real chloride penetration and preventive maintenance are essential for a more controlled durability and service life.

References

Key references are listed here. For any other publication details, please contact the author.

Aarstein, R., Rindarøy, O.E., Liodden, O. and Jenssen, B.W. (1998) 'Effect of coatings on chloride penetration into offshore concrete structures', *Proceedings, Second International Conference on Concrete under Severe Conditions – Environment and Loading*, Eds by O.E. Gjørv, K. Sakai and N. Banthia, London: E. & F.N. Spon, pp. 921–929.

American Association of State Highway and Transportation Officials, (1986) *Strategic Highway Research Program Research Plans*, Transportation Research Board.

American Concrete Institute (1972) 'Ad hoc board committee on concrete – Year 2000', *ACI Journal, Proceedings*, Vol. 68, pp. 581–589.

Atwood, W.G. and Johnson, A.A. (1924) 'The disintegration of cement in sea water', *Transactions, ASCE*, Vol. 87, pp. 204–230.

Beslac, J., Hranilovic, M., Maric, Z. and Sesar, P. (1997) 'The Krk Bridge: chloride corrosion and protection', *Proceedings, International Conference on Repair of Concrete Structures – From Theory to Practice in a Marine Environment*, Ed. by A. Blankvoll, Norwegian Public Roads Administration, Oslo, pp. 501–506.

CEB, (1997) 'New approach to durability design', CEB Bulletin 238, Lausanne.

Comptroller General of the United States, (1979) *Solving Corrosion Problems of Bridge Surfaces Could Save Billions*, United States Accounting Office PSAD – 79–10.

Cramer, S.D., Covino, B.S., Bullard, S.J., Holcomb, G.R., Russel, J.H., Nelson, F.J., Laylor, H.M. & Stoltesz, S.M. (2002) 'Corrosion prevention and remediation strategies for reinforced concrete coastal bridges', *Cement & Concrete Composites* Vol. 24, pp.101–117.

Det Norske Veritas (1976) *Rules for the Design, Construction and Inspection of Fixed Offshore Structures*, Oslo.

Elf Petroleum Norge A.S (Private communication).

EN 206-1, (2000) *Concrete – Part 1: Specification, performance, production and conformity*, CEN, Brussels.

Espelid, B. & Nilsen, N. (1988) 'A field study of the corrosion behavior on dynamically loaded marine concrete structures', ACI SP-109, Ed. by V.M. Malhotra, pp. 85–104.

Féderation Internationale de la Précontrainte (FIP) (1973) *Recommendations for the Design and Construction of Concrete Sea Structures*, London.

FIP Concrete Sea Structures Commission (1994) *Durability of Concrete Structures in the North Sea, 1994, State-of-the-Art-Report*, London.

Fjeld, S. & Røland, B. (1982) 'Experience from in-service inspection and monitoring of 11 North Sea structures', Offshore Technology Conference, Paper No. 4358, Houston, USA, 9 p.

Fluge, F. (1997) 'Environmental loads on coastal bridges', *Proceedings, International Conference on Repair of Concrete Structures – From Theory to Practice in a Marine Environment*, (Ed.) by A. Blankvoll, Norwegian Public Roads Administration, Oslo, pp. 89–98.

Gjørv, O.E. (1968) *Durability of Reinforced Concrete Wharves in Norwegian Harbours*, Ingeniørforlaget AS, Oslo, 208 p.

Gjørv, O.E. (1970) 'Thin underwater concrete structures', *Journal of the Construction Division, ASCE*, Vol. 96, No. CO1, Proceedings, Paper No. 7326, pp. 9–17.

Gjørv, O.E. (1975) 'Concrete in the oceans', *Marine Science Communications*, Vol. 1, No. 1, pp. 51–74.

Gjørv, O.E. & Vennesland, Ø. (1982) 'A new probe for monitoring steel corrosion in offshore concrete platforms', *Materials Performance*, Vol. 21, pp. 33–35.

Gjørv, O.E. & Kashino, N. (1986) 'Durability of a 60 year old reinforced concrete pier in Oslo Harbour', *Materials Performance*, Vol. 25, No. 2, pp. 18–26.

Gjørv, O.E. (1989) 'Mechanisms of steel corrosion in concrete structures', *Proceedings, Vol. 2, EVALMAT 89, International Conference on Evaluation of Materials Performance in Severe Environments, Kobe, Japan*, pp. 565–578.

Gjørv, O.E., Sakai, K. and Banthia, N. (Eds), (1998) *Proceedings, Second International Conference on Concrete Under Severe Conditions – Environment and Loading*, London: E. & F.N. Spon.

Gjørv, O.E. (2000) 'Controlled service life of concrete structures and environmental consciousness', *Concrete Technology for a Sustainable Development in the 21st Century*, (Eds) by O.E. Gjørv and K. Sakai, London: E. and F.N. Spon, pp. 1–13.

Gjørv, O.E. (2002) 'Durability and service life of concrete structures', *Proceedings, The First fib Congress 2002*, Session 8, V. 6, Japan Prestressed Concrete Engineering Association, Tokyo, pp. 1–16.

Gjørv, O.E. (2004) 'Durability design and construction quality of concrete structures', *Proceedings, Fourth International Conference on Concrete under Severe Conditions – Environment and Loading*, (Eds) by B.H. Oh, K. Sakai, O.E. Gjørv and N. Banthia, Seoul National University, Korea Concrete Institute, Seoul, pp. 44–55.

Hasselø, J.A. (1997), 'Ullasundet Bridge – the life cycle of a concrete structure', Course on Life Cycle Management of Concrete Structures, Department of Building Materials, Norwegian University of Science and Technology, Trondheim, 15 p. (in Norwegian).

Hølaas, H. (1992) 'Condition of the concrete structures at the Statfjord and Gullfaks oil fields', Report OD 92/87, The Norwegian Petroleum Directorate, Stavanger, (in Norwegian).

Kompen, R. (1994) 'New specifications for securing the concrete cover', *Betongindustrien*, Vol. 26, No. 2, pp. 28–32, (in Norwegian).

Lahus, O., Gussiås, A. and Gjørv, O.E. (1998) 'Condition, operation and maintenance of Norwegian concrete wharves', Report BML 98008, Department of Building Materials, Norwegian University of Science and Technology, Trondheim, (in Norwegian).

Lahus, O. (1999) 'An analysis of the condition and condition development of concrete wharves in Norwegian fishing harbours', Dr. ing. Thesis 1999:23, Department of Building Materials, Norwegian University of Science and Technology, Trondheim, (in Norwegian).

Malhotra, V.M. (Ed.), (1980) 'Concrete in marine environment', *Proceedings, First International Conference, St Andrews-By-The-Sea, Canada*, ACI SP-65.

Malhotra, V.M. (Ed.), (1988) 'Concrete in marine environment', *Proceedings, Second International Conference, St Andrews-By-The-Sea, Canada*, ACI SP-109.

Malhotra, V.M. (Ed.), (1996) 'Performance of concrete in marine environment', *Proceedings, Third International Conference, St Andrews-By-The-Sea, Canada*, ACI SP-163.

Matta, Z.G. (1993) 'Deterioration of concrete structures in the Arabian Gulf', *Concrete International*, Vol. 15, Nov., pp. 33–36.

Metha, P.K. (Ed.), (1996) *Odd E. Gjørv Symposium on Concrete for Marine Structures, Proceedings*, CANMET/ACI, Ottawa.

Nilsson, I. (1991) 'Repairs of the Øland Bridge, experience and results of the first six columns carried out in 1990', Report, NCC, Malmø, Sweden, (in Swedish).

Norwegian Association for Harbour Engineers (2004) *Recommended Specifications for Increased Durability of New Concrete Harbour Structures*, Oslo, 17 p. (in Norwegian).

Norwegian Association for Harbour Engineers (2004) *Practical Guidelines for Service Life Design of New Concrete Harbour Structures*, Oslo, 43 p. (in Norwegian).

Ohta, T., Sakai K., Obi, M. and Ono, S. (1992) 'Deterioration in a rehabilitated prestressed concrete bridge', *ACI Materials Journal*, Vol. 89, pp. 328–336.

Østmoen, T., Liestøl, G., Grefstad, K.A., Sand, B.T. and Farstad, T. (1993) 'Chloride durability of coastal concrete bridges', Report, Norwegian Public Roads Administration, Oslo, (in Norwegian).

Østmoen, T. (1998) 'Field tests with cathodic protection on the Oseberg A Platform', *Ingeniørnytt*, Vol. 34, No. 6, pp. 16–17, (in Norwegian).

Sakai, K., Banthia, N. and Gjørv, O.E. (Eds), (1995) *Proceedings, First International Conference on Concrete Under Severe Conditions – Environment and Loading*, London: E. & F.N. Spon.

Sandvik, M. & Wick, S.O. (1993) 'Chloride penetration into concrete platforms in the North Sea', *Proceedings, Workshop on Chloride Penetration into Concrete Structures, Division of Building Materials, Chalmers University of Technology, Gothenburg*, (Ed.) by L.O. Nilsson, 7 p.

Sandvik, M. (1994) 'Mix design, optimization of high-performance concrete', Course on Life Cycle Management of Concrete Structures, Department of Building Materials, Norwegian University of Science and Technology, Trondheim, 25 p. (in Norwegian).

Sandvik, M., Haug, A.K. and Erlien, O. (1994) 'Chloride permeability of high-strength concrete platforms in the North Sea', ACI SP-145, Ed. by V.M. Malhotra, Detroit, pp. 121–130.

Smeaton, J. (1791) *A Narrative of the Building and a Description of the Construction of the Eddystone Lighthouse*, H. Hughs, London, 7 p.

Strait Crossings (1986) *Proceedings, First Symposium on Strait Crossings, Kristiansund*, Norwegian Public Roads Administration, Oslo.

Strait Crossings (1990) *Proceedings, Second Symposium on Strait Crossings, Trondheim*, Norwegian Public Roads Administration, Oslo.

Strait Crossings (1994) *Proceedings, Third Symposium on Strait Crossings, Ålesund*, Norwegian Public Roads Administration, Oslo.

Stoltzner, E. & Sørensen, B. (1994) 'Investigation of chloride penetration into the Farø Bridges', *Dansk Beton*, Vol. 11, No. 1, pp. 16–18, (in Danish).

The Norwegian Petroleum Directorate (1976) *Regulations for the Structural Design of Fixed Offshore Structures*, Stavanger.

Vennesland, Ø. & Gjørv, O.E. (1981) 'Effect of cracks in submerged concrete sea structures on steel corrosion', *Materials Performance*, Vol. 20, Aug., pp. 49–51.

Wig, R.J. & Ferguson, L.R. (1917) 'What is the trouble with concrete in sea water?' *Engineering News Record*, Vol. 79, pp. 532, 641, 689, 737 and 794.

Wood, J.G.M. & Crerar, J. (1997) 'Tay Road Bridge: analysis of chloride ingress, variability & prediction of long term deterioration', *Construction and Building Materials*, Vol. 11, No. 4, pp. 249–254.

5 The Durability of Steel Structures in Different Environments

David H Deacon

It was stated by the late Ken Chandler, formerly Head of the British Steel Corrosion Advice Bureau, that engineers claim that: 'Corrosion protection of steel structures amounts to 10% of the construction costs but involves 90% of the time and problems.' It is hoped that this chapter will go some way to redressing that balance.

5.1 Introduction

The chief factor influencing the durability of steel structures is corrosion. Steel corrodes because, in its natural and lowest energy form, it exists not as steel, but in one of the oxide forms of iron. Iron and steel are artificial substances and large amounts of energy need to be put into their manufacture. An electric arc blastfurnace consumes a huge amount of energy in converting iron ore sources into useable engineering materials, such as one of the forms of iron or steel.

Iron and steel are, therefore, both thermodynamically unstable and will tend to revert to their lowest energy state, unless prevented from doing so by some physical, chemical or electrochemical means. This uninhibited process is commonly referred to as corrosion, or more often as rusting, and is an electrochemical reaction. A steel surface will consist of grains, the presence of which are sufficient to set up small differences of electric potential over the surface. In the presence of moisture, small cells may be established on the surface, with anodic and cathodic reactions proceeding, the resulting product of both reactions being an oxide layer, visibly familiar as rust.

The oxide layer formed on steel is friable, porous and has only moderate adhesion to the parent metal. Stainless steels, on the other hand, are produced with their composition modified by the addition of less reactive metals such as chromium, nickel and molybdenum. These high cost additions provide steels where the basic resistance to oxidation is increased, due in part to their relative positions in the reactivity series of metals. However, this is not the only reason why they are 'stainless'.

Improvements to acid resistance begin with just a 2% addition of chromium to iron, but become significant with around a 10% addition when ferritic stainless steels are formed. Further additions enhance the resistance until at 18% chromium with 8% nickel the popular 18–8 austenitic stainless steel is produced – with increasing corrosion resistance achieved through 25–12 and 25–20 combinations, post-treated by an annealing process. All three are characterised by being non-magnetic and possessing a low carbon content.

Other elements such as copper, manganese and molybdenum may be used, or additions made which incorporate high cost or high performance elements that easily exceed the original steel content (to the extent that there is little iron or steel left in the composition). This may proceed until sophisticated alloys such as Inconel (60–80% nickel with up to 20% chromium) or Monel (the most frequently used industrial composition which is 70% nickel and 30% copper) are reached. Stainless steels appear to resist the rusting process because the alloying elements prevent oxidation due to their position in the reactivity series for metals, and because they provide a different type of surface oxide layer.

The formation of an oxide layer comprising chromium oxide, alone or with oxides of other less reactive metals, provides a modified surface oxide layer which is protective by being very tough and adherent, electrochemically sound, and a very good barrier to the further ingress of oxidising agents.

Before the methods of preventing corrosion are examined, it is necessary to consider briefly the differing environments to which steel structures are exposed. This is because the mechanisms of corrosion involved are different, depending upon the particular climate local to the steel surface.

5.1.1 Atmospheric corrosion

Steel will remain relatively free from rust if left inside a dry building but will corrode more rapidly if exposed to a moist environment – either internally or externally. This is quite obvious to most people, but the fundamental reason is often not quite as obvious, with perhaps the term 'oxidation' being to blame.

There is plenty of oxygen in the air in each of the three internal and external environments listed yet the rate and extent of rusting may be very different in each case. From this it follows that oxygen supply is not the rate-determining property in the oxidation, or corrosion process. It is the moisture content, known as the relative humidity, of the environment, which is the driving force in atmospheric corrosion where there is an unlimited supply of oxygen.

Many studies have taken place to show that in pure air, little corrosion will occur below about 99% relative humidity. Pure air contaminated with the merest presence of impurities may, however, begin the corrosion process once the relative humidity climbs above 50% and will proceed rapidly above 70% relative humidity. These relative humidity values are common throughout northern Europe and other temperate zones. They also occur through overnight condensation in otherwise arid regions and may be near constant in tropical regions.

Proximity to the sea, coastal estuaries and tidal rivers will further aggravate corrosion. Studies have shown that the corrosion rate of steel 50 metres from the sea can be accelerated by over two orders of magnitude as a result of sea salt, compared to steel located 1 kilometre from the sea.

Improvements to the atmospheric environment, brought about by legislation have reduced the level of pollutants present in the air of most industrialised countries, but little can be done regarding salt-containing airborne sea spray. To this is added the frequent use, during winter months, of de-icing salts on roads and bridges, often producing local microclimates where corrosion rates rise above that generally anticipated for the area.

Design features and even orientation, play a part in atmospheric corrosion and these features will be mentioned later in this chapter.

5.1.2 Corrosion in water

In water it is the availability of oxygen, which is dominant in the corrosion process. Deep down in cold waters the rate of corrosion may be low, but in tidal areas and especially in the splash zone, with abrasive sand and large volumes of moving water, accelerated rates of corrosion occur. The composition of the water itself will have a marked and often complex role to play. There will be many components dissolved in naturally occurring waters, with some being particularly aggressive to steel.

The hardness and conductivity are a measure of dissolved salts whilst acidity and particularly the presence of organic matter and dissolved gases all play a part in possible reactions affecting the steel. For this reason seawater in the tidal zone is often most severe, possessing an abundance of salt in a wet and airy position.

5.1.3 Corrosion in soil

Additional complications are likely to arise should steel, generally in the form of piles or pipes, be buried in soil. The nature of the soil, be it sandy or alluvial, the position of the water table and consistency of moisture, all come into play in determining the nature of the electrochemical effects on the buried steel. These effects will often cause pitting, rather than an even rusting, in which case a local corrosion rate many times the general rate may occur. A soil survey may be the prudent and necessary action required to ensure the longevity of the buried components.

5.1.4 Bacteriological corrosion

It has previously been recognised that oxygen was necessary for the corrosion, or oxidation, of steel. With bacterial corrosion in certain immersed or buried conditions, this is not the case.

Sulphate-reducing bacteria is the most common form of bacteria to cause corrosion under oxygen-depleted conditions, characterised by a 'bad eggs' smell and blackening of the steel. A bright orange bacterial effect, accelerating corrosion below the low-tide level, has also been reported, as has the attack of aerobic and sulphate-oxidising bacteria, also capable of showing a surprisingly severe corrosive action to vulnerable steel. The effects of these differing types of corrosive forces on steel may be hindered, reduced or suspended for long periods of time, provided adequate attention to detail and correct actions are taken.

It is important to bring many disciplines together to prevent nature from degrading steel structures and the iron reverting to its oxide form.

5.2 Design

In the design of steel structures, corrosion control plays an unassuming part in comparison to appearance, structural integrity and cost. The ease of maintenance of the structure in later years is not always foremost in the mind of the designer, due to other more immediate requirements, even with the change to recent thinking and the advances made in considering 'whole life costs' and the inconvenience and cost of maintenance.

The 'fit and forget' philosophy for deep-water manifolds may be appropriate at the bottom of the sea with little free oxygen present, but is rarely possible or cost-effective for land-based steel structures.

5.2.1 Fasteners and fixings

Structural steel has become a commodity product and in most cases it will be obtained from various supply sources. The first challenge in preventing corrosion arises when joining the structure together. The avoidance of crevices, those small gaps between two steel surfaces which are inaccessible for preparing, cleaning or painting, yet allow moisture or condensation to enter, is the first task of the engineer.

The solution may not always be the obvious one. Filling with weld or blocking with a long-lasting sealant or filler might be the soft option but may not necessarily be the best option. Increasing the gap to allow free passage of water, or pre-painting the faces, may also be worthy of consideration. The fastening and fixing together of the components of the structure will receive attention, but some aspects are overlooked in the need to satisfy other eventualities. Although structurally strong, these areas are frequently the weakest point in the battle against corrosion.

Damage will occur in the fixing process and in areas where further protection, for instance by paint, may be difficult. Zinc-coated nuts and bolts may enhance the protection and reduce thread corrosion. It is important to not only specify hot-dip galvanised fixings (not the cheaper zinc plated option) but also to ensure they are actually used. Galvanised wall ties to exterior brickwork have, however, tended to cause embarrassment through rapid consumption of the protective layer and subsequent staining. Stainless steel fixings are used with success, but areas of ponding rainwater, or frequent splashing of salt water from vehicles, may give rise to corrosion of the base metal due to the bimetal effect. This may be particularly severe in cases where contaminants through sea salt or acid rain increase the corrosivity of the ponding water remaining on the two surfaces. Here again, attention to the need to ensure complete drainage from steel surfaces helps to avoid corrosion traps.

Embedding steel in an adequate covering of concrete will, after an initial thin layer of rust forms, generally prevent further corrosion, provided the concrete cover is sufficiently thick and will not be reduced in alkalinity.

It may be a sensible precaution to consider a convex concrete plinth at ground level to prevent constant, or even intermittent, immersion of steel members. If concrete is used around the base to protect the steel, attention is needed to ensure a full and permanent bond and seal between the concrete and the steel. Corrosion caused by the inappropriate joining of different metals may be quite basic, but joining other dissimilar construction materials also needs consideration.

Bricks which may stay wet because of local environmental conditions should not be directly in contact with steel frames, nor should wood or other absorbent materials without taking some other form of action to protect the steel. The avoidance of sharp edges and angles, protrusions and complex shapes will add to the life of the steel. Stitch welding will be less expensive than continuous welding and, where structurally appropriate, is quite sound in internal areas not subject to moisture or the formation of condensation. Due consideration of the formation of crevices, previously mentioned, should be taken if these weld runs are liable to be exposed to moisture. For the same reason, butt joints are preferred to lap joints. Spatter should always be removed from adjacent surroundings to prevent staining and the onset of pits in moist environments.

It may be obvious that external areas protected from the harshest environmental conditions experience less corrosion. However, it is interesting that areas frequently washed by rain which drains well from the surface, also share lower levels of corrosion, providing further potential benefits to corrosion prevention from the earnest consideration of all design features.

5.3 Control methods

There is an enormous range of methods and materials that can be used to slow down the natural process of the iron in steel reverting to its natural oxide. Consideration must be given to many factors, the essential majority of which will be covered in the following pages of this chapter.

5.3.1 Specifications and surveys

One of the first considerations for new work is the provision of an effective specification. The specification is also necessary for remedial or planned preventive maintenance, but in a maintenance situation, the condition of the existing paint coating can only reliably be confirmed after a suitable site survey has been commissioned.

The site survey needs to be carried out by an experienced and qualified paint technologist, knowledgeable in general engineering, but particularly in the control aspects employed on the structure. A comprehensive survey, to be of value in deriving an adequate specification, is unlikely to be satisfactorily achieved even by an experienced paint application inspector. To achieve the whole gambit of information necessary for even an audit survey, a higher level of understanding, knowledge and responsibility is required than may be demonstrated from paint application experience alone.

The paint survey itself should cover many aspects. A detailed examination of the paint film on the structure is required using magnification, solvent tests and limited micro-destructive subjective tests, to determine adhesion/intercoat adhesion and cohesion of the various layers as well as other physical properties. Where necessary, the presence of salts should be determined, and where metal coatings have been used, these should be examined to determine the extent of loss by corrosion and to estimate the remaining life of the coating.

The need for maintenance of a structure will be governed by various factors, but generally either appearance or the need to ensure structural stability will be the determining factors. There are various ways of approaching maintenance. Too often the maintenance programme is the first casualty where cutbacks in expenditure have proved necessary. This may have short-term financial advantages, but inevitably leads to a significant increase in long-term costs.

Although it may sometimes be necessary to drastically cut budgets, including maintenance, account should be taken of critical areas where a delay will result in accelerating breakdown of coatings or an increased loss of steel by corrosion. These areas should be given priority and even in circumstances of financial restraint, they should be treated to bring them to a satisfactory standard. One of the problems with this approach is that maintenance planning is often carried out on an ad hoc basis, or alternatively is planned for certain fixed periods. Consequently, it is often difficult to know the actual state of the structure, so even if a limited budget were to be available, it would be difficult to know how to spend it to achieve optimum benefit.

Although maintenance re-painting is often delayed to a point where eventually considerable expense will be incurred, it is not necessarily economic to regularly re-paint structures at fixed intervals, irrespective of their condition.

Examinations of older structures have shown that on some areas as many as 30 coats of paint can be detected on detached paint films, whereas in other areas there is little paint because corrosion has regularly occurred causing breakdown and flaking of the paint. If the money 'wasted' on this painting technique had been used to ensure that the more susceptible areas had been treated properly, the overall costs would have been much reduced. This balance of expenditure can really be achieved only by proper surveys of structures and buildings before maintenance is planned. A survey is also of benefit in the planning of future planned maintenance functions, especially with respect to re-painting, where a correctly conducted survey will yield the most cost-efficient specification.

The performance of a coating is significantly influenced by the state of the steel surface to which it is applied. This is equally true for maintenance re-painting. Consequently if the paint is allowed to break down to a stage where the underlying steel is rusting, a choice will be necessary. Either expensive blast cleaning methods will be required to clean the steelwork to a suitable standard, or alternatively loose rust and paint will be removed manually and the paint will be applied to a poor surface. In the latter situation, the coating will last a comparatively short time. In some situations, of course, blast cleaning is not practicable so manual cleaning will be the only choice. Clearly it is advantageous to avoid such a situation by ensuring that maintenance painting is carried out before widespread surface preparation of the steel becomes necessary.

The natural deterioration of a paint film, originally applied to a sound surface, will be from the air-to-paint interface. Surface chalking followed by surface crazing, if allowed to develop beyond a critical point, will eventually provide an unsound base for the application of additional coats of paint. A careful survey will include tests on the existing paint, which will establish categorically, its suitability to receive fresh coats. Modern paints are far more highly stressed in the curing and ageing stages than the previous generation of air-drying oil-based paints which can lead to embarrassing and costly, premature paint failures such as flaking and delamination.

It is not always appreciated that paint breakdown rarely follows a linear relationship with time. Typically, a coating lasts for a number of years with only minor breakdown and surface degradation, in say 5–8 years, in the next year there may be 1% breakdown, but if no action is taken this may rise to 5% in the following year. In such a situation a delay of a year can double the cost of cleaning and painting, even discounting any effects of inflation.

Surveys prior to re-painting cover a detailed, close examination of the physical properties of the degraded coating system to determine the extent and type of maintenance required. A proper report should be prepared, which can act as a detailed 'log book' of work. The survey also provides an opportunity to examine and make recommendations concerning particular design details, which may be promoting corrosion and coating breakdown, e.g. expansion joints and water traps. Determining the adhesion to the substrate and inter-coat adhesion by subjective destructive standard methods and by magnification followed by removal of small flakes for laboratory analysis and identification, i.e. measurement of the number of coats and the thickness of individual coats, all play their part in a thorough and meaningful site survey.

The UK Highways Agency, formerly the Department of Transport, set down certain procedures in order to achieve maximum economy with maintenance painting of highway bridges. A paint survey is required first to establish the extent and type of any coating failure and to establish the condition of the paint system. Various categories of failure have been listed by the Highways Agency and are the criteria for determining the extent of the maintenance required:

Category I	Local failures only
Category II	Normal weathering of finishing coat
Category III	General failure of the finishing coat
Category IV	General failure of system, with substrate corrosion

For all categories of failure a pre-specification, overall survey is recommended. In the case of Categories III and IV a qualified paint technologist who has experience as a paint surveyor is appointed from a list maintained by the BES (Bridge Engineering Structures) Division of the Highways Agency. Once the survey has been completed a draft specification is initially prepared to ensure that the re-painting provides a sound, long-life compatible protective system. A feasibility trial is then carried out on larger structures to confirm the practicality of the draft specification. To write an effective specification for painting, the following five main objectives must be satisfied:

- as part of the Contract, to state the whole of the detail by which the required life of the protective system is to be achieved;
- to serve as a basis for accurate pricing and tendering;
- to be a complete reference document for the suppliers of materials, the contractors, the sub-contractors and all other parties to the Contract;
- to provide guidelines and authority for the painting inspector;
- to provide the basis from which any subsequent disputes, failure investigations and possibly arbitration, can be resolved.

The specification

The basic scope of the work needs to be covered in the specification, including type and location of structure, location of coating operations, the general environment where the coating is to be carried out and all areas requiring special attention, i.e. problems of access and timing of operations, if maintenance painting. Additional to the scope, the specification needs to address the documents and relevant standards, surface preparation, materials, application, control of coating materials, their testing, inspection, handling and transport, plus any remedial work.

Writing effective specifications is an important factor in achieving a successful painting or re-painting contract. The specification should be helpful to all parties. It must be remembered that it should 'say what it means and mean what it says' and be appropriate for the expectations of all concerned. Indeed, the well-written and complete specification may form the basis for a long-term insurance-backed guarantee for the works.

5.3.2 Surface preparation

For a paint system to perform well on a steel substrate, the surface must be chemically clean and have a sufficiently rough profile to provide an anchor pattern for the applied coating. When steel is blast cleaned, a surface profile of tiny peaks and valleys is established. This profile is sometimes referred to as the anchor pattern. This anchor pattern increases the surface area of the substrate and improves the mechanical bonding of the priming coat to the steel surface. Not all methods of surface preparation impart a profile to the steel surface. For example, high pressure and ultra-high pressure water jetting do not impart a profile, although it is a superior method for removing soluble chemical contaminants when compared with dry methods of surface preparation. They may, however, reveal the original surface profile. The following list provides a summary of the most common methods of surface preparation:

1. wheelabrator blast cleaning;
2. dry air/abrasive open blast cleaning;
3. vacuum blast cleaning
4. low pressure, wet/abrasive blast cleaning;
5. wet blast cleaning with soluble abrasives;
6. high pressure water jetting;
7. ultra-high pressure (UHP) water jetting (above 25 000 psi/1700 bar)
8. mechanical hand and power-tool abrading.

It remains essential to ensure the surface of the steel is free from oil and grease before carrying out one of the full surface preparation methods listed above. This is a two-step process. Firstly removing, with the aid of solvents, any excessive amounts of grease and secondly washing with a mild emulsifying agent and with clean water until the surface is 'grease free'.

The first type of enclosed system of blast cleaning is used only in fabricator's works or painting shops and is generally most suitable for sheet steel or piling. This method is often referred to as a 'wheelabrator'. It comprises a series of rotating impeller wheels which centrifugally 'throw' the abrasive at the steel surface at high velocity thus removing most of the visible contaminants and creating a profile. The second method may be used in or adjacent to a paint shop, or on-site, for new or maintenance work since the only items of equipment required are a compressor and blast-pot. The third system employs an enclosed blasting and recovery system, reducing waste and dust by an effective series of containment skirts. This system operates best on large flat areas and may be employed on flooring. All three of these methods are capable of effective removal of millscale from hot rolled steel. Removal of millscale is an essential part of the cleaning process prior to any type of paint application. The other five methods are employed exclusively on-site for maintenance work.

For the dry methods, the abrasive and the debris must be separated if the abrasive is to be recycled. Non-metallic abrasives such as copper slag are seldom worth recovering. Recovery of the abrasive media is not viable for the 'wet' methods employed in site maintenance cleaning. The type and grade of abrasive employed control the visual cleanliness of the steel. Blast cleaning is invariably the preferred method of surface preparation for new steel and may be specified according to recognised International Organization for Standardization (ISO) Standards. Automatic blasting with metallic shot or grit, followed by the automatic application of a pre-fabrication primer is an available and good quality source for cleaning construction steel and structural steel straight from the steel mill.

Environmental issues are leading to a reduction in the frequency of on-site dry blasting. This reduction can be reversed if major attention is paid to containment issues. The cost of the disposal of spent abrasive becomes significant where there are significant amounts of lead contamination. The lead content of new paint is zero.

Surface preparation and cleaning in cases where toxic pigments are being removed must be extremely well controlled, with very efficient and effective containment and regular medical checks on the operatives concerned. ISO Standards are being drafted for wet abrasive blasting and the methods proposed are finding favour for many maintenance projects. They offer advantages for structures which have suffered salt contamination, or where any airborne contamination by the abrasive (e.g. into gearboxes), must be avoided. It is also possible to raise a profile on the structural steel substrate with this method of cleaning.

Ultra-high pressure (UHP) hydro-blasting is an alternative to on-site maintenance that uses grit blasting. The terms hydro-blasting, hydro-jetting and water jetting all describe the same process, but there may be confusion between water washing and hydro-blasting. The following terms are listed in the SSPC/NACE Standards. These Standards have been prepared jointly by a working party of members of the North American Steel Structures Painting Council and National Association of Corrosion Engineers.

Low pressure water washing – operates at pressures less than 68 bar (1000 psi)
High pressure water washing – operates at pressures between 68–680 bar (1000–10 000 psi)
High pressure hydro-blasting – operates at pressures of 680–1700 bar (10 000–25 000 psi)
Ultra-high pressure hydro-blasting – operates at pressures above 1700 bar (25 000 psi).

UHP hydro-blasting operates at pressures in excess of 1700 Bar (25 000 psi), but the majority of the new UHP machines operate in the 2000–2500 bar range (30 000–36 000 psi). Whilst there is a tendency for surfaces to 'flash rust' following hydro-blasting, the amounts of energy involved are usually sufficient to heat the substrate locally, which helps to evaporate residual water, thus reducing the formation of flash rusting.

Analysis has also shown that the flash rusting left by UHP hydro-blasting, is an inert ferric oxide and as long as the deposits are not too heavy or loosely adherent, performance of compatible coatings may not be adversely affected. New developments of UHP with pressures up to 2000 bar, but using just 22 litres of water per minute, have proved to be effective in removing complete coating systems, including thermally sprayed metal systems.

These developments in the safety and the efficiency of UHP water jetting systems have made them an attractive alternative, both technically and environmentally. The complete removal of a coating system reveals the original surface profile or steel condition. Lower pressures and decreased water volumes are even capable of removing, when used by an experienced operator, selectively, single or multi-coat layers of a paint system. It is probable that in the future, engineers and specifying authorities, because of the heightening of international environmental awareness and legislation, will increasingly be looking for an alternative to open dry grit blasting.

The latest developments in UHP equipment and a greater understanding of the process, will allow end-users to take advantage of the overall benefits of UHP hydro-blasting. These benefits are summarised as follows:

- no dust produced, thereby improving health and safety and reducing contamination of machinery and freshly painted surfaces;
- no grit disposal costs;
- other trades persons able to work in close vicinity;
- reduced surface salt levels compared to dry grit blasting;
- individual paint layers can be stripped;
- reduced clean-up costs and clean-up time;
- the ability to continue blasting when raining; and
- an overall reduced environmental impact.

Where safety issues preclude the use of either abrasive blasting or wet methods (e.g. where electrical switchgear may not be isolated), modern power-tool cleaning may be the only employable method. Although this is a slower and less efficient method, especially around bracings and stiffeners, modern pneumatic tools have greatly reduced vibration levels when in use, thus improving conditions for operators.

However, mechanical tools and equipment remain tiring for the operator to use for long periods and a vacuum hose can make it more difficult to reach into tight corners, thus placing additional stress on the operator. With the continuing development of power-tool technology and the promotion of 'surface tolerant' paint systems, a resurgence of power-tool cleaning in maintenance contracts will happen over the next few years.

5.3.3 Coatings

Coatings, of one form or another, are generally acknowledged to be the most cost-effective method of arresting, or at least slowing down, the ravages of corrosion and retarding what is commonly referred to as rusting. Control of corrosion on steel structures to increase their durability can be split into two general classes as follows: protection afforded by metal coatings and protection afforded by paint and allied materials.

Metal coatings

Metal coatings are used widely to protect steel from corrosion; galvanised sheet and copper plated pipes are examples of their use. In this chapter only metals used for the protection of structural steel and steel components used for construction will be considered. In this category there are only two metals to be considered, namely zinc and aluminium. Of these, zinc is used to a much greater extent than aluminium.

Although aluminium has been employed as a protective coating for structural steel, its use is limited. Paints containing a high percentage of zinc, known as zinc-rich, are sometimes incorrectly, called metal coatings, but as the influence of the paint binder is of considerable significance to the overall properties of the coating, these paints are more correctly considered as a priming paint and as such are considered later.

There are five basic methods of applying metal coatings to steel, although not each method is applicable to all coatings:

1. hot-dipping;
2. spraying;
3. diffusion;
4. electro-deposition; and
5. cladding.

Generally, only hot-dipping and spraying are used for coating structural steel. Diffusion and electro-deposited coatings are used for fixings.

Hot-dipping

Essentially, this consists of carefully cleaning and then dipping the steel into a bath of molten metal. The molten metal reacts with the steel, forming an alloy of the coating metal, which is metallurgically bonded to the steel at the interface producing an almost pure metal coating at the outer surface.

The process using molten zinc is called hot-dip galvanising and consists of three main stages. The steel to be coated is degreased and cleaned of scale and rust in a pickling solution of dilute acid, usually hydrochloric acid. The steel is then given a pre-fluxing treatment and finally immersed in a molten zinc bath, maintained at a temperature of about 450°C. The coating thickness depends upon a number of factors, the thickness of steel, the composition of steel and the withdrawal rate from the bath.

On heavy structural steel members the coating thickness is usually between 85 and 125 micrometres. Generally, coatings are designated as the weight of zinc per unit area (e.g. 610 g/m^2), which is equivalent to 85 micrometres on each face of the steel substrate. Thicker coatings of zinc can be obtained by grit-blasting the steel before galvanising. Steels containing more than about 0.3% silicon will provide thicker coatings because the zinc–iron alloy adheres by a different mechanism, so that the coating thickness is proportional to the time of immersion in the molten zinc bath.

At the same time, this changes the composition of the zinc layer so that the iron–zinc alloy layer becomes thicker and the pure zinc layer near the surface becomes thinner. If the iron–zinc alloy layer is very close to the top surface, the zinc tends to have a dark grey appearance without the characteristic spangle. This layer can also stain an iron-rust colour and give the appearance that the galvanising is corroding. In fact, although it might be considered unsightly, it can be an indication of an exceptionally thick protective zinc layer.

Steel sheet is often galvanised by a continuous process rather than the individual immersion of small sheets in a bath. The process is basically the same as for heavier steel, but coatings are generally thinner. Aluminium metal coatings can be applied by hot-dipping, in a manner similar to zinc. Generally, it is a more complex process in which the steel is heated before being immersed in molten aluminium. Large steel sections are not coated by this method in the United Kingdom.

Hot spraying

In this method, an atomised stream of molten metal is projected from a special gun, fed by either wire or powder onto a grit-blasted steel surface. Although any metal available in a suitable form and capable of being fused in an oxy-acetylene or similar type of flame can be sprayed onto the steel, in practice only zinc, aluminium and alloys of these materials are applied in this way. The metal droplets from the spray gun are blown by compressed air onto the steel surface. These drops are not molten when they hit the surface but are flattened to some extent and, as spraying continues, overlapping layers are built up to the required thickness.

The applied coating can be very porous, unlike hot-dipped coatings, there is no alloying between the steel and the coating, adhesion being dependent upon mechanical bonding. Thermally sprayed coatings by gas or by arc produce porous films, which are not, therefore, metallurgically bonded to the steel surface. With arc-sprayed metal coatings, in addition to the porosity within the layer, a far rougher surface is produced over which the paint has to be applied and care needs to be taken with de-nibbing to avoid polishing the surface, which may cause adhesion difficulties with subsequent coatings.

Whichever type of hot spraying is used, the zinc or the aluminium coating must be sealed as soon as possible after application, otherwise, moisture will enter the porous film and cause breakdown within the metal layer. There has been an increasing practice to adopt the use of thermally sprayed aluminium or zinc, or hot-dip galvanising , any of which may or may not be over-coated with a paint system. The maintenance of coating systems on a non-ferrous metal base is more difficult to implement than on steel.

Electro-deposition

Often termed electroplating, this method is used to apply many different metal coatings to steel. In this section, only zinc will be considered. The term 'electro-galvanising' is sometimes used to refer to electroplating by zinc. In this process, the steel to be plated becomes the cathode and is immersed in an electrolyte containing salts of the metal to be deposited, generally in a solution with various agents to provide the desired coating properties such as brightness and hardness. The coating thickness is controlled by the current density of the solution and the length of time of plating.

Generally, this method produces very thin coatings (e.g. 2–5 micrometres), although coating thicknesses of 25 micrometres are sometimes provided on components for service in aggressive conditions. The plated piece-parts are then generally 'passivated' by further chemical treatment, often giving a bluish or yellow hue to the surface of the plated article. Steel sheet is electroplated in a continuous process to provide a thin zinc coating, termed 'Zintec' which is then generally over-coated with paint coats as appropriate to the anticipated service conditions. Alloy layers are not formed in this process, but coatings of zinc are usually uniform and quite dense.

Diffusion coatings

This method is sometimes called 'cementation'. Generally the steel is heated to a temperature near to, but below, the melting point of the coating metal and is then covered by a metal coating powder.

In 'sheradising', the term used for zinc coatings, the articles to be coated are packed in a box in contact with the metal coating powder. Heat is applied and diffusion occurs to form an alloy coating. Often the components are tumbled during the process to improve the contact between the steel and powder. Other methods using a gaseous form of the coating metal are available (e.g. for chromium diffusion).

Sherardising is widely used for nuts and bolts, since the process has the advantage that very small dimensional changes occur. Unlike other coating methods, the coating is virtually all alloy. This means that some rust staining occurs at an early stage of exposure even though the basic steel is protected. The thickness of the alloy coating depends upon time and temperature but is generally in the range 10–125 micrometres.

'Calorising' is the term used for aluminium diffusion coatings and is generally used for heat-resistant purposes.

Cladding

In this process a sandwich of steel between layers of the metal to be coated is heated and rolled to thickness. Good bonding between the steel and coating metal is essential. Stainless steel and aluminium alloys are the metals most commonly applied, but cladding of this type is not generally used for ordinary structural steelwork.

Other methods

Methods such as vapour or vacuum deposition and chemical treatments such as 'electroless plating', are not generally used for steel components in structural or construction work.

5.3.4 Corrosion resistance of metal coatings

Zinc

Zinc coatings are anodic to steel and under most circumstances afford protection by sacrificial corrosion at cut edges and damaged areas. This is an advantage, particularly as it reduces or eliminates rust staining on the coating, although such areas do have a finite life unless additionally protected by coatings. The protection afforded to steel by zinc tends to follow an S-shaped graph; that is, good for a length of time, but then degrading quite rapidly once a certain level of zinc loss has been reached.

The lower corrosion rate of zinc results from the formation of thin films of surface corrosion products. These films provide added protection from further attack. The rate of corrosion is governed largely by the exact nature of these protective films, which in turn depend very much on the exposure environment. For example, in acidic conditions such as industrial areas, the zinc will be attacked at a much greater rate than in clean rural areas. The effect of acidity is well illustrated by tests carried out in a railway tunnel many years ago when steam engines were used. The acidic exhaust gases and steam resulted in zinc corroding at about the same rate as steel.

Recently, a 'corrosion map' for the UK has been drawn up by MAFF (the Ministry of Fisheries and Food). This indicates the environmental improvements made in the UK over the past decades, but it still shows areas of severe potential corrosion to galvanising. In non-industrial inland areas the corrosion products on zinc tend to be basic in nature and of low solubility. Zinc coatings will last many years without further protection in truly rural areas, whereas in an industrial or coastal environment, the same coating would last a much shorter period of time.

Aluminium

The method by which aluminium protects steel is generally by providing a barrier coating of a more corrosion-resistant metal, not by being sacrificial. Aluminium corrosion products are less soluble than those of zinc and tend to form good protective layers in many situations. Tests on sprayed aluminium coatings of 80 micrometres thickness have shown that they can give protection to steel for 20 years or more in industrial and marine environments. It has been shown that on a surfing beach (a very corrosive environment) aluminium coatings are much better than zinc. Tests have shown that a long service life can be attained from aluminium metal fully immersed in seawater. Results from extended external testing of aluminium show that exposure to rain is beneficial to aluminium and that the gases often found in an industrial environment have little effect in accelerating the corrosion of aluminium.

It will be of interest to engineers to know that in order to achieve a good adhesion of aluminium metal spray to a structural steel substrate, the highest possible level of cleanliness is necessary. This highest level of surface cleanliness is known and identified by the standard ISO 8501-1 (Specification for rust grades and preparation grades of uncoated steel substrates after overall removal of previous coatings). This grade is known as Sa3. It is both difficult to initially achieve and to keep as 'white metal', for there is a natural tendency of the pure, clean steel surface to oxidise, even slightly. The lower or commercial grade of 'near white metal' is referred to as Sa1 1/2 by ISO 8501-1 and as SP10 by SSPC (the North American Standard). The ISO 8501-1 Sa1 1/2 level of cleanliness is quite satisfactory for zinc, but is insufficient for aluminium; a further indication of why aluminium metal spray is used less for structural steelwork than zinc.

5.3.5 Paint coatings

Paint has been around a long time and has been protecting and decorating steel structures since the industrial revolution. Paint remains the most common method of preventing deterioration of steel structures yet paint is a subject with which few engineers feel comfortable – whether they have a civil, design, maintenance or structural background. Perhaps this is due to the differing levels and philosophy of the technology involved in each discipline.

The paint chemists must learn to grow their 'living' organic polymer through adaptation of their science, rather than follow the more inanimate and precise metallurgical tools of the engineer's vocation. The fact that the parameters of paint remain a mystery to all but a few engineers may go some way to explaining the number of unsuitable paint systems which have, in the past, resulted in premature failures.

Engineers and specifiers do not need an in-depth knowledge of paint technology, but a broad understanding of the nature of the different types of the most usual forms of protection of steel structures is useful. In this chapter, it is intended to clear away some of the confusion regarding the use of paint to protect steel.

A basic insight into the make-up of the common types of paint used on structural steelwork, will help to avoid the adoption of unsuitable paint systems and provide realistic expectations for modern paint systems, examining how they may be used successfully to delay the deterioration of steel structures.

Paint is an extremely complex engineering material and is a complicated blend of organic and inorganic materials. Complex detail is not included in this chapter. The intention is to provide an overview of the essential components of paint, whilst correcting possible erroneous beliefs and demystifying some of the common jargon.

The types of paints referred to in the following pages are limited to those encountered in the protection of steel structures. Materials employed for decoration alone, such as emulsion paints, will not be covered. It is beneficial to regard the types of paint used for protection as coatings, rather than paint.

The differentiation between paints used here, is that coatings offer an inherent level of protection from corrosion. Additionally, a protective coating may be anticipated to have a thickness of dry film of not less than 50 micrometres. This thickness is just two-thousandths of an inch, or approximately the thickness of a normal human hair, hence one may imagine it does have to be rather special to afford the level of protection demanded. Some protective coatings, and a full coating system, may have a recommended dry film thickness an order of magnitude above this.

It remains interesting to consider that a protective coating is seldom impervious to either moisture or oxygen, so, again, the mechanism by which coatings protect steel has to be rather special. Paint is formulated to meet a specific set of design characteristics, derived from market intelligence, customer preferences and sales demands.

It has also become a prime requirement to satisfy legislative and cost constraints. It must be compatible with other coatings, with which it will be used to form the necessary protective coating system, and comply with design, local or national standard requirements set by third parties.

The 'recipe', or formulation, for a protective coating will, typically, contain some 30 different ingredients. For simplicity, they may be classified under the four following headings:

- resins or binders;
- pigments and extenders;

- solvents and diluents; and
- additives.

Each class of material imparts its unique properties to the make-up of the coating. In order to gain an appreciation of the general nature of the types of protective paint coatings found as deterrents to the degradation of steel structures, the characteristics of each class needs to be investigated.

It is sensible to start with the resin component. This is because the majority of the final properties of the coating will, to a large extent, depend upon this 'binder' material. 'Binder' is an excellent name, as it is the resin material, which 'binds' the remaining ingredients together into a homogeneous coating. Indeed, the majority of paint coatings used to protect steel structures are actually classified and referred to under the name of the main resin employed in the coating formulation.

A broad list of the resin, or binder, components which lend their names to the generic type of coatings currently be found on structures, is set out in Table 5.1 below.

The scope of the variety may be exemplified by the fact that many in the above list are available in a number of forms. Whilst the liquid form is the most common type in use on steel structures, epoxies, polyurethanes and polyesters are frequently encountered as having been applied by powder coating.

Table 5.1 Main resin types

Oil based	**Chlorinated rubber**
Oleo-resinous	Acrylated rubber
Alkyd	Vinyl
Epoxy & coal-tar epoxy	Acrylic
Polyurethane	Bitumen
Zinc silicate and silicones	Poly-siloxane

In the case of powder coatings, application is made in a factory or contractor's shop by an electrostatic spray or by the fluidised bed technique, followed by fusing in an oven. The original misnomer 'organic vitreous enamel' is, fortunately, rarely met these days and the items so coated are generally piece parts, particularly finding favour in shop fittings.

The liquid forms will be one of the following types; solvent based, solvent borne, high build, high solids, water based, water borne, solvent-less, solvent-free, plastisol, or organasol. The types of solvents will be reviewed later.

It should be noted that whilst 'water based' and 'water borne' indicate that water is the main solvent used, significant levels of organic solvent may be contained within the formulation of the protective coating to aid coalescence, or film formation.

Most generic types of paint coatings are capable of being formulated as a primer, an undercoat or build coat, and as a finish coat, in many of the solvent classes noted above. To take primers as an example, further specialisation occurs when one considers that primer types may be classified according to function, such as etch primer, wash primer, blast primer, weldable primer, prefabrication primer, post fabrication primer, primer-sealer or surface tolerant primer. Added to this is the further classification of a primer according to, not just generic type, solvent category and function, but to mode, e.g. zinc-rich primer, zinc phosphate primer, aluminium (mastic). Clearly the naming issue can be thought to be getting a little out of hand!

The variations given above give rise to a feeling of some trepidation. Hence many engineers consider it a wise precaution to employ consultants, who are specialists in the field of paint, to prepare, or at least audit, specifications for major steelwork projects.

The purpose of the next few pages is to show that there is an order to protective coatings and to give guidance on how to navigate this potential maze, by reducing paint coatings to their basic essentials. Previously it was found sensible to first consider the types of generic resin or binder that are used.

These resins may be conveniently split into two types as follows:

1. Type-one resins are 'non-convertible'. They are exactly the same substance in liquid form in the original container, as when they have been applied, lost their solvent portion and are fully dry on the steel. These have been listed on the right-hand side of Table 5.1 with the exception of the polysiloxane.
2. Type-two resins are 'convertible'. They undergo some form of chemical change between being the liquid in the container and the dried (or cured) film on the steel. The change is not simply due to the loss of solvent. These have been listed on the left hand side of Table 5.1.

Type-one products dry wholly and solely by evaporation of the solvent used in their composition and are thermoplastic by nature. Type-two products, change, or 'cure' rather than simply dry. They may be single pack or multi-pack materials, but they tend to be thermoset by nature.

Each type of resin will now be considered, with the characteristics given by the resin to the coating.

Oil-based paints

These were the earliest type of paint and are based on treated natural oils like linseed, sunflower and tung. These single pack coatings are very good at wetting steel surfaces, particularly those prepared by mechanical hand tools; they dry by oxidation and reactions of the fatty acid content of the oil. This means the drying characteristics are slow, as is the rate of application, as brushing is the preferred method for all oil-based coatings.

'Red lead in oil' paints tended to be the standard by which others were measured, but acute lead toxicity has meant that the application of this classic primer material has virtually ceased. However, we now have the problem of removing the residue of previously applied red lead on many existing structures. This must be carried out safely, with great care and expertise. Similar environmental problems have been caused by the use of calcium plumbate primers, particularly on old galvanised window surrounds, where the high lead content makes containment and licensed disposal, absolutely essential.

The use of oil-based paints on steel structures is now limited to the restoration work of structures important to the national heritage.

Oleo-resinous paints

These modified oil-based paints were originally developed to maintain the ease of brush application and the durability of single pack oil paints, but with improved drying times. This was achieved by blending the oil with either natural or synthetic resins. Water resistance, particularly with phenolic types, is good, but whites and pastel shades tend to yellow, so where used, they are mainly restricted to primers and undercoats.

Alkyds

These paints incorporate a huge range of synthetic resins. They are single pack, convertible, thermoset materials, which are still in use today, although mainly for internal steelwork. The protective coating types are derivations of the type used in homes to paint wooden doors, surrounds and skirting boards. Alkyd paints are, generally applied by brush or roller. They are preferred by applicators, who have their roots in painting and decorating, rather than those with an industrial paint contracting background.

Oils, or their fatty acids, are employed in the manufacture of alkyds and industrial coatings, and with chemical modifications they became established as 'quick-drying synthetics'. Alkyd paints have reasonable temperature and water resistance, but should not be used for frequently wet conditions. Silicone alkyds have improved colour and gloss retention and urethane alkyds have the capability of being formulated to enhance their durability.

Epoxy esters

This single pack epoxy is a thermoset material, best regarded as a modified alkyd, which has improved chemical resistance. Alkyds and epoxy esters must not be used on zinc surfaces, as a reaction termed 'saponification' will occur in moist conditions, leading to failure due to a severe loss of adhesion. Since the advent of water-miscible types, alkyds have enjoyed something of a resurgence as candidates for steel structures, even to the extent that weldable prefabrication primers are now available.

Chlorinated rubber paints (CRP)

Chlorinated rubber paints are of the non-convertible, thermoplastic variety and enjoyed a position of eminence as single pack coatings for steel structures between the 1960s and the 1980s. Chlorinated rubber coatings have many benefits; they are resilient to damage, but easily repairable should transit or mechanical damage occur. Their availability in a range of practically non-yellowing colours, in gloss or satin finish, each possessing good chemical resistance and infinitely over-coatable with themselves made the chlorinated rubber range the first choice for coating structural steelwork.

In addition to these advantages, it was possible to formulate chlorinated rubber coatings for use in leisure pools and to achieve the highest value for nuclear decontamination ever recorded. The fall from a position of common use to one of rare occurrence was dramatic and was forced on the paint industry by resin manufacture. The trigger was in 1988, when the United Kingdom (UK) was one of the many governments that signed the Montreal Protocol, agreeing to cease the use of ozone-depleting materials. Chlorinated rubber uses an ozone-depleting material in its manufacture. The replacement of this material was found to be both technically and commercially impossible. The small residue remaining in the paint, also proved unacceptable.

The final act, which excluded chlorinated rubber was the UK Environmental Protection Act (1990). With the aim of reducing the quantity of solvents emitted to the atmosphere, this Act limited the amount of solvents used in all paint. The large quantity of highly aromatic hydrocarbon solvent used to dissolve the solid chlorinated rubber was impossible to reduce, consequently the material was not compliant with this Act either. Global paint manufacturers removed the range from their lists and although a derived material remains available from two sources, the material is no longer used on structural steel.

Acrylated rubber (AR) was at one time considered as a replacement for chlorinated rubber, because it possesses some of the features of its predecessor. Acrylated rubber used no ozone-depleting substances in its manufacture and required less aromatic solvents to prepare a liquid form. However, the solvent content was too high to become compliant and although it was used in concrete construction, where it was formulated to allow the passage of moisture, but not carbon dioxide, its use on structural steel is now scarce.

Vinyls are a further group of single pack, non-convertible, thermoplastic coatings with high potential for use in aggressive chemical environments. Requirements for strong solvents in large amounts have limited the use of this range, which may further shrink with the ever tightening reductions sought in the use of volatile organic compounds (VOCs) in paint coatings.

Acrylics are a large group of resinous materials, related to vinyls and may be produced as non-convertible or convertible coatings. It is the second class, which are employed most often in structural steelwork. Their excellent pale colour helps to produce topcoats with good gloss and colour retention.

Acrylic modified polyurethanes, or the newer non-isocyanate types are becoming established as the top coat of preference for structural steelwork exposed to aggressive industrial and coastal environments.

Two-pack epoxies form an enormous range of products marketed with the 'base' component in one package and the 'hardener' (or 'curative' or 'reactor') contained in a separate, smaller pack. The predominant classes of curing agent for epoxies are polyamide, for general duties, or polyamine for enhanced chemical resistance.

Epoxies provide tough, hard chemically resistant coatings, but this was one of their drawbacks. Age embrittlement and the difficulty in over-coating such hard films have now largely been formulated out of third generation epoxies that are used for structural steelwork. Added to this, the slowness of cure of pure epoxies has been overcome by modern variations, which cure below 0°C and are now infinitely over-coatable. Negative features of epoxies are limited to concern for their skin sensitisation characteristics in liquid form and the 'chalking' of the surface of all epoxies subjected to ultraviolet light, a subject covered later.

Polyurethanes are a large group, often just referred to as PU. Coatings prepared from polyurethanes may be single- or two-pack, but are always classified as convertible. It is possible to produce liquid polyurethanes, which dry to films as thin as 25 micrometres or as thick as 1000 micrometres in a single coat. Added to this, two-pack polyurethanes can continue to cure in very cold conditions, even claimed to be as low as −17°C. They are a truly versatile group. Aromatic polyurethanes have the best resistance properties, but they react with aliphatic polyisocyanates to enhance their overcoatability, they provide a higher order of gloss and colour retention and are often used as a finish coat to top coat epoxies on steel structures.

Single-pack polyurethanes, which cure by reaction with atmospheric moisture, have been used on structural steel, but have properties closer to an alkyd than to the two-pack protective polyurethane coatings generally employed as finishing coats. Despite the use of di-isocyanates (regarded as a compound toxic to people and capable of causing severe respiratory problems) in their manufacture, coatings based on polyurethanes may be used as safely as any other coating, provided sensible precautions, as detailed on the label of the container, are followed.

As with all potentially harmful materials, a thorough understanding of the Product Data Sheet and the Material Safety Data Sheet, are necessary before adoption and use of liquid coating materials are considered.

Zinc silicates are inorganic, convertible materials, which find use as thin, 18–22 micrometre coats of weldable pre-construction primer, or 75 micrometres of post-construction primer. The former allows the production of high integrity welds by automatic and manual means, with negligible burn-back of the coating and a remarkable absence of weld spatter. The latter have enjoyed success on steel structures subject to severe coastal or off-shore environments due to their tough, tenacious nature and have also been used, over-coated with silicones, where resistance to high temperature is required.

Other silicones, silicates and siloxanes have only limited use in specialised areas of structural steelwork corrosion control, except for the advent of epoxy-polysiloxanes, which are a further alternative to isocyanate-free, colour stable top coats.

Bituminous coatings and coal-tar epoxies (CTE)

These are somewhat old-fashioned, black substances. The former is non-convertible and always soft whilst the latter is a convertible, two- or three-pack material. Their use is restricted to buried steel, or water immersed steelwork, or that suffering near constant condensation. The latter is also potentially carcinogenic in liquid form and is virtually banned on mainland Europe, but not in the UK, where its use remains the most cost-effective protection of semi-submerged structural steelwork, despite coal-tar free versions being available. However, their use is rapidly diminishing.

An ultra high-build elastomeric polyurethane version has had some success as a liquid applied retro-fit damp-proof membrane on, for instance, car park decking.

Unsaturated polyesters are only employed where extreme abrasion or chemical corrosive attack is likely. The only use of thermally cured phenolics in steel structures is within the tubes of heat exchangers.

Having dealt with the main component, the resin or binder, at some length, we need to deal briefly with the other three classes of ingredients for coatings. These groups of materials are covered below.

Pigments and extenders

This group of raw materials gives the coating its 'body' and provides, with the resin, the bulk of the dry coating. The difference between a pigment and an extender is essentially one of refractive index. Pigments are generally synthetic and expensive, providing both opacity and colour to the paint coating. Extenders tend to be considerably less expensive and impart neither colour nor true opacity to the dry paint film. It was stated earlier that the use of red lead and calcium plumbate have ceased due to toxicity issues. Previously popular corrosion-inhibiting pigments such as zinc chromate, zinc tetroxy-chromate and strontium chromate are no longer used because of known carcinogenic effects.

Following the lead set by North America, the same fate has befallen the brightly coloured, high opacity yellow and orange lead chromates and lead molybdo-chromates, which are now, classified as potentially toxic to people. Red oxide is renowned as a pigment, with its use particularly prevalent in primers. The material is cheap, has excellent opacity and a pleasant brick red colour. It frequently comes as a surprise when people learn it has no value as a rust inhibiting pigment. Indeed, being red iron oxide, it may be considered essentially as a form of rust itself!

A special pigmentary form of iron oxide, much used in Europe but seldom in North America, is worthy of mention. This is micaceous iron oxide (MIO), a lamellar and inert form of iron oxide. The pigment, when correctly treated, lies as discrete platelets in the paint film, and increases the difficulty for the aggressive corrosive ions in penetrating the dry paint film. In this way, the effective film thickness to permeation is increased. Glass flake pigments have a similar function in decreasing permeability and thus assist in reducing the passage of corrosive ions through paint films.

Titanium dioxide is the most widely used white pigment. Coloured pigments used in top, or finishing coats must have sufficient light fastness in full colour and when 'diluted' with white into pastel shades, to retain the hue throughout the life of the system. For this reason, many bright colours used in printing inks are usually unsuitable for use in paint coatings. Suitable coloured pigments used may be inorganic, such as chromium oxide green, or organic, such as phthalocyanine blue. Examples of common extenders, which are simply used to add bulk, are blanc fixe (inert barium sulphate) and chalk. Extenders, which do have an effect on the paint coating are talc, which imparts a nice smooth feel and 'slip' on sanding and kaolin, which helps to provide a form of 'false body' or thixotropy to the paint coating.

Both of these extenders also help to ease the settlement of the pigmentation of a coating, on prolonged 'shelf' storage. Solvents and diluents help to dissolve or dilute the resin. Typical solvents are Xylene and MiBK (methyl isobutyl ketone). Water and white spirit are not generally found as solvents for coatings for structural steelwork, but the former may well become a significant solvent issue in the future.

Solvents play a part in the manufacture of the coating, the application of the coating and even the film-forming nature of the coating. They are not present in the dry film; hence they may be ignored for the purpose of this chapter. Diluents are not true solvents, but are used to impart enhanced rheology, drying or commercial characteristics to the coating. In the new millennium, both solvents and diluents must be reduced or used sparingly, to have the lowest achievable environmental impact.

Additives are generally used in small, but essential amounts, for a variety of reasons and do impart a range of benefits to the wet and dry paint coating. At the manufacturing stage, additives are used to aid dispersion of the pigments into the resin and to reduce foam, the possibility of settlement and the possibility of flocculation. At the application stage, additives provide rheological benefits by imparting thixotropy and false body, flow properties, combined with sag resistance and aid drying and adhesion to the substrate, with dry films improved in gloss and mar resistance by appropriate additives.

5.3.6 Paint and coating testing

All paint coatings are thoroughly tested during their research and development (R&D) stages to ensure compliance to their design and concept criteria. During production the manufacturer relies on simple quality control testing to ensure the coating meets set tolerances. In today's global markets, paint coatings can be manufactured in a country remote from the land of use. Manufacturing costs in Third World countries and China are very low, as are the costs for local raw materials. Transport to the Americas and Europe from these sources adds little to cost, but much to the bottom line profit of global paint manufacturers.

Even paints of known European manufacture are subject to commercial forces to reassess the use and content of expensive pigments, additives and resin blends, such that cost-optimisation is as vital as development and more commercially lucrative, in the short-term anyway, than research.

For this reason, major users insist on samples being taken from unopened tins on-site for audit checking by independent test houses. The surprisingly high number of 'failures' has ensured that this system of control is continued, with several potentially catastrophic premature failures on infrastructure steelwork in the UK having been prevented on a number of major trunk road and motorway bridges.

On-site testing and inspection is important, with independent inspectors ensuring that: environmental conditions, substrate suitability and cleanliness, mixing, the temperature of paint, application of coats of wet paint, assessment of dry film thickness and the integrity of the coating systems are all within the parameters which will lead to the completion of a coating system that will meet the lifetime requirements determined at the design stage.

5.3.7 Cathodic protection

Cathodic protection is an important method of corrosion control, but is limited in its application to structural steel. This is because cathodic protection is only used effectively where steel is buried or immersed. There are two methods of protecting steel by this means.

Firstly, where sacrificial anodes made of a more reactive metal are fixed to the submerged steelwork, the anode material, being at a more negative potential than the steel, will corrode in preference to the steel. Sacrificial anodes need to maintain good electrical contact throughout their life and require inspection and replacement at appropriate intervals. Sacrificial anodes may also be used as 'just in case' protection on submerged steelwork, which has been protected with a coating, but which is considered liable to become damaged in service.

Coatings also reduce the amount of steel in contact with the corrosive environment, hence reducing the area of steel to be protected by sacrificial anodes and thereby reducing costs. Ships, off-shore structures and some lock gates have employed sacrificial anodes with success.

The second method of cathodic protection is by the use of an impressed direct electrical current. In this system, current is supplied through an inert anode, so that the whole of the buried or submerged steel structure becomes the cathode of an electrical circuit. In this way the electrochemical reactions at both anode and cathode are suppressed and the steel is protected from corrosion.

Some success has been claimed with this system on, for instance, steel reinforcing bars embedded in concrete, but with conventional paint coatings, any over-potential within the circuit may lead to delamination or blistering of the coating, particularly if susceptible to cathodic disbondment.

5.3.8 Tapes, wraps and waxes

Wrapping steel items with tape, whether self-adhesive or on top of an adhesive or solvent-free coating, is generally restricted to pipes and tubular sections. The wrapping is carried out by machine prior to burying and gives good resistance in the case of angular back-fill. It is possible to hand-wrap structural steel components, but this method is time consuming and it is difficult to achieve high quality work.

The composition of the tape may be petroleum based wax or grease, reinforced with polypropylene, natural fabric, bituminous or polyvinyl chloride (PVC) wrapping. Cementitious coatings, sometimes with the addition of a water dispersible polymer, have been used over petrolatum tapes, to provide a degree of physical toughness and protection. As with wax based coatings, these materials are seldom found on steel structures, unless used in inaccessible areas, but they may be traced to supplemental pipe work, associated with the utility supply to structures.

5.3.9 Environmental conditioning

Another method of providing protection from corrosion is to remove the elements, which are corrosive to steel structures. By using air-conditioning or dehumidifiers, the internals of steel structures may be protected, provided a constant level of less than 50% relative humidity is maintained and that this is acceptable to any personnel working within the environment.

Attempts have been made to purge vessels with dry nitrogen or to enclose silica gel in sealed box members, but these have been found wanting for long-term assurance of freedom from corrosion and are of limited use in most steel structures.

5.4 Coating failures

Having examined the nature of corrosion, paint, specifications and surveys, it is appropriate to consider the nature and type of coating failures which reduce the life of structural steel.

Coating failure may be anticipated and may even be the preferred mode of ageing; or it may be unexpected and regarded as premature. The nomenclature, which includes industry jargon, is referenced below.

5.4.1 Types of failure

Blistering

Blistering of coatings is generally caused either by solvents, which are trapped within or under the paint film, or by water which is drawn through the paint film by the osmotic forces exerted by hygroscopic or water soluble salts at the paint–substrate interface. The gas or the liquid will then exert a pressure, stronger than the adhesive strength of the paint. If the pressure becomes greater than the cohesive strength of the paint film, the blister breaks.

Solvent entrapment can be due to 'skin curing' of the top layer of the coating film, by too thick a film, or by over coating, heating or immersing too quickly. The measurement or evaluation of the degree and size of blistering can be carried out by the use of photographs or diagrams in Standards such as in ISO 4628-1 (photographic standard for the measurement of blisters in paint films).

Corrosion blistering

Coatings applied to iron and steel generally fail by disruption of the paint film caused by the formation and accumulation of a large volume of corrosion products at the coating–metallic substrate interface. If the thickness of the paint coating is inadequate, then the corrosion will be of an overall nature. But when the corrosion is due to water and aggressive ions being drawn through the film by the osmotic action of the soluble iron corrosion products, then the attack will start from corrosion pits. This can be overall or confined to isolated areas.

Characteristically, paints on iron and steel often fail in the same areas from maintenance period to maintenance period. Corrosion does not automatically spread under paint films if the adhesion is good. The blistering of paint films on steel in seawater involves the operation of corrosion cells on the metal surface. Iron dissolves in anodic areas and hydrogen evolves in cathodic areas leaving an accumulation of sodium hydroxide in the water of the cathode blisters. Anode blisters that develop before the appearance of rust are small and filled with acid liquid. These readily fracture and become the seat of anodic pits. Paints based on linseed and other drying oils, including alkyds, are attacked by alkali and therefore these paints are especially prone to blistering under immersed conditions.

Filiform corrosion

Filiform corrosion is a characteristic form of corrosion, which normally appears on painted cold rolled steel, for example car finishes and office steel furniture. In appearance it looks like thin snail-like trails of corrosion under the paint film. The paths are comparatively straight but change direction when they approach the cathodic liquid of another path.

Loss of adhesion (flaking)

Loss of adhesion is generally visible as the lifting of the paint from the underlying surface in the form of flakes or scales. If the cohesive strength of the film is strong, then the coating may form large shallow blisters. Excessive mechanical damage during handling can be a symptomatic of poor adhesion.

Other causes include:

- loose, friable or powdery materials on the surface before painting (e.g. rust, dirt, salts, dry spray, mill scale, chalking, etc);
- contaminants that prevent the paint, from 'wetting' the surface (e.g. oil, grease, silicones, amine bloom, etc);
- too smooth a surface to provide mechanical bonding (e.g. new galvanising, aged polyurethane or epoxy coatings, too shallow blast profile, etc);
- application of catalysed materials in a too-advanced state of their curing (e.g. in excess of the material's pot life).

Chalking

Chalking is the formation of a friable, powdery coating on the surface of a paint film caused by the disintegration of the binder due to the effect of weathering, particularly exposure to the actinic (photochemically active) rays of the sun and to condensation from dew.

Different binders react at different rates, for example, epoxies react quite quickly whereas acrylics and polyurethanes can remain unchanged for long periods. Chalking of epoxies however is considered only a surface phenomenon and therefore not harmful except to the decorative appearance.

Oil-based paints are thin enough to be eroded away so that eventually the undercoat is 'grinning' through. In all cases it is considered the most acceptable form of failure since the surface preparation for subsequent maintenance consists only of removing loose, powdery material and then applying further compatible coatings. It is not necessary to blast clean to the substrate.

Cracking

Cracking is visible in several different forms and degrees:

- 'Hair cracking' consists of fine cracks which do not penetrate the top coat and occur erratically and at random over the surface.
- Fine cracks which do not penetrate the top coat and are distributed over the surface giving the semblance of a small pattern are known as 'checking'.
- 'Cracking' resembles checking but the cracks are deeper and broader.
- 'Crocodiling' or 'alligatoring' is a severe type of crazing that produces a pattern resembling the hide of a crocodile or an alligator. This may be caused, by applying a hard paint film, over a soft layer, for instance an epoxy over an acrylated rubber.
- There is a special form of cracking called 'mud cracking', which resembles the appearance of dried out mud, but on a smaller scale. This effect tends to be specific to applications of zinc silicates, which are well above their recommended dry film thickness.

All paint films, but especially coatings containing solvents, are subjected to internal stresses on curing or drying. This may have the effect of reducing their tolerance to external stresses. Thus if a coating cracks spontaneously, it is because the internal stress has increased to a value greater than the tensile strength. Internal stresses are dependent upon plasticisation, pigmentation, ageing and the conditions under which the coating was applied and cured.

Some paint films can be sufficiently flexible for most purposes, but not when applied over a softer or more flexible undercoating (e.g. alkyd paints over bitumen generally crack and in the particular form known as crocodiling).

Paint systems can also fail by cracking when the substrate is affected by thermal expansion and swelling due to variations in the temperature and the relative humidity of the atmosphere. This particularly applies to paint coatings on wooden components abutting steel structures.

Wrinkling

Wrinkling of a paint film occurs when the surface expands more rapidly during the drying than the body of the paint. This can occur with oil paints, which normally contain additions of metallic driers, if they are applied too thickly.

Runs and sags

A run is a downward movement of a paint film over an otherwise flat surface, which is often caused by the collection of excess quantities of paint at irregularities on the surface, i.e. cracks, holes, etc., the excess material continues to flow after the surrounding surface has set. It is unwise to consider such defects as affecting only the appearance and therefore unimportant for industrial applications, since due to surface tension effects, the perimeters of these areas can be accompanied by 'pinholes' or 'holidays'.

Runs and sags in a paint film application can be symptomatic of more than just over-thick application or excess thinners. With two-pack catalysed materials it can also indicate a failure to add the catalyst correctly or using paint beyond its shelf life.

Pinholes

Pinholes are minute holes formed in a paint film during application and drying. They are often caused by air or gas bubbles, perhaps from a porous substrate (e.g. metal spray coatings, or zinc silicates), which burst, forming small craters in the wet paint film and then fail to flow out before the paint has set. The term 'holidays', is used to define either skipped or missed areas, left uncoated with paint, or to mean pinholes.

Cissing, crawling, fish-eyeing

These are all forms of surface defects where the paint has not wetted the surface correctly and the wet paint recedes from small areas leaving either holes of various shapes and sizes, or attenuated films. The cause is generally oil, grease or silicones on the surface to be painted, or water or oil in the paint during application. Two typical sources of this type of contamination are the air supply to the spray gun and the use of the wrong solvent.

Blooming or blushing

Bloom (often called blushing in the USA, although in the UK this term is confined to this phenomenon on lacquers that cure by rapid evaporation), is a deposit like the bloom on a grape which sometimes forms on glossy finishes, causing loss of gloss and dulling of colour. It is generally caused by high humidity, or condensation during the curing period. This occurs with amine, cured epoxies in particular, when applied with the relative humidity above 85%.

Blooming is not generally considered detrimental to the coating, but can reduce the adhesion of subsequent overcoats. The bloom can normally be removed by careful wiping of the surface with a rag saturated in the appropriate solvent, but this needs to be carried out safely. In the UK this would mean after a full Control of Substances Hazardous to Health (COSHH) assessment had been undertaken.

Lifting or pulling-up

This is the softening and expansion of a previously applied paint when a new paint with strong solvent is applied over it. The effect is similar to that obtained with paint removers and is sometimes termed 'pickling'. It can be prevented by a careful choice of over-coating paints to ensure that their solvents are compatible, or in some instances, allowing the previous coat to dry and harden to a more complete extent.

Complete or partial failure to cure of two-part materials

Two-pack materials such as epoxies and polyurethanes can sometimes dry without the addition of the curing agent. However, such a film is incorrectly cured and will not give the service intended. The paint film will also be softer than the fully cured coating and may have the tendency to sag. Even when the two materials are added together they must be mixed adequately and in the correct ratio, otherwise the polymer formed will be less durable.

There might also be a tendency for the mixed coating to have a shorter pot life than expected and may even 'set up' in the pot or the spray lines. With the exception of the isocyanate cured epoxies, most of the other two-pack epoxies will not cure at ambient temperatures of below 5°C, unless they are of the modified and infinitely recoatable, third generation type.

Dry spray

This occurs when spraying paint and when the particles hitting the surface are insufficiently fluid to flow together to form a uniform coating. The result is a powdery cohesively weak layer. The integrity of the coating is, therefore, not attained and this will lead to a loss in the properties of the system as a whole. The defect is generally caused when rapid drying materials such as zinc silicates or two-pack epoxies are sprayed, with the gun too far away from the surface, or when there are very strong wind cross currents.

Bittiness

A 'bitt' finish is one with particles of skin, gel, flocculated materials or foreign matter projecting from the surface of an applied paint film. It may also have been caused by insufficient dispersion during manufacture, or, more likely, by the filter splitting after a sample had been taken for quality control, but before the batch has been filled.

The term 'seedy' specifically denotes bits, which have developed in a paint during storage and is frequently initiated by the type of thixotrope used.

Bleeding

This is the diffusion of a soluble coloured substance, such as bitumen, wood preservative, etc., from, into and through, a paint from beneath. It may also occur when strong solvents are used in a subsequent paint coat when applied to a coating which has only semi-resistant pigment dyestuffs as colourants.

Blocking

This is the unwanted and undesirable sticking together of painted items when stacked for storage. Over-thick applications of chlorinated rubber paints and under-cured two-pack epoxies are typical examples. The painted items have to be separated with considerable difficulty and generally considerable damage occurs to the coating contact points on the top and the bottom faces.

Cheesyness

This is the character of a film of paint, which although dry, is mechanically weak and rather soft. It may be due to the retention of a diluent within the film. Two-pack paints where the hardener has been omitted will exhibit the appearance of being 'semi-hard' like a piece of cheese.

Efflorescence

This is the development of a crystalline deposit on the surface of brick, cement, etc., due to water containing soluble salts, coming to the surface and evaporating, so that the salts are deposited. In some cases the deposit may be formed on the top of the paint film, but usually the paint film is pushed up and broken.

Fattening

An increase in consistency of paint during storage. This increase is not necessarily to such an extent as to make it unusable, but the paint becomes more viscous, indicating that some initiation of a polymeric reaction of the resin or binder has taken place, increasing its molecular weight.

Feeding

An increase in consistency of the paint, to such an extent, that it is unusable except by undue and excessive thinning. This is generally due to the chemical reaction between its constituents and in most cases the polymer is in an advanced state of reaction. In such cases, the paint should not be used and if it is still within its 'shelf-life', it should be returned to the manufacturer with a complaint, rather than simply be discarded.

Floating or flocculation

This is the variegated colour effect in either wet paint or applied film due to the partial separation of one or more of the pigments. It is generally a fault in the composition of the paint caused by aggregation of the pigment particles.

Grinning through

This is where patches of the underlying surface show through the topcoat, due to inadequate opacity of the paint film. This may be due to too thin an application that is below the recommended level of wet or dry film thickness, or to pigmentation. The use of lead-free pigments to match bright, full colours often require an extra coat of paint to achieve an adequate level of obliteration of previous coats.

Orange peel or pock marking

The pock-marked appearance of a sprayed film resembling the skin of an orange is due to an incorrect spraying technique, too high a pressure or, rarely, inadequate flow from the paint coating, which may not have been formulated for spray application.

Pulling

The behaviour of a very quick-drying paint, which during application by brush, becomes so sticky that the resulting film is thick and uneven.

5.4.2 Causes of coating failure

These may be separated into seven distinct categories.

1. Faults due to inadequate specification:
   a) Use of incompatible materials, for example, a hard coating over a soft coating, or mixing convertible and non-convertible coatings.
   b) Failure to specify stripe coating for edges, nuts, bolts, threads, etc.
   c) Specifying the wrong curing agents or the wrong type of materials for ambient conditions, during application or in service.
   d) Specifying a finish coat to be applied at site, but taking no account of the maximum over-coating time–temperature relationship of the preceding, shop or works applied coating.
2. Faults due to coating materials:
   a) Variations between batches of paint.
   b) Use of cheap and inferior raw materials in response to market competition.
   c) New products tested under accelerated conditions in the laboratory, not performing the same way in practice.
   d) Paint suppliers, not being given the full details of the operating conditions and then recommending, or supplying the wrong type of material.
3. Faults due to surface preparation:
   a) Inadequate cleaning of a surface to be painted so that it remains contaminated or in an unstable condition.
   b) Failure to abrade or 'key' a chemically cured coating surface that has exceeded its maximum over-coating time.
   c) Contamination of surrounding newly applied paint with dirt or abrasive particles, often caused by blowing spent abrasive with an air-line, instead of vacuuming the whole area.
   d) Use of steel that has uncorrected manufacturing surface defects such as laminations.
   e) Failure to clean and dress welds adequately, leaving sharp weld areas instead of a smooth ripple (or preferably a radiused weld for arduous conditions) and non-removal of residual weld spatter.
   f) Damage of surrounding paint, by spot blasting procedures.

4. Faults due to application:
   a) Inadequate mixing of paint so that the final product has inferior properties.
   b) Incorrect or careless spraying technique so that the coating is dry sprayed.
   c) Misses, or thin areas.
   d) Use of two-pack materials beyond their pot life.
   e) Ignorance of the induction time requirements of some two-pack materials.
   f) Failure to 'condition' heavy duty protective coating paints, by not leaving them in a warm environment the day before intended use.
   g) Use of poorly maintained spray apparatus, such as incorrect 'set up', badly worn and overlarge fluid tips or nozzles, the fluid hose being insufficiently cleaned so that the flow is restricted and there is incomplete atomisation of the liquid coating.
   h) Application of incorrect wet film thickness, being either too thick or too thin, and failure to check this on a regular basis.
   i) Inadequate access for the painter to carry out the work satisfactorily.

5. Faults due to ambient conditions:
   a) Too hot, causing poor flow of paint across the surface.
   b) Too cold, causing incomplete or very slow, cure of paint.
   c) Too windy, causing dry spray, over spray and dirt contamination.
   d) Rain or condensation on the prepared surfaces or on the surfaces just painted.
   e) Humidity too high for the materials which are sensitive to such conditions.
   f) Humidity too low for the materials that require a minimum humidity.

6. Faults due to poor inspection:
   a) Incomplete observation and monitoring of the full coating procedure.
   b) Ignorance of the properties and requirements of modem coating systems.
   c) Incorrect use of instruments.
   d) Indiscriminate use of destructive tests.
   e) Use of faulty or non-calibrated instruments or those with an incorrect range.
   f) Lack of recognised inspector training and qualification (Institute of Corrosion).

7. Faults due to transport and storage:
   a) Rough handling of painted items.
   b) Storage too close to ground or on unsuitable surfaces.
   c) Stacking of materials, not completely cured (e.g. acrylated rubber coated items).
   d) Contamination during the incomplete paint system on sites or in fabrication shops.

5.5 Conclusion

This chapter has given an insight into some of the methods and materials employed to prevent the corrosion of steel structures. The intention has been to outline the range of materials available and explain the nomenclature used, so that it is easier to understand the use of such terms (jargon). The desire has been to assist in the assimilation of the basic data for protecting steel structures from deterioration due to corrosive forces.

It is hoped that the guidance offered has achieved, in part, this mission. If so, it will reduce the risk of premature coating failures from occurring and help the reader to obtain the optimum performance from the correct coating, correctly applied and gain the benefit of improved whole life costing.

Further reading

Bayliss, D.A. & Deacon, D.H. (2002) *Steelwork Corrosion Control*, London: Spon Press.

See also www.steel-protection.co.uk and the websites of other organisations mentioned in this chapter.

6 The Durability of Aluminium Structures

Dr John W Bull

This chapter describes the use of aluminium in a wide variety of engineering applications and addresses aluminium, its alloys, design, fabrication, performance, products and welding. The chapter concentrates specifically on the durability of aluminium, its natural high corrosion resistance and how this resistance and its durability in corrosive environments can be enhanced by design and corrosion protection. Examples are given of the wide range of products where aluminium is used. Specific examples of award winning structural designs are included.

6.1 Introduction[1,2,3]

6.1.1 General

Aluminium is the most abundant metal in the earth. It does not occur in the free state, but is found in a very stable combination (bauxite) with oxygen and other elements, but it requires high temperatures and large amounts of energy to reduce the combination to aluminium.

Aluminium was first isolated in the early nineteenth century and remained a scientific curiosity until the 1850s when attempts were made to develop it commercially using a chemical process. Until the 1880s it remained a high cost metal used for such things as spoons and statues when an electrolytic process for the industrial production of aluminium was developed independently in both France and in the USA. By the end of the nineteenth century aluminium was used for housing, panelling, railway wagons, roofing and torpedo boats.

As a structural material, aluminium first came to prominence in the early twentieth century with its use for airships and again in the middle of the twentieth century for its use in military aircraft. Since then aluminium has been used for bridge construction, buildings, hulls of multi-hull ships, modules on offshore oil platforms, railway carriages, roofs, transmission towers, the superstructure of ships and vehicle bodies.

The major constraint on the exploitation of aluminium is customer conservatism. Aluminium's metallurgy, properties, fabrication and service behaviour in all environments has been well studied and reported. The highly specialised aviation industry has been the driving force behind many of the significant developments in aluminium alloys, fabrication techniques and the testing of aluminium's performance. The commercial development of aluminium into the civil engineering construction industry, naval construction and surface transport has added to the understanding and confidence in the development of standards covering fabrication, mechanical design and metallurgy.

Aluminium's ease of fabrication, long reliable life and advantageous weight saving in weight-sensitive structures has long been recognised. For example, using aluminium instead of steel gives a 40% weight reduction if yield strength is the limiting factor and up to 70% in less critical areas. An aluminium beam is one-third of the weight of an identical steel beam. For the aluminium beam to deflect the same as the steel beam, requires a 50% increase in web depth, but the aluminium beam will still be only 50% of the weight of the steel beam. Similarly, a flat aluminium plate will be 50% thicker and only 50% of the weight of a steel plate of equal stiffness. If, in both cases, stiffeners were added the difference in thickness disappears and the aluminium construction will be even lighter. Aluminium's strength increases as the temperature reduces. Because aluminium is easy to extrude to complex and accurate sections, even in quite small amounts, the shape of the extruded section can be optimised to place the metal precisely where it is required thus giving a significant weight saving. One of the economic advantages of fabricating in aluminium is that handling equipment designed for steel fabrication can accommodate larger sections. Less expensive facilities are required for conventional-sized aluminium structures.

Aluminium has a longer life and lower maintenance requirements than steel. By using the most appropriate aluminium alloys, no painting or other forms of protection are required against atmospheric corrosion because unprotected aluminium forms a protective oxide coating.

In passenger ship construction the use of an aluminium superstructure allows the incorporation of extra decks thus greatly increasing the ship's revenue-earning capacity. Aluminium has proved itself in the North Sea environment. More than one hundred helidecks of interlocking aluminium sections are in service and aluminium is used for acoustic enclosures, cable ladders, cladding, filtration enclosures, explosive relief panels, gantries, lifeboats, louvres, mini accommodation modules, mud mats, railings, reflective heat shields, stairways, telescopic bridges, walkways and many other applications. The first all-aluminium offshore platform was built in 1957 and included a 220-person accommodation module, designed entirely in aluminium was built with a weight saving of 50%.

6.1.2 The cost of aluminium structures

Aluminium's high initial cost per tonne is offset when life cycle costs are considered. Because of aluminium's low weight, a significantly reduced tonnage is used, savings are made in supporting the structure's weight, plus the use of larger sub-assemblies and specially designed sections simplifies assembly procedures, construction and design. Lighter weight structures mean reduced transportation costs and simplified handling. Structural welding can be reduced by 50% and the majority of weld bevels can be accurately pre-formed in the structural extrusions. The use of level aluminium floors eliminates the need for screeding. Aluminium requires no painting. Detailed studies have shown that capital costs for typical aluminium structures can be as low as 70% of that of other materials. Even in a worst case where the capital cost is 20% higher, the additional cost can be justified if maintenance costs are considered. All aluminium has a high scrap value, which is a significant factor for components designed for a limited life – as in transportation.

6.1.3 Sustainable development[4]

The UK aluminium industry is committed to sustainable development. For example, the aluminium industry looks at ways to minimise waste, to be more energy efficient and to recycle aluminium because of its high scrap value. The aluminium industry has developed descriptions of best practices to benchmark UK emission levels against Europe and the rest of the world. Over 50% of the energy used in the primary production of aluminium is derived from hydropower, an energy source that does not contribute to air pollution or global climate change. Recycling aluminium scrap requires only 5% of the energy used for primary aluminium production. Aluminium has the advantage that it can be repeatedly recycled with the quality being as good as for the primary metal. The recycling rate for the scrap collected during fabrication is 100%, while that for recovered aluminium is 90% in transport, 70% in building and 36% for beverage cans. The 1999 Blue Peter TV appeal for used aluminium beverage cans raised enough funds to build three primary schools in Mozambique! Because of its lightness, a 40% weight saving can be made when aluminium is used instead of steel in a car body. This results in greater vehicle fuel economy.

The environmental and social issues raised by bauxite mining are a special consideration for sustainable development. Globally, the aluminium industry produces 20 million tonnes of primary aluminium per year and, with current reserves, could do so for three hundred years. The majority of bauxite deposits occur in forest areas, with only 11% being mined from areas in the tropical rain forest. In the forest areas 80% of the mined sites are restored to indigenous forest, with the remaining 20% being returned to commercial forest or agriculture. These mine restoration programmes have gained a United Nations Environmental Programme Award.

6.1.4 Some advantages and disadvantages of aluminium

Once the surface of aluminium has oxidised that surface remains stable. Further, aluminium's mechanical properties improve as its temperatures decreases. Aluminium section shapes are almost limitless as they are produced by extrusion, giving a very high strength-to-weight ratio especially when used in bridges, buildings and transportation. Extruded shapes can be designed to simplify assembly. The majority of aluminium alloys can be welded with adhesive bonding being an acceptable method for jointing structural aluminium. Aluminium has excellent workability and can be processed by bending, drilling, embossing, forging, pressing, rolling, and other metal working methods. Aluminium is non-magnetic and does not receive magnetic induction from magnetic fields. Aluminium has high strength; it is resistant to corrosion and is durable in extremely hot or cold weather. Aluminium has high electrical conductivity, high heat conductivity and is used for electrical distribution.

Certain very strong aluminium alloys corrode in hostile environments and therefore need protection. Aluminium has a reduced buckling failure load when compared with steel, but changing the shape of the aluminium section compensates for this. Aluminium weakens as the temperature increases. It expands and contracts more than steel and the heat-affected zone (HAZ) due to welding can have a significant strength loss. Aluminium's fatigue strength is lower than for steel and aluminium can corrode where it is jointed to other metals.

6.2 Engineering applications of aluminium

Aluminium has many uses including: air conditioning units, billboards, body panels, bridges, bridge railings, bus shelters, cabinets, camera casings, cans, carports, conductor rails, conservatories, copiers, covers, curtain walls, fencing, computers, decking, engines, entranceways, facades, furniture, greenhouses, guard railings, helicopter platforms, highway signs, honeycomb panels, kitchen appliances, marinas, marine containers, partitions, portable seating systems, prefabricated housing, railings, reflectors, roofing systems, scaffold platforms, scaffolding, seating, shelters, shop fronts, signs, solar panels, soundproof walls, sports equipment, stairs, stairways, structural units, tanks, traffic signs, wheels, windows and wire fencing.

A developing field for aluminium is in transportation, which includes airline cargo containers, automotive components, automotive railway stock, bicycle parts, buses, catamaran boat decking, containers, electric cars, mobile homes, portable showrooms, sailboat accessories, space frames, superstructure for ships, train seats, truck container accessories, ultra-light aircraft and vehicle suspension parts.

6.3 Aluminium Awards[5]

The 1999 the Aluminium Imagination Architectural Awards had a record number of entries and they were of the highest standards yet achieved. The overall winner by Future Systems was the Media Centre at Lord's Cricket Ground in London. The Media Centre (shown in Figs 6.1 and 6.2) was considered as one of the finest and most innovative aluminium architectural designs anywhere in the world. The aluminium structure was of semi-monocoque construction involving a structural skin, which worked in conjunction with a boat-like grid of ribs curving in three dimensions. The structure was assembled on site from twenty-six sections prefabricated in Cornwall by Pendennis Shipyard. Internally, the lower level contains the press benches. The press mezzanine immediately above is suspended from the upper part of the structure using concealed hangers.

Stratford railway station, by Chris Wilkinson Architects, uses aluminium and glass in an elliptically curved shelter above the station. There is a standing-seam roof, composite aluminium sheet cladding, subway cladding and a 30 000 m^2 ceiling made from specially designed and extruded aluminium tongue-and-grooved planks that have a silver metallic finish giving a light-reflecting surface for the lighting.

The great aluminium roof at North Greenwich Transport Interchange, by Foster & Partners is a section taken from a 1 km diameter sphere and is shaped like a bird's wing in flight. Aluminium was chosen because of its low maintenance, speed of erection and ability to accommodate bi-directional geometry. Aerofoil shaped eaves shed rainwater

Figure 6.1 *The world's first all-aluminium building. A front view of the media centre at Lords Cricket Ground. Reproduced with kind permission of the Aluminium Federation.*

Figure 6.2 *A view of the rear of the media centre at Lords Cricket Ground. Reproduced with kind permission of the Aluminium Federation.*

and smooth airflow across the roof. The suspended light fittings are also in aluminium and are designed to bounce light off the reflective aluminium ceiling panels.

6.4 Aluminium products

6.4.1 Flat-rolled products

Flat-rolled products such as plate, sheet and strip are used for aircraft, cans, car bodies, computer discs, curtain walls, household appliances, nameplates, packaging, printing plates, railway trains, roofing and ships. Roll forming is used to achieve extra close tolerances and to make special geometric shapes.

6.4.2 Extrusions

Aluminium extrusions include bars, pipes, wires and an infinite number of hollow or solid shapes to improve strength and torsional rigidity. Extrusions are used for aerospace equipment, aircraft, buildings, cars, condensers, drums, electrical machinery, evaporators, radiators and recreational goods. Extensive research and development into welding has increased the use of extrusions for motor vehicles and railway vehicles. In general, extrusions maximise the moment of inertia of a structural section. Extrusions are designed to avoid additional fabrication, to allow for welding by adding metal to the welding zone and to keep welded connections away from areas of high stress.

6.4.3 Casting and forging

Forging technology developed rapidly following its initial use in the manufacture of aircraft propellers and engine cases. It then developed into heavy forging, precision stamp forging and ring forging. These latter forging processes have resulted in high-quality forging for aircraft, impellers, railway stock, ships, wheels and windows.

6.4.4 Painted aluminium

Painted aluminium enhances aluminium's natural resistance to corrosion and is used on antenna, electrical equipment, heat insulation and roofing where highly corrosive environments may be encountered.

6.5 The alloys of aluminium[6,7,8,9]

6.5.1 General

Detailed information on the specifications for cast and wrought aluminium and aluminium alloys for general engineering purposes can be found in various British Standards. However, many countries have their own designations and BS 8118 lists the nearest foreign equivalent complying with the relevant British Standards[6]. The country designating wrought aluminium alloys is the USA, where the Aluminium Association's system is almost universally accepted as the international standard[10]. The system consists of a four-digit number, which defines the aluminium alloy, followed by letters and/or numbers that define the condition or temper of the alloy. BS 8118 uses the alternative alloy temper designation of ISO 2107, with aluminium alloy castings being designated in accordance with BS 1490[6,7,8,9,10,11]. Care must be taken when using the British Standards, as they will be superseded by Eurocodes.

Aluminium of 99% purity and above (alloy group 1) has considerable corrosion resistance, but does not have high strength; consequently, alloying elements are introduced to improve the strength of aluminium. The main alloying elements are copper (alloy group 2), manganese (alloy group 3), silicon (alloy group 4), magnesium (alloy group 5), magnesium and silicon (alloy group 6) and zinc (alloy group 7). Alloy group 8 is used for all other alloying elements and alloy group 9 is an unused group. The first of the four digits in the designation indicates the alloy group. In group 1, the last two of the four digits indicate the minimum aluminium percentage. For example, 1080 indicates aluminium with a minimum purity of 99.80%. If the second digit is not zero, then it indicates modifications in the impurity limits or the addition of alloying elements. For the remaining groups, 2 to 8 inclusive, the last two of the four digits serve only to identify the different alloys in the group, while the second digit, if it is not zero, indicates alloy modifications. Aluminium alloys are divided into two groups, the heat-treatable alloys and the non-heat-treatable alloys.

6.5.2 Heat-treatable alloys

The heat-treatable alloys are divided into three groups, 2***, 6*** and 7***. The strength of the heat-treatable alloys is enhanced by solution heat treatment and precipitation heat treatment sometimes called ageing. With some heat-treatable alloys, ageing occurs naturally and the second formal heat treatment may not be needed. Suffix letters and numbers (e.g. O, F, T4, T5, T6 or T8) identify the heat-treatable condition. T6 for example identifies solution heat-treatment and then artificial ageing.

The 2*** group has high strength but low corrosion resistance. This group is used widely in the aerospace industry but is not used generally in engineering situations due to its low weldability.

The 6*** group is the most used of the heat-treatable alloys and is the most popular for extruded alloys. This group combines medium to high strength and good corrosion resistance. The group is easily formed, machined, welded and has excellent finishing qualities. It is admirable for extrusion depending upon whether it has been manufactured to have an increased or a weaker strength. It is used for architectural sections, bridges, cranes, heavy structures, rail transport and road transport. It is used in bolted, riveted and welded connections.

The 7*** group, like the 6*** group can be manufactured to have an increased or a weaker strength. The increased strength alloys in this group includes the strongest aluminium alloys. These alloys have reduced corrosion resistance, reduced extrudability and reduced weldability. To increase their corrosion resistance, they are usually plated or painted. The weaker strength alloys are corrosion resistant and used in welded structures as their HAZ softening is less severe than for the 6*** group. The 7*** group is heat treatable and has excellent welding characteristics. The 7*** group is used where high strength is required as in aircraft, formwork for concrete, freight containers and vehicle wheels.

6.5.3 Non-heat-treatable alloys

The non-heat-treatable alloys are divided into four groups, 1***, 3***, 4*** and 5*** and arc usually produced in plate and sheet form. The strength of these alloys is improved by strain hardening, which is defined by an H-number (e.g. H1, H2, H3). Suffix letters such as F and O are also used to indicate the 'as fabricated' condition or the annealed condition respectively. Heating decreases the strength of the non-heat-treatable alloys.

The 1*** group is used where high corrosion resistance is required. As the purity of the aluminium increases so does its corrosion resistance, but its strength reduces. The 1*** alloys have excellent finishing capability, excellent formability, high corrosion resistance, high electrical conductivity, high thermal conductivity, but low strength and poor merchantability. Typically this group is used for thermal or electrical conductors, profile sheeting and tanks.

The 3*** group has increased strength over that of the 1*** group but retains good resistance to corrosion. This group has excellent thermal conductivity, good corrosion resistance, good formability and good weldability but poor machinability and medium to low strength. Typically this group is used for chemical equipment, coolers, heat exchangers, profile sheeting and radiators.

The 4*** group is used for castings and welding wire and has properties similar to the 3*** group.

The 5*** group is the major non-heat-treatable alloy. It has a range of ductility, strengths and excellent corrosion resistance, especially in the marine environment, but less so in hot environments. It is used mostly as plate or sheet for sheet-metal fabrication and welded plate construction.

6.6 Design principles[6,9]

6.6.1 General

Aluminium structures must be designed by considering the limit states at which they become unfit for their intended use. These limit states are the ultimate limit state of static strength, the serviceability limit state of deformation and the limit states of durability, fatigue and vibration.

6.6.2 Loading

The aluminium structure must be designed to resist, within reason, all loads to which it will be subject. These loads are dead loads, imposed loads, wind loads and temperature effects. Fluctuating loads need to be considered, as do loads due to explosions, fire, seismic activity, structure assembly, support settlement and vehicular impact. Dynamic effects are particularly important for aluminium structures, due to aluminium's ability to resonate, its high flexibility and low mass. To reduce dynamic effects, structural damping and prototype testing may be required.

6.6.3 Nominal loads

The nominal loads are those that the structure may reasonably be expected to carry in normal service and are used for checking the limit states of deformation, fatigue and vibration.

6.6.4 Factored loads

Factored loads are the nominal loads multiplied by the overall load factor and are used to check the limit state of static strength. The overall load factor allows for accidental overloading, variability in loading, etc.

6.6.5 Dynamic effects

Dynamic effects are treated as imposed loads. As aluminium has a low weight, its response to dynamic effects means that resonance must be considered and structural damping applied as required.

6.6.6 Static strength

An aluminium structure is acceptable in terms of static strength if the action effects under factored loading are less than the structures factored resistance.

6.6.7 Factored resistance

The factored resistance is the calculated resistance divided by the material factor. The material factor takes into account the differences between the strengths of the test specimens and those of the actual structure.

6.6.8 Deformation

The structural deformation is acceptable if the elastic deflection due to the nominal loading is less than the allowable limiting deflection. Aluminium is three times more flexible than steel and BS 8118 suggests deflection limits between span/360 and span/100. BS 8118 assumes that where a structure has been designed in accordance with the static design of members, there will be no significant permanent deformation under the action of the nominal loads. BS 8118 accepts that structures which are repeatedly assembled and disassembled are subject to a gradual build-up of deformation.

6.6.9 Durability

BS 8118 is quite limited in its interpretation of durability. Alloy groups are given durability ratings. Provided a durable alloy is used and protected in accordance with the BS 8118, taking into account the degree of exposure and the design life, the structure is considered satisfactory.

6.6.10 Vibration

Due to aluminium's low weight, undesirable vibrations may occur under normal service loads. Machine-, wave- and wind-induced vibrations can occur, but they can be reduced by limiting displacements, adding vibration dampers or ensuring that resonance of the structure does not occur. If vibration continues to be a potential problem then fatigue failure must be assessed.

6.6.11 Fatigue

If a structure or a component is subject to significant changes in loading, then fatigue should be considered. There are two limit states of fatigue, total collapse under the unfactored design load spectrum and stable crack growth. Fatigue occurs due to the repeated loading of localised details, which causes increased stress. This increase in stress may, under static loading, be of little significance, but under repeated loading considerable increases in stress may occur causing a microscopic crack to form.

The crack then propagates each time the load is applied. Initially the rate of propagation is small, but as the number of cycles increases, the rate of propagation also increases, later causing catastrophic failure[3].

Fatigue failure may also result from a stress raiser at the root, the toe or in the rippled effect of a weld. Cracks may form at a hole for a bolt, at a sudden change of cross-section or at the surface roughness of a section. Consequently, high standards of fabrication and inspection are required to reduce the possibilities of fatigue failure.

The choice of the aluminium alloy has little effect on fatigue performance as N the number of cycles to failure at a specified detail is related to f_r the stress range. The stress range is the difference between the maximum and the minimum stress at each cycle. Essentially the fatigue strength of aluminium is one-third that of steel, but BS 8118 gives far more accurate results than this. BS 8118 uses Miner's rule for variable amplitude loading and for parts of the detail classification scheme[6]. The S–N curves of BS 8118 were established using statistical data analysis and engineering judgement. Regression analysis was carried out on individual data sets and a series of best-fit slopes was determined. The constant amplitude cut-off stress was located at 10^7 cycles, below which the constant amplitude stress cycles were assumed to be non-damaging. However, even if occasional cycling occurred above this level, they will cause propagation which as the crack extends, will cause lower amplitude cycles to become damaging. For this reason the stress-range-cycles-to-failure curves of BS 8118 changes between 5×10^6 and 10^8 cycles for the general spectrum of loading conditions[6]. Any stress cycles below the variable amplitude cut-off stress, which occurs at 10^8 cycles, are assumed non-damaging. These results may be conservative for some spectra and testing may be used to give increased fatigue cycles. The trend of the stress range-cycles-to-failure curves is to flatten out with increasing detail strength. In the endurance range between 10^3 and 10^5 care must be taken to ensure that the maximum tensile stress in the design stress range does not exceed the static design stress[6]. Where adequate data did not exist, fatigue test data from steel specimens was used, by reducing the strength by a factor of 3. This assumes that the total fatigue life consists solely of crack propagation and that the rate of crack growth can be correlated on the basis of elastic modulus only.

6.6.12 Total collapse

For the limit state of total collapse, the designer must check that the design life of the structure is less than its predicted life. To determine the predicted life, it is necessary to identify all potential fatigue sites, then to determine the number of loading cycles to failure at each of these sites and ensure the number of loading cycles to failure is more than the required design life.

6.6.13 Safety

An important criterion in the application of aluminium relates to the safety issues of fire performance and thermite sparking. Aluminium alloys are non-combustible and do not burn. Further, aluminium striking aluminium will not produce a spark. However, research into the thermic reaction of aluminium and rusty steel shows that incendive sparking with gas ignition will occur only under very specific and remote conditions. The correct methane–air mixture must be present, as must unpainted steel with a layer of rust and a smear of aluminium. The thermite spark will occur, either at the initial impact or when a previously deposited aluminium smear is subsequently struck either by a steel or an aluminium impactor. Thermite sparking can be eliminated by anodising the aluminium or by ensuring the steel surfaces are either painted or galvanised. Fortunately, aluminium smears on rusty steel corrode within three days, thus removing the thermite-sparking hazard.

6.6.14 Fire

In fire, aluminium neither burns nor gives off an inflammable vapour when heated. Where there is a need for fire-resistant materials, conventional fireproofing methods and materials can be utilised. Aluminium's melting point is much lower than that of steel and in fire aluminium loses strength and distorts. Aluminium does not burn nor does it support combustion. In practice the reflectivity and high thermal conductivity of aluminium reduces the rise in temperature and the formation of hot spots.

6.7 Fabrication

6.7.1 General

Many of the processes used for the fabrication and jointing of aluminium are similar to those used for steel, however, aluminium does require special measures in a number of areas. For example, a dedicated, segregated workshop for fabrication is required as are approved welders and staff trained in the handling of aluminium. Aluminium requires less machining force and can accept higher machining speeds. Aluminium is not a fire hazard and when molten must be kept apart from water. If hot, aluminium must be handled carefully as it shows no colour change when heated.

6.7.2 Health and toxicity

Aluminium is poorly absorbed by the body and has never been found to be toxic except under quite unusual and specific conditions of exposure.

6.7.3 Drilling

Drilling can be carried out at high speed. The drilled hole will be larger than in steel and excessive heat can be generated so a coolant is required.

6.7.4 Bending

Extrusions are easily bent and only complex sections requiring special tooling.

6.7.5 Adhesives

Adhesives are used to bond aluminium to aluminium or to other materials. Adhesive bonding is used in the aerospace and high-tech electronics industries and for truck bodies. Pre-treatment of the aluminium is essential for effective gluing and unsealed anodising gives the best results for achieving a strong adhesive joint.

6.7.6 Punching

To avoid burrs, the punching tool must be ground regularly and always kept sharp. The clearance between the upper and lower punching tool must be changed for different alloys, tempers and wall thicknesses.

6.7.7 Riveting

In aluminium, cold rivets produce little impact stresses. For high strength, the rivet must fill the entire hole and the hole must be drilled and trimmed.

6.8 Durability and corrosion protection[3,6,12]

6.8.1 General

Aluminium alloys are categorised into a descending order of durability, A, B and C which is used to determine the required protection. If more than one alloy is used, the protection must be in accordance with the alloy with the lowest durability rating. Normally, the standard alloys listed in Tables 2.1, 2.2, 2.3 and 2.4 of BS 8118 do not need surface protection[12]. The corrosion resistance of these alloys is due to the self-sealing film that forms on the surface of the alloy, which in mild environments allows the aluminium's surface to retain its original appearance. As the environment becomes more aggressive, the surface may first darken and roughen, then the roughness increases and increasing amounts of visible white powdery surface oxide forms. In the latter case, added protection is necessary as in coastal and marine environments.

Where surface corrosion does occur, the corrosion–time curves have an exponential form. There may be a rapid loss of reflectivity over six months to three years, followed by little change over perhaps the next eighty years, except in extremes of acidity or alkalinity. If the surface oxide film is damaged, it immediately reforms and it is this ability to repair surface damage that gives aluminium its durability. In general, tropical environments are no more harmful than temperate environments.

Aluminium normally requires no protection against surface corrosion, but if corrosion protection is required, the precise conditions at the actual site must be studied. BS 8118 categorises aluminium alloys into three durability ratings, A, B and C and goes on to determine the degree of required protection. One of six protection procedures can be specified for metal-to-metal contact surfaces, with 0 representing no protection and 1 through 5 increasing in protection. Corrosion resistance is enhanced by anodising, metal spraying, painting, plastic coating or by surface treatment. Anodising is an electrolytic process that produces a hard thick film of aluminium oxide on the surface of the aluminium. Aluminium may be painted with either a bituminous or a non-bituminous system. Pre-coated aluminium and factory applied finishes may be used, but care must be taken over any heating that the process requires. Extra protection may be necessary at joints. When aluminium is in contact with building materials, the contact surface must be coated with bituminous paint. Further consideration must be given when aluminium is in contact with water; sea-dredged aggregates or with chemicals normally used in the building industry.

The effects of corrosion are related to the chemical composition of aluminium, the fabrication process, the heat treatment and the stress field. Atmospheric corrosion causes concentrated corrosion, exfoliation, delamination corrosion, inter-granular corrosion and surface corrosion where the surface is denied oxygen to form its usual oxide layer. Stress corrosion occurs in the tension zones of a stress field when a corrosive agent is present.

6.8.2 Forms of corrosion

The principal forms of corrosion are corrosion fatigue, crevice corrosion, galvanic action, pitting and stress corrosion. These forms of corrosion are not unique to aluminium, but understanding them leads to the redesigning of joint details to avoid them.

6.8.3 Galvanic corrosion

If two dissimilar metals are in electrical contact with each other in the presence of an electrolyte, an electric current will flow between them resulting in the corrosion of one of the metals. The rate of corrosion depends on the type and concentration of the electrolyte, the resistance of the metallic contact and the relative areas of each metal. Aluminium and its alloys will corrode in contact with copper, copper alloys, iron, mild steel, nickel and stainless steel. Aluminium will not corrode when in contact with zinc and magnesium alloys, which corrode sacrificially and protect the aluminium. To prevent galvanic corrosion it is necessary to ensure that dissimilar metals are electrically isolated from each other by means of non-metallic surfaces. Steel bolts or other steel components must be zinc galvanised or aluminised when they are in contact with aluminium.

6.8.4 Crevice corrosion

Crevice corrosion occurs where water is trapped in a confined space, e.g. at joints or in grooves between the threads of nuts and bolts. When a corrosion deposit is formed in a confined space distortion can take place.

6.8.5 Pitting corrosion

Pitting corrosion produces a cavity, which is deeper than its width. If the aluminium is in the form of a thin sheet, perforation of the sheet may occur.

6.8.6 Stress corrosion cracking and corrosion fatigue

A small number of aluminium alloys are susceptible to stress corrosion cracking if a corrosive environment exists and the alloy is under tensile stress. Removing any one of the three criteria will eliminate stress corrosion cracking. The initiation of fatigue cracks is facilitated by corrosion such as pitting. To control corrosion fatigue the corrosion and or the amplitude of the local cyclic stress must remain within the design restraints.

6.8.7 Other causes of corrosion

Corrosion of aluminium alloys can also be caused by direct chemical attack from acids and alkalis. If the surface of the aluminium is denied oxygen it continues to corrode as an oxide layer cannot be formed. This form of corrosion is called 'poultice action'. Poultice corrosion occurs when insulating materials such as felt washers, fire protection or foam stop oxygen getting to the surface of the aluminium. The chemical attack of aluminium alloys arises from caustic soda, strong acids and washing soda. Any areas of aluminium exposed to such spillage must be protected. When aluminium alloys are exposed to marine environments, only slight weathering of the surface will occur. If corrosion attack does take place it is confined to the first few years and then it will reduce. Where additional corrosion protection is required, then anodising, metal spraying and painting should be considered.

6.9 Corrosion protection systems

6.9.1 Design to minimise corrosion

Although many aluminium alloys, when freely exposed in a marine environment, have a good resistance to corrosion, in practice, it may be necessary to apply some form of corrosion protection treatment. Thoughtful structural design can avoid potential sites for corrosion by the use of correct detailing and the choice of appropriate aluminium alloys.

When considering corrosion protection, it is necessary to be aware of the localised environment, the required life of the component, the level of inspection, the level of maintenance and the thickness of the metal. For example, thick components are unlikely to require protection because no significant loss of mechanical strength will occur. Corrosion protection is required where crevices exist, such as at mechanical joints, wherever aluminium is in contact with other materials and wherever aluminium is in contact with a corrosive agent for any extended period of time, e.g. acids having pH <5 and alkalis having pH >8. Less seriously vulnerable areas are joints between dissimilar alloys of aluminium. These include welded connections where inter-metallic compounds may be precipitated out of the solid solution and give rise to localised galvanic action leading to pitting corrosion. Porous materials such as gaskets and washers in permanent contact with aluminium can become corrosion sites if they absorb water and remain wet. The designer should ensure there is no possibility for corrosion by correct design and by suitable localised corrosion protection.

6.9.2 Corrosion protection of offshore structures

Usually aluminium members will just weather with their mechanical properties and structural integrity remaining intact. Certain localised areas such as metal-to-metal joints will require protection. Louvres and similar components will need anodising. An inert layer must be placed between aluminium and any fire protection layers. Organic coatings must cover drill pipes, floors, tank linings, etc. which are exposed to corrosive liquids. To prevent pitting corrosion and to provide a sterile surface to fresh water tanks, chlorinated rubber coatings are necessary.

6.9.3 Organic coatings

Prior to painting, the surface of an aluminium alloy must be pre-treated to remove its oxide film. This pre-treatment is carried out by using etch primers or conversion coatings. Etch primers are more suitable for site-applied coatings. Conversion coatings are used for factory-applied coatings as they require the aluminium component to be dipped into hot solutions and then dried.

Lead-free paints, which are suitable for protecting steel, are suitable for aluminium. The paint has to be lead free to avoid the galvanic attack of the aluminium. Paint systems last longer on aluminium because of aluminium's inherent durability. In the marine environment, an anti-fouling paint applied to aluminium must not contain copper or tin. Paints, such as stoved-polyurethane modified with polyamide for a scuff-resistant finish, are available.

6.9.4 Anodising

Anodising consists of increasing the thickness of the hard, dense, protective oxide coating on the surface of the aluminium. Provided a thick coating is applied, there will be no tendency for pitting or for a general attack of corrosion. Anodising increases the coating thickness by up to five times the thickness of the natural oxide film. Hard anodising is a variant of the normal anodising process and provides a coating with better abrasion resistance and wear resistance. The hard-anodised surface is not usually sealed but a sealing coat may be applied to increase corrosion resistance but there will be some reduction in abrasion resistance. Colours are often introduced into the anodising process.

6.9.5 Coating

For decorative purposes, a single coat of paint or lacquer may be applied. For additional corrosion protection, insulation or resistance to chemical effects, high mechanical stresses, moisture effects or temperature effects, multi-coat finishes are used.

6.9.6 Mechanical surface treatments

Special surface treatments can be achieved by brushing and polishing. Brushing provides a decorative non-reflective finish, while polishing gives a bright, reflective high gloss finish.

6.9.7 Metal spraying

Aluminium alloy plates, sections, and welded joints which are located in inaccessible areas can be given increased corrosion protection by metal spraying using a 99.5% purity aluminium wire. Similar sealed coatings on steel provide both cathodic protection and corrosion protection. Similar layers give increased wear resistance and provide an anti-slip coating.

6.9.8 Aluminium-to-aluminium joints

Mechanical joints are made by bolting, riveting and interlocking specially designed extruded sections. Adhesive bonded joints are not usually used for primary structural connections, although they are used in building structures and for secondary structural connections, such as web stiffener attachments. The adhesive bonding of aluminium produces joints with a lower stress concentration and a higher fatigue life than the same joint constructed using mechanical fasteners or by welding. Adhesive joints do not have heat-affected zones and have better corrosion resistance and sealing properties. Adhesive bonding is usually used in conjunction with thin materials and provides a separation layer between the aluminium and the different metals or non-metallic materials joined to the aluminium.

In general, adhesive joints are used to carry a shear load, improve fatigue strength, to prevent fretting and to provide a seal. Bonded joints should not be used to carry tension or where the load will tend to open the joint. Loads should be carried over as large an area as possible. The joint width has more influence on increasing the joint strength than does increasing the joint length. Valid data on the resistance of the bonded joint is required and will necessitate testing, as there are no accurate analytical methods for fatigue loading. Special attention needs to be given to bond integrity, damage to the joint while in service, loading rate sensitivity, service humidity, service temperature and surface preparation. Specialist advice should be sought when adhesive bonding is being considered, as the factors of safety for bonded joints are higher than for other types of joints.

The corrosion protection of bolted joints and riveted joints require them be sealed at the time of assembly by the application of protective coatings, such as gaskets, priming paints or sealants. The contact surfaces must be assembled with the paint still wet so that all crevices, into which water might penetrate, are filled. Bolt holes, fasteners and rivet holes must also be coated. It may be necessary to avoid galvanic corrosion with non-aluminium bolts electrically isolated from the aluminium by the use of washers made from inert material. For conditions of severe corrosion the bolt shank should also be insulated by an inert bush inserted into the bolt hole. If the joint is highly stressed the bushes and the washers must be made from stronger inert materials.

If steel rivets are used, galvanised washers should be inserted between the head of the rivet and the aluminium surface. This is in addition to the standard corrosion protection required at such joints such as cleaned surfaces pre-coated with primer and assembled wet, or separated by tape or suitable gaskets, etc. Following assembly, the fasteners and surrounding area must be painted. Aluminium rivets used to join aluminium components do not require additional protection. Corrosion protection at joints between extruded sections which interlock must be sealed at assembly using a sealant injected into the cavity, or by the use of plastic tape. Crevices must be closed off to stop water penetration.

Welded joints in aluminium may suffer from galvanic action and pitting corrosion if the filler wire or the welding technique is incorrect. The weld metal has a mixture of filler wire and parent metal while the HAZ will contain the alloying elements of the original metal. The result can create galvanic attack in the form of pitting. The corrosion resistance of welds is therefore ensured by the correct selection of filler wire and welding technique. Filler wires containing magnesium produce welds having the maximum corrosion resistance.

Service experience with welded aluminium structures has shown that low, in-service temperatures increase the fatigue strength of welded aluminium alloys.

6.9.9 Aluminium-to-steel joints

If an electrolyte is present, steel will cause corrosion of any aluminium alloy with which it is in contact. Thus there must be effective electrical isolation of the aluminium from all steel components. Gaskets and sleeves must be used in bolted steel-to-aluminium connections. Bolted joints and riveted joints of either aluminium or steel must be sealed with gaskets, paint, sealant or tape to ensure no crevices exist.

With rivets it is not possible to use inert gaskets for insulation. To protect aluminium components it is necessary to have galvanised washers under the head of an aluminium rivet on the steel surface and under the head of a steel rivet on an aluminium surface. All the edges of the joint should be over-painted and the heads of all rivets should be coated for additional protection. Before painting, steel rivets heads and surrounding steel areas should be sprayed with aluminium or zinc, but if the steel surfaces are already galvanised it is only necessary to protect the rivet heads. The joint design should ensure all water drains away from the connection areas.

6.9.10 Aluminium to concrete

Water draining from fresh concrete surfaces will cause corrosion or straining of aluminium components, therefore protection of the aluminium is required, e.g. the use of a bitumen-based paint.

6.9.11 Aluminium in contact with other materials

Metals such as copper which are joined to aluminium for reasons of electrical contact must be completely enclosed by paint or other organic coatings. Porous materials permanently in contact with aluminium, such as gaskets and insulation materials set up poultice corrosion if they become wet. A film of polythene between the insulation and aluminium should be provided.

6.9.12 Remedial measures

Where corrosion of the aluminium has occurred already and will continue into the future, protection against further attack is required. All corrosion products must be removed by abrasive blast cleaning, brushing and high-pressure water jets. Small holes caused by pitting can be drilled out and filled with a screwed plug. If the corrosion is extensive, then the defective aluminium must be replaced with a more suitable alloy bolted, riveted or welded in place. The reasons for the corrosion must be thoroughly investigated.

6.9.13 Inspection

Aluminium is often selected because of its low maintenance requirements. To ensure the long-term performance of the aluminium, preventative maintenance programmes require scheduled cleaning and inspections. The inspections should include:

- a check on unexpected chemical spillage;
- a check of all highly stressed and fatigue-loaded joints and welds for cracking and damage;
- a check of all highly stressed and fatigue-loaded hollow sections for corrosion or seam cracking;

- the inspection of sealed hollow members for signs of water ingress and freezing; and
- a check for signs of erosion, joint deterioration and wear.

6.9.14 Cleaning

The need to clean aluminium and aluminium alloys may be required to ensure an acceptable 'cosmetic' appearance. Types of cleaning include acid cleaning, alkaline cleaning, chemical cleaning, dry abrasive blast cleaning, high pressure washing, solvent cleaning, steam cleaning, washing and wet blast cleaning.

6.10 Welding

6.10.1 General

Although the assembly operations and fabrication for aluminium are broadly the same as those used for steel, there are a number of special requirements, which are easily accommodated by fabricators. It is essential that the following phases of welding be given equal attention:

- assembly methods, fixtures and jigs;
- detailed joint design;
- preparation of materials for welding;
- selection of consumables and materials;
- welding process selection;
- welding procedure quantification and specification; and
- welder approval and training.

The majority of welding techniques can be used to join aluminium and its alloys. Fusion welding of aluminium and its alloys by the inert-gas shielded arc welding (MIG and TIG) processes requires different techniques to that used for the welding of ferrous metals. Trained and approved welders of aluminium are essential and they must meet code requirements. The MIG and TIG processes require calm, draught-free conditions otherwise defective welds may be produced. Aluminium is prone to weld porosity if it is contaminated with moisture; consequently, aluminium is welded in a separate shop from other materials. Aluminium assemblies should not be exposed to dust particles from grinding operations of copper alloy or ferrous metals as these can contaminate the prepared joints awaiting welding and pose corrosion and weld quality hazards. The welding of aluminium alloys on site is not recommended, as it requires draught- and weather-proof shelters to enclose totally the welding operations.

6.10.2 Process selection

The welding of aluminium falls into three major categories, fusion welding, resistance welding and solid phase bonding. For general engineering fabrication, the welding processes which give repeatable high-quality welds in stressed structural aluminium, at economically acceptable metal deposition rates are the MIG and TIG processes. The use of aluminium-to-steel transition pieces produced by solid phase bonding allows aluminium components to be welded to steel. Most aluminium alloys can be welded by resistance welding and solid phase bonding, but fusion welding processes are influenced by metallurgical factors with some alloys being easier to weld than others. MIG and TIG welding are the most suitable for the greatest number of alloys.

The strength of non-heat-treatable wrought alloys is derived from solution hardening and by cold working, so that the work-hardened tempers are appreciably stronger though less ductile than the annealed alloy. The effect of fusion welding on the work-hardened alloy is to produce partial annealing in the HAZ, the extent of which is dependent upon the heat input. Therefore, the joint strength is based on the annealed strength of the parent metal, irrespective of the original temper of the parent metal.

The heat-treatable wrought alloys derive their mechanical properties from solution heat treatment and natural ageing. The effect of the heat treatment is partially offset by fusion welding, with the loss of strength of about 20% to 40% occurring within 25 mm of the weld. Thus it is essential that the welding heat input be carefully controlled.

Most aluminium casting alloys can be fusion welded using the MIG and TIG processes.

6.10.3 Safety requirements

When arc welding aluminium, additional attention is required as the higher current density arcs produce higher levels of ultraviolet light. Further, the highly reflective surface of the aluminium produces additional reflections to the arc's rays. Ozone gas is also generated and effective extraction and ventilation are required.

6.10.4 Preheat and interpass temperature

Preheating is not normally required for welding aluminium alloys, but it can reduce cracking and provide balanced heat input when welding components of different thickness. Preheating of the metal reduces the stress in the solidifying metal; however the mechanical properties and metallurgical characteristics of the alloy will be adversely affected by elevated temperatures.

With MIG and TIG welding, preheating of the metal is used to prevent weld defects associated with starts and restarts, including lack of fusion, overlaps and localised porosity. Preheating also prevents welding defects associated with the welding of sections thicker than 25 mm and is used to dry the metal when welding under damp conditions. Preheat temperatures are restricted to 250°C for the non-heat-treatable alloys and 100°C for the heat-treatable alloys.

6.10.5 Approval and training of welders

All welders must have passed the relevant welder approval tests in accordance with the welding code being used. Welders welding non-critical structures, where welding procedure approval is not required, should also be tested. The welding characteristics of aluminium demand that welders understand the special requirements of welding aluminium. They should understand the higher welding speeds, higher currents and different melting characteristics. Training courses should be provided in MIG or TIG welding, and be based on the required production requirements. A welder who is skilled in arc welding ferrous materials should be able to achieve the relevant standard of competence in aluminium with five days of training.

6.10.6 Solid phase bonding

Solid phase bonding is a process where a weld is made between two materials by sufficient pressure to cause plastic flow of the surface to be joined. The surfaces may require to be heated. Although surface contamination may prevent a continuous bond, it is possible to achieve full joint strength efficiency. Dissimilar metals can be joined and as no melting occurs, the formation of brittle inter-metallic compounds is largely avoided. The absence of a liquid metal pool gives significant metallurgical advantages over traditional fusion welding processes. Consequently, solid phase bonded joints are strong and ductile. The HAZ associated with fusion welding is of much less significance. Transition joints can also be made between aluminium and steel structures that prevent galvanic corrosion, as corrosion that occurs at the edges is unable to penetrate into the interface between the metals.

6.10.7 Pressure welding

Pressure welding requires the local deformation of the components to be joined and is achieved by pressure rolling. Applying heat will reduce the pressure necessary to achieve bonding.

6.10.8 Explosive welding

Explosive or high energy welding uses a controlled high velocity impact to produce a metallic bond between the two metals. The process is highly stable, as all that is required is the necessary explosive charge plus the ignition and detonating equipment. The process can be used to spot weld similar or dissimilar metals and is widely used to produce transition joints. Transition plates produced by explosive welding are used to weld aluminium deckhouses to steel decks. The transition joint is welded to the aluminium and then welded to the steel. This sequence allows the heat developed in welding the steel to be dissipated to the aluminium. The temperature at the aluminium–steel interface must remain below 315°C to minimise the possibility of the formation of brittle inter-metallic compounds. The transition joints are in plate form and cold-cut to a range of dimensions. They are also supplied in tubular form to permit leakproof joints.

6.10.9 Friction welding

Friction welding is achieved by producing enough frictional heat at the interface of the pieces to be joined to make the interface metal become plastic. Mashing and stirring occurs and the pieces are welded together. The welded joint is flat on both sides.

6.11 The influence of weld defects on weld performance

6.11.1 Static properties

The loss of static strength due to weld defects is proportional to the loss of effective cross-sectional area of the weld. Weldable aluminium alloys do not have notch brittle weld metal and the effect of notches and stress concentrations produced by defects do not have a significant effect on the static strength properties of the joint. However, welds in heat-treatable alloys which have been given full post-weld heat treatment are more notch-sensitive and it is essential that stress-raising defects such as cracks and lack of fusion are avoided. Stray arc strikes can be associated with solidification cracks and present serious stress raisers. Isolated or uniformly distributed porosity does not have a significant effect on the tensile properties of the joint unless gross porosity is present.

6.11.2 Dynamic properties

Fillet welds have lower fatigue strengths than butt welds. The fatigue strength of a welded joint is governed by any stress concentrations arising from the joint geometry, thus cracks and notch-like defects, etc. are to be avoided. Defects that are internal to the weld can be ignored, but defects at the toe or the root of a fillet weld will cause failure. Removing the weld reinforcement bead from a butt weld very much reduces stress concentrations. Where fatigue loading is expected, holes should be drilled undersize and reamed out.

6.12 Conclusions

This chapter has described the use of aluminium in a wide variety of engineering applications and has discussed aluminium, its alloys, design, fabrication, performance, products and welding. The chapter has concentrated on the durability of aluminium, its naturally high corrosion resistance and how this resistance and durability in corrosive environments is enhanced by thoughtful design and corrosion protection. Examples have been given of a wide range of aluminium products and specific examples of award winning structural designs in aluminium.

Acknowledgements

The support of the following organisation and companies is gratefully acknowledged: Aluminium Federation Ltd, Birmingham; Council for Aluminium in Building, Cheltenham; Hydro Aluminium Extrusion Group and SECO Aluminium Limited, Witham.

References

[1] Sharp, M.L., *Behaviour and Design of Aluminium Structures*, New York: McGraw-Hill, 1992.
[2] Mazola, F.M., *Aluminium Alloy Structures*, London: E. & F.N. Spon, 1995.
[3] Dwight, J., *Aluminium Design and Construction*, London: E. & F.N. Spon, 1998.
[4] Alfed, *Aluminium for future generations, The UK aluminium industry: Towards a sustainable development strategy*, Birmingham, UK: The Aluminium Federation Ltd, 1999.
[5] Aluminium Extruders Association, *Aluminium Imagination, 1999 Aluminium Imagination Architectural Awards*, Birmingham, UK: The Aluminium Extruders Association, 1999.
[6] British Standards Institution, BS 8118:1991, *Structural use of aluminium – Part 1: Code of practice for design*, London: BSI.

[7] ISO 2107, Aluminium, magnesium and their alloys – Temper designations.

[8] Aluminium Federation, *The Properties of Aluminium and its Alloys*, Ninth edition, 1993.

[9] Bull, J.W., *The Structural Design of Structural Elements in Aluminium*, Aldershot: Gower Publishers, 1994.

[10] The Aluminium Association, *Registration record of international alloys designations and chemical composition limits for wrought aluminum and wrought aluminum alloys*, Washington, USA, updated regularly.

[11] British Standards Institution, BS 1490:1988, *Specification for aluminium and aluminium alloy ingots and castings for general engineering purposes, ingots and castings*, London: BSI.

[12] British Standards Institution, BS 8118:1991, *Structural use of aluminium – Part 2: Specification for materials, workmanship and protection*, London: BSI.

Bibliography

Alcan Offshore and Wimpey Offshore, 'Aluminium in offshore structures', *Aluminium Design Guide*, Gerrards Cross: Alcan Offshore, 1993.

Sharp, M.L., Nordmark, G.E. and Menzemer, C.C., *Fatigue Design of Aluminium Components and Structures*, New York: McGraw-Hill, 1996.

7 The Durability of Masonry Construction – an Overview

Prof Emeritus AW Hendry

Given time, the natural agencies of sun, rain, wind and frost will remove mountains. In much less time, they will remove buildings and it is a measure of the inherent resistance of masonry materials to these elements that there are many buildings still in use several centuries after their construction. Their survival, of course, has not been unaided and has been the result of continual maintenance and repair, sometimes amounting to partial reconstruction. Durability could indeed be defined as the ability of the material or structure to remain serviceable without excessive or unexpected maintenance. This of course raises the question of the required or expected service life of the building so that there can be no unique definition of durability.

Nevertheless, the survival of very old buildings must have some lessons for contemporary construction. Thus, the original concepts and details must have been sound, the materials carefully selected and the workmanship of a high standard. Where any of these fell short, the building will have been the subject of endless remedial work and in the end may have fallen into ruin or been completely rebuilt. Although it is not expected that many present day buildings will last for centuries, most are expected to last for many decades without excessive (and expensive) maintenance. Therefore we must apply the same essential criteria of careful design and construction as were so evidently applied in the case of these old buildings which have survived.

Fundamentally, deterioration results from either physical or chemical causes. Physical causes include foundation movement, thermal and moisture movements and frost damage. We have nowadays a greater appreciation of foundation design than our predecessors and therefore we should be able to avoid serious difficulties from this cause which, however, is not specifically related to masonry structures. Thermal and moisture movements pose greater problems in present day construction because of our use of different materials in proximity to one another. Problems arise particularly when concrete and masonry elements with different characteristics of shrinkage, creep and thermal movement are used in parallel and where the movement of one may interfere with the movement of the other. In such circumstances large forces can be generated, sufficient to cause cracks in one or the other, opening the way for the penetration of water and leading to further damage.

Even if there are no cracks in external masonry, penetration of water can cause damage by repeated freezing. This was well understood in the past and buildings were provided with overhanging eaves, cornices, string courses and other details which, in addition to contributing to appearance had the function of throwing water off the masonry. It is an unfortunate feature of many contemporary buildings that these features have been discarded and, as a result, the masonry walls have become much wetter and more liable to frost damage and other potentially deleterious effects.

In persistently wet walls damage by chemical attack also becomes more likely. Thus if the masonry units contain soluble sulphates, these may go into solution and cause failure of the mortar, especially if this is of a lean mix and made with a fine sand. Sub-surface crystallisation, although not essentially a chemical effect, can cause damage to certain types of masonry where it occurs.

All these problems point to the need for careful detailing of buildings, based on a proper understanding of the materials being used.

Another feature of modern buildings is the widespread use of metal components in masonry walls, for example for wall ties in cavity walls and for ties between masonry cladding panels and a concrete or steel structure. Wall ties are generally of galvanised mild steel the durability of which may be limited in exposed conditions. Steel is protected from corrosion when embedded in mortar only so long as an alkaline environment persists around it. However, mortar changes chemically over a period of years as a result of the penetration of carbon dioxide from the atmosphere. When the process of 'carbonation' is complete, the alkaline protection no longer exists and the wall ties will be open to corrosion in the presence of water and oxygen, unless protected by galvanising. The lifespan of galvanised coatings is uncertain but a rather heavy coating of zinc is necessary to ensure a life of, say, 60 years. Under conditions of severe exposure it is advisable to use ties of austenitic stainless steel which are likely to last indefinitely.

With the use of reinforced or prestressed masonry the durability of metals embedded in mortar and grout become of even greater importance. It would appear that no practical amount of cover will protect a carbon steel bar if embedded in mortar and that such reinforcement should be heavily galvanised or preferably of a suitable stainless steel. Experience of mild steel in bed joint reinforcement in masonry has been variable but no difficulties appear to have arisen with such reinforcement in hollow blockwork where the bars are surrounded by a suitable thickness of concrete. Grouted cavity construction where again sufficient concrete cover can be provided should also be immune from corrosion. Experience with prestressed masonry is more limited but if concrete practice is followed satisfactory protection of tendons should be achieved.

Water penetration of masonry is another source of trouble and although this may not lead to deterioration of the masonry itself, it can certainly lead to severe damage to the interior fabric of the building and if caused by bad workmanship in the construction of the masonry may require it to be re-built. The type and thickness of the masonry must of course be selected with regard to exposure conditions as defined in the relevant code of practice, such as BS 5628 Part 3, 1999. The protection of walls by overhangs and other details has already been mentioned as a means of keeping walls relatively dry and thus reducing the risk of rain penetration.

The outer leaf of a cavity wall, however, will have a limited capacity for absorbing and re-evaporating water and is thus likely to remain very wet in exposed positions for considerable periods. In such a situation the potential for frost damage is likely to be increased, especially in heavily insulated walls which will not be dried out by heat escaping from inside the building.

It would appear then that although masonry materials are inherently very durable there are a number of features in modern masonry construction which can, if not taken into account in the design of buildings, reduce their durability.

The factors outlined in this section and the characteristics of the various materials used are discussed in some detail in the following two chapters. Chapter 8 reviews these problems in terms of mechanisms which may result in deterioration and includes a list of references whilst Chapter 9 deals particularly with brickwork and blockwork.

Bibliography

Grimm, C.T. (1999) *Design for Masonry Volume Change*, Boulder, Colorado: The Masonry Society.

Harding, J.R. and Smith, R.A. (1986) 'Brickwork Durability', BDA Design Note 7, Windsor: Brick Development Association.

Hendry, A.W. and Khalaf, F.M. (2000) *Masonry Wall Construction,* London: Spon Press.

Richardson, B.A. (1991) Defects and Deterioration in Buildings, London: E. & F.N. Spon.

Sowden, A.M. (1990) *The Maintenance of Brick and Stone Masonry Structures*, London: E. & F.N. Spon.

Thomas, K. (1996) *Masonry Walls*, Oxford: Butterworth Heinemann.

8 The Durability of Masonry in Aggressive Environments and Techniques for its Conservation and Protection

Luigia Binda, Giulia Baronio

8.1 Introduction

Numerous examples of building material deterioration that occurred during the second half of the 20th century involving traditional (reinforced concrete, stones, renders, etc.) and new (synthetic materials, consolidants, insulation materials, etc.) materials, have pointed out the effects of poor design and of lack of maintenance. In the recent past, the design of new construction based on accurate calculations with highly refined structural models, has frequently, totally ignored problems concerning the durability of the building and of its components.

In the past the choice of the materials was made on the basis of empirical rules. This was the result of long experience. Today this choice is often left to the contractor and to manufacturers, and is generally made without proper consideration of durability.

In many countries, national and international standards and codes of practice do not pay special attention to the durability of materials and to their behaviour in aggressive environments – or to possible incompatibility between the materials chosen. Frequently, the only exception is represented by facing materials and floors, which have to show durability to freeze–thaw action, and to fatigue.

The buildings of the past were constructed to last and many of them stood (or still stand) for centuries and millennia. Ancient treatises and manuals contain suggestions and proposals for the choice of durable materials. Long-term experience of using the same types of materials (stones, bricks, mortars, timber and steel) would suggest to the 'designers', and to the 'constructor' which materials were compatible materials for use in masonry walls. This choice would avoid physical and chemical post-construction decay.

- In order to protect their earth-based constructions, the Assyrians (in 5000 BC) used bitumen and asphalt for the exterior and gypsum plasters for the interior.
- To choose durable stones, Vitruvius proposed (in 25 AD) an on-site durability test. He suggested that the stones should be left in open space at the site of the construction and that any showing deterioration after two years should be discarded.
- In the following centuries treatises and manuals described recipes for special mortars and renders to be used for walls immersed in water, or for exposed

surfaces. The choice of mortar (based on lime and pozzolana, lime and brick dust, or the use of natural hydraulic mortars) took durability into consideration.

- The research which resulted in the discovery of Portland cement in the 18th century was also undertaken in order to find a durable material for aggressive climates. Only later, was the cement used for reinforced structures with strength in mind.
- The need to make wall paintings brighter and more durable in 19th century Germany led to the use of mineral colours.
- In 1880 Dr. A. Geikie presented a paper to the Royal Society in Edinburgh entitled 'Rock weathering as illustrated in Edinburgh Churchyards', based on an investigation of tomb stones.
- Air pollution which, greatly increased in the 20th century, was already affecting London at the end of the 13th century[1].
- St Paul's Cathedral was heavily damaged by black crusts due to the use of coal for heating in the streets[2].

In the past, Europe's historic buildings were continuously subject to maintenance by cleaning and the substitution of decayed bricks and stones, mortar repointing and render repair. However, by the second half of the 20th century many historic centres in Italy, Spain, Greece and elsewhere fell into disrepair. In fact after 1945 many architecturally important historic centres were abandoned and left to decay[3].

As for contemporary construction, facing concrete was widely used without considering the necessity of protecting it against fast carbonation and steel corrosion (Fig. 8.1). New products obtained through advanced technology (organic synthetic materials, etc.) were declared durable and employed without adequate experimental research into their durability and their compatibility with traditional materials. Decay was often detected shortly after construction was completed[4].

Figure 8.1 *Detail of a reinforced concrete structure subject to steel corrosion.*

The first International Conference on Durability of Building Materials and Components (DBMC) took place in Ottawa, Canada in 1978[5]. This was after a conference on durability had taken place in Boston in 1958. The 9th DBMC conference was organised in Vancouver, Canada in 1999[6]. Since 1978 research has taken place on the durability of building materials, especially concrete, with the results presented at conferences, workshops, seminars and courses.

Section 8.2 considers the debate on definitions of durability and service life as developed by national and international committees. In Section 8.3, the most common aspects of deterioration in masonry, their causes, and strategies to increase the service life of masonry are discussed. In Section 8.4, the moisture movements in masonry and their effects are briefly presented. In Section 8.5, a brief description of the state of the art of research on masonry durability is given. In Section 8.6, a systematic approach to the study of masonry durability, as proposed by the authors, is described.

8.2 Durability and service lifetime: concepts and definitions

Two fundamental concepts of research on the durability of materials need to be defined: durability and service life (or lifetime).

Fronsdorff and Master in 1981[7] defined 'durability' as an essential attribute of construction materials. The dictionary definition: 'quality of being durable, ability of duration under a continuous use' does not seem to fit. In fact, material durability is not an absolute quantity, but a function of the environment.

According to ASTM 'Recommended Practise for increasing Durability of Building Constructions Against Water-Induced Damages' two definitions are given:

- durability – service, in terms of safety, of a structure or its parts for the design lifetime;
- durability – capability of a product, component, assemblage or construction to maintain its serviceability. Serviceability is defined as the capability of the material, component, assemblage or construction to offer the designed services.

Both of these definitions of durability imply satisfaction of the rules given by the design.

According to Sentler[8], the durability of a structure can be defined as its ability to resist, with success, severe deterioration for a defined number of years.

Kreijer[9] defines durability as the loss of serviceability as a function of time. He also considered the building as composed of a number of elements and components built with assembled materials. Considering the building as an assemblage of materials differently connected, he introduced the concept of compatibility between materials. He also defines the concept of serviceability as the specific behaviour of the building and its parts in a specific environment.

Sijöström[10], using the terminology of CIBW80/RILEM71-PSL joint TC, defines durability as the resistance of a material or component to environmental factors or agents, i.e. the strength of the materials to deterioration.

Eurin[11] and S.C. Saunders[12], at the conclusion of session 3 of the first International RILEM Conference on 'Durability of Construction Materials' Paris,

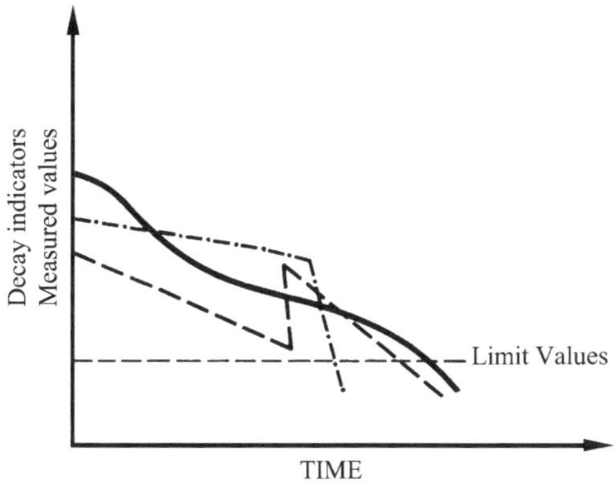

Figure 8.2 *Examples of functions representing the hypothetic serviceability of a material[10].*

1987, do not change Sijostrom's definition, and connect durability to the ageing of the materials and to the reduction of serviceability due to ageing. They say that the 'science of ageing' is based on a multidisciplinary approach to the physics and chemistry of materials, which takes into account the combined effects of all the ageing agents. Therefore, it is necessary to go from material science to material engineering.

According to the last updated definition, given by CIBW80/RILEM71-PSL joint TC[13], durability is the capability of the building, assemblage, product or construction to preserve its functionality or serviceability within the time required by the design.

It can be seen that the definitions of durability are many and not identical; new definitions can be expected in the future.

Master and Fronsdorff[7], define serviceability as the capability of a building, component, etc. to satisfy the functions for which it is designed.

For Sentler[8], the lifetime is the time interval within which the structure can give serviceability without maintenance.

Kreijger[9], defines the concept of serviceability as the behaviour of the building or its parts with reference to durability.

Sijöström in[10], defines performance as a function describing the properties that indicate decay which vary with time (Fig. 8.2).

According to the definition given by CIBW80/RILEM71-PSL joint TC[13], the service life time is the period of time during which all the properties are greater than or equal to the minimum acceptable values when the building is subject to ordinary maintenance.

In conclusion, it is possible to say that, even with apparent differences, the concepts of durability and service life for materials and structures have not changed over time and that the updated definitions given by CIBW80/RILEM71-PSL joint TC can be accepted.

8.2.1 Deterioration of porous materials

The deterioration of porous materials due to aggressive agents appears as decohesion, fissuration, detachment, delamination, disintegration (Figs 8.3 and 8.4), etc.

The deterioration of masonry is generally a result of exposure to the environment and is connected to:

a) the chemical and physical properties of the material;
b) climatic condition (temperature, relative humidity, presence of pollutants, etc.);
c) water critical saturation.

Nevertheless, the causes of decay in building materials are frequently connected to other factors such as[8]:

a) the design procedure, and
b) the building process.

Figure 8.3 *Deterioration of a capitol of Angera stone.*

Figure 8.4 *Delamination of bricks in a natural environment.*

Decay due to the environment

When finished, and during construction, the building is subjected to environmental action. This action can have damaging effects on the materials. Its influence, together with the characteristics of the materials and of the structure, is outlined in Fig. 8.5.

The lifetime of a building is dependent upon the following factors: aggressiveness of the environment, the quality of the material, construction detailing external actions, and structural behaviour. The deterioration processes can be of both a physical and a chemical nature.

Physical decay is a consequence of: erosion, freeze–thaw action, salt crystallisation and cracking due to mechanical action (Figs 8.6 and 8.7). Chemical decay is a

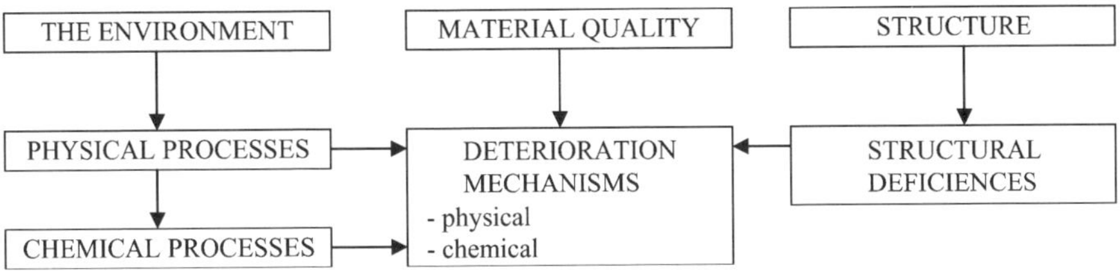

Figure 8.5 *Influence of environment, material quality and structure on deterioration mechanisms (Sentler 1983)* [8].

Figure 8.6 *Reduction of cross section of a wall due to environmental action.*

Figure 8.7 *Combined effects of load and chemical–physical agents.*

consequence of: dissolution, sulphation, reaction with pollutant deposition (Fig. 8.8), carbonation of mortar and corrosion of the steel reinforcement, etc. In the case of brick masonry for example, the chemical decay is usually related to the durability of the mortar. Nevertheless, lack of maintenance certainly accelerates these processes (Fig. 8.9).

Deterioration due to the design method

Within the design process, attention to durability is usually limited to the choice of materials compatible with the structure's environment. However, it is important to consider specific aspects within the design process such as:

- the behaviour of the material in service and over time;
- the construction details of the structure.

These details are frequently excluded from the structural analyses and also from the design drawings. Also, they are usually rarely addressed in codes of practice and standards.

Deterioration due to the execution process

Problems can occur during the construction process that compromise the durability.

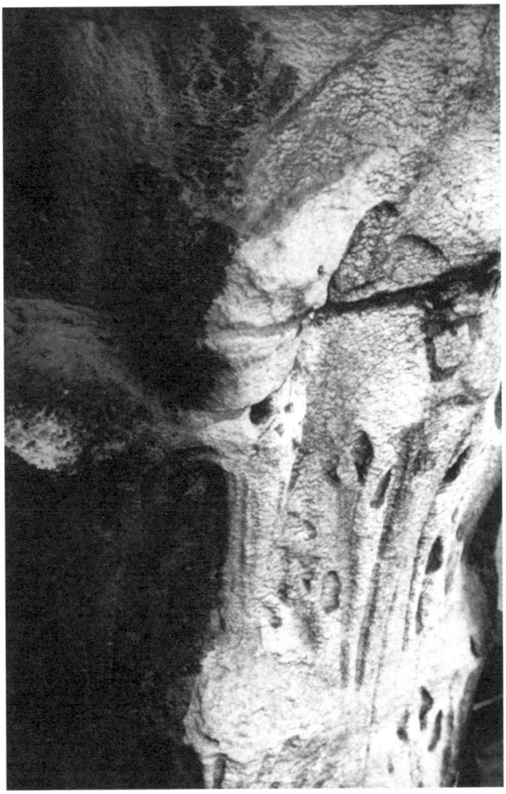

Figure 8.8 *Deterioration due to an aggressive environment (Milan, detail of the Roman colonnade of S. Lorenzo).*

Figure 8.9 *Decay of a building abandoned for thirty years (Milan, Cascina Rosa, 19th century).*

In the case of brick masonry the following defects will cause problems:

- lack of bond between the mortar and the bricks due to poor laying of joints;
- lack of verticality of the walls;
- lack of connection between the walls and the floors;
- the wrong choice of mortar type due to not taking into account brick absorption.

8.2.2 A strategy for the durability of buildings

The most important concepts for achieving a satisfactory service life are:

1. That the proved durability is based on:
   - knowledge of aggressive agents;
   - knowledge of the deterioration processes;
   - choice of materials appropriate to the expected exposure conditions;
   - execution of appropriate construction details to prevent decay; and
   - inspection of the building during construction.
2. Maintenance must be carried out according to the defined schedules by the designer, in order to increase the service life of the building. The designer must have adequate experience and skill in order to determine these schedules. It is important therefore, to define the acceptable level of deterioration in case of the lack of maintenance. In seismic areas the prevention of earthquake damage must be considered.
3. Prevention. When maintenance is adequate, repair should only be necessary in case of mistakes during construction, in case of exceptional events (earthquakes, fire, flood, etc.) or when the function of the building is changed.

8.3 Causes and decay processes of masonry components

8.3.1 Brick deterioration

Bricks are ceramic materials made with a porous coloured paste. They have a regular geometric shape and are obtained by firing clay (or clay mixed with other materials which improve their plasticity). Together with stone and timber, bricks are the oldest materials used for building construction[14]. Solid bricks have a prismatic shape and their dimensions vary from country to country. For example, in Italy, the average dimensions are 5.5 × 12 × 25 cm.

Factors affecting their durability can be subdivided into internal and external factors.

Internal factors

Clays

Under a mineralogical and chemical profile, clays can be divided into two categories:

- siliceous clays, and
- calcareous clays.

Siliceous clays are found in glacial lake deposits. Calcareous clays are more common and can be found in alluvial deposits.

Partially admissible impurities

Clays can be based on pure clay minerals and on non-clay minerals or impurities. The most frequent impurities are: quartz, calcium or magnesium carbonate, mica, sulphur or iron oxides, sulphates, etc.

Special attention must be given to calcium carbonate and sulphates. Calcium carbonate, which is present as calcite, is usually found in very variable quantities, normally between 19% and a maximum of 30%. Such high percentages can be accepted provided the following conditions are satisfied:

- the calcium carbonate particles must be finer than the clay particles;
- the calcium carbonate is uniformly distributed in the mass.

During firing, free calcium oxide is obtained according to the equation:

$$CaCO_3 \Rightarrow CaO + CO_2$$

The calcium oxide may react with the amorphous silica and alumina coming from the broken lattice of the clay minerals. This reaction creates new formation products in crystalline phases, as calcium silicate and silico aluminate[15].

When large grains of calcium carbonate are present, even small percentages cannot be accepted. In fact, in that case, the free calcium oxide will not react with the silica and alumina. It will hydrate since it is in contact with the humidity of the environment. The subsequent volume increase will cause high pressure and push out the material around it. Clays with a high content of calcium carbonate will create weak and easily damaged bricks.

The Italian Code of Practice accepts a maximum of 0.05% of alkaline sulphates in fired bricks. This very low value is determined by the high risk that these soluble salts, in certain conditions of temperature and relative humidity, will cause efflorescence and delamination of the exposed surfaces of the materials. Therefore, bricks cannot have a high content of soluble salts.

Influence of clay grain size on the resistance to freeze–thaw action

Clays consists of lamellar particles with an average diameter that ranges between fractions of a micron (10^{-6} m) to 1 nanometre (10^{-9} m). Since the average diameter of the pores is directly connected to the average diameter of the particles, if the clay composition is controlled, it is possible to obtain products sufficiently resistant to frost action.

According to Ravaglioli[16], the optimum grain size distribution should be chosen according to the following instructions. Residuals in a sieve of dimension of 0.06 mm should be between 17–22% of which:

- the fraction from 1 to 3 mm = 8%; and
- the fraction less than 1 mm = 92%.

The size of the grains passing through the 0.06 mm sieve:

- the fraction greater than 20 nm = 42%,
- the fraction from 2 to 20 nm = 42% and
- the fraction less than 2 nm = 16%.

Influence of the production process

Bricks were handmade from ancient times until the 19th century. Their inhomogeneity was due to:

- the heterogeneity of clay;
- the type of mix – frequently clay was mixed with chamotte (Fig. 8.10); and
- the lack of uniform pressure.

Therefore, when choosing new bricks for a repair it is better to use bricks designated as 'handmade' rather than bricks that have been extruded. In fact in the complicated process of the production of bricks, the machine production is the most delicate step. Due to the production technology, defects can be found in the formation of brick (Fig. 8.11).

These machine-made defects can be due to:

- the wire-drawing machine not being correctly calibrated for the assigned plasticity of the basic paste;

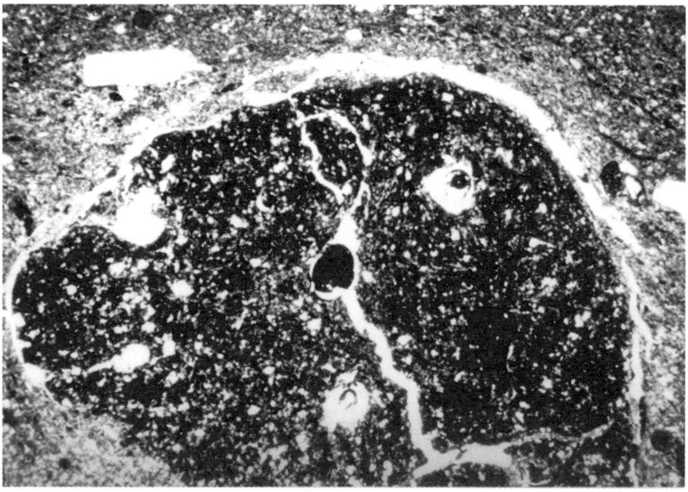

Figure 8.10 *Thin section of a 'Chamotte' fragment.*

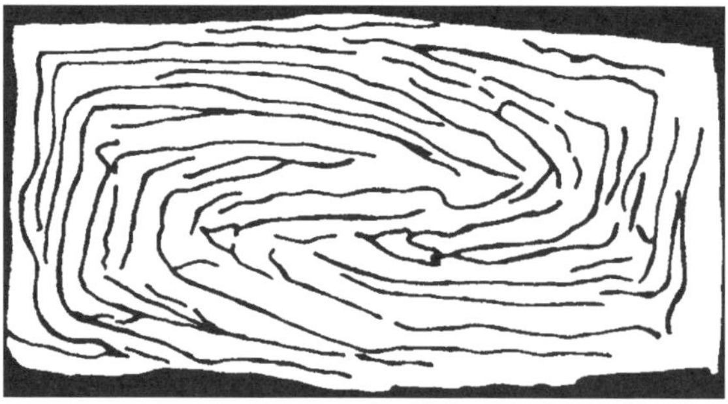

Figure 8.11 *Typical S-structures in an extruded brick* [17].

- the differential speed of the clay going through the machine; and
- the formation of smooth surfaces created by the rotating helix as these surfaces will not bond well.

Micro-cracks may then occur during the processes of desiccation and firing. Such bricks must not be used for the restoration of masonry surfaces exposed to the environment.

Some specimens of brick sampled as being defective, were subjected to accelerated ageing tests, such as freeze–thaw and crystallisation tests. In comparison to handmade bricks (obtained by an industrial process, but such that the handmade brick characteristics are maintained), the defective bricks failed under freeze–thaw conditions after only 7 cycles, while the handmade bricks managed over 20 cycles and also maintained a compressive stress of 75% of the undamaged bricks.

When subjected to salt crystallisation tests with a saturated solution of sodium sulphate, the defective bricks failed with large fissuration along the already cracked surfaces (Fig. 8.12), the handmade bricks suffered a thin exfoliation and pulverisation after 7–8 cycles (Fig. 8.13). The same tests were performed on ancient bricks and the inhomogeneity of the material was clearly shown (Fig. 8.14).

Influence of the firing temperature

The chemical and mineralogical composition of the clay and the firing cycle are fundamental parameters in determining the durability of bricks. In order to obtain a well-compacted brick with a low porosity and a high strength, the firing temperature

Figure 8.12 *An extruded brick with S-shaped structure after a number of crystallisation cycles with* Na_2SO_4.

Figure 8.13 *The deterioration of a new 'handmade' brick after salt crystallisation tests. Two mechanisms of failure occurred: a) exfoliation, b) crumbling of the sanded surface.*

Figure 8.14 *Inhomogeneity of the structure shown by ageing tests (old Venetian brick).*

should be such that the clay's viscosity should ensure that all the pores are filled without the high deformation in the firing product.

During firing, important irreversible chemical and physical transformations take place within the material. Between 400°C and 600°C, loss of the chemically bonded water takes place, with consequent demolition of the crystal lattice; amorphous silica and alumina are formed between 600°C and 900°C, also the dissociation of calcium carbonate occurs with the elimination of CO_2 and the formation of free CaO. This reacts with the previously obtained amorphous materials producing calcium silicates and aluminates[17]. Beyond 900°C a new formation takes place; the mullite ($3Al_2O_3$ $2SiO_3$). The silica which still remains free can enter a liquid phase due to the alkali present, or it may be transformed in different allotropic phases.

Due to the large dimensions of the present kilns, the temperature cannot be uniformly distributed and, as a consequence, low-fired or over-fired bricks can result. This situation, together with the chemical–physical composition of the original clay, influences the brick's composition and porosity. The porosity tends to decrease as the firing temperature increases. Fig. 8.15 shows the water absorption as a percentage against the firing temperature for two brick types, A and B.

Fig. 8.16 shows that the compressive strength increases with firing temperature. Nevertheless, bricks fired at too high a temperature, and with a very low porosity, can have a very low bond with mortars.

External factors

The external factors causing decay in bricks are all related to the environment around the building (e.g. climate, atmosphere composition, water, etc.). The deterioration caused by external factors can be: chemical, physical and biological.

Chemical decay

Chemical decay can be due to a polluted atmosphere, acid rain or simply rain running down a wall. Hydrocarbon combustion produces gaseous substances (e.g. CO_2, SO_2,

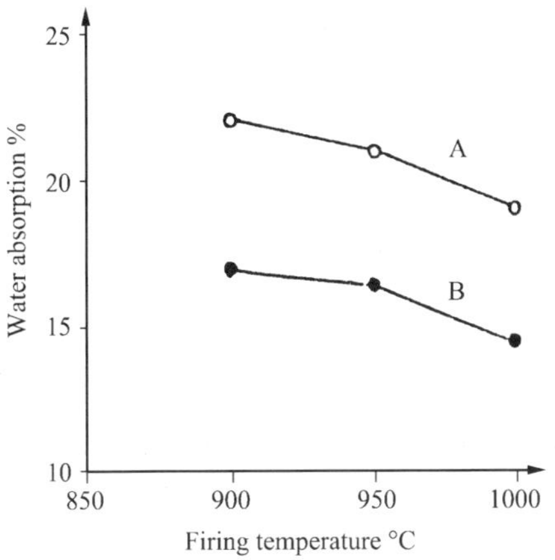

Figure 8.15 *Water absorption against firing temperature for two different clays.*

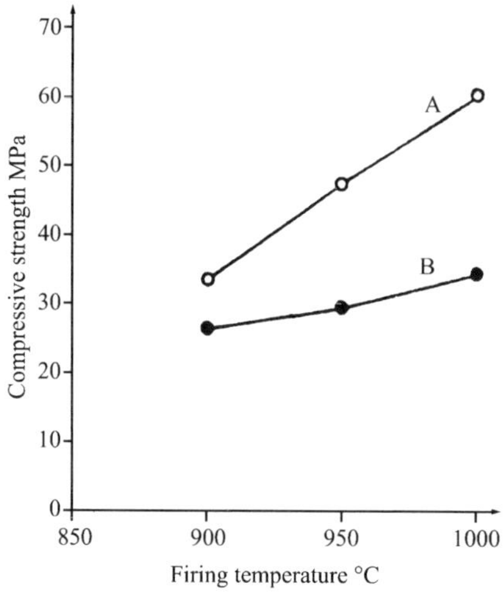

Figure 8.16 *Compressive strength against firing temperature for two different clays.*

NO_x, etc.) which are aggressive to masonry surfaces. These gases are usually found in urban and industrial atmospheres with increased amounts in winter. Some of the chemical actions will be briefly explained.

CO_2 is water soluble (rain, fog) and its concentration increases as the temperature decreases. The water becomes acid and aggressive towards masonry materials (mortars more than bricks).

Sulphuric anhydrite is also soluble and by catalytic oxidation, is transformed into SO_3 and then into sulphuric acid. Among all the pollutants, sulphuric acid is the most aggressive because:

- it is constantly produced as industrial waste;
- it does not evaporate; and
- has a high affinity with many ions such as Fe, and Ca contained in the vitreous matrix of the bricks.

The most common bricks are, in fact, mainly composed of quartz (which cannot be etched by acid solutions). Nevertheless, the single quartz grains are bonded among themselves, and with other minerals of neo-formation, by a vitreous mass composed of alkaline silicates and alkaline earth silicates which can be etched[18].

Physical deterioration

It has been demonstrated that the highest percentage of decay of brick–stone masonry walls is caused by the crystallisation of soluble salts (Na_2SO_4, $MgSO_4$, $CaSO_4$, NaCl)[19]. In masonry, the rate of evaporation is greater than the rate of migration of water from the internal part of the masonry. The salts dissolve into water concentrate and crystallise at the surface of the masonry. These salts can crystallise into an anhydrous or a hydrated state depending upon the external thermodynamic conditions (temperature and relative humidity).

The word 'efflorescence' indicates the loss of water from a hydrated salt when the vapour pressure of the hydrated system is higher than the water vapour pressure of the air. Dehydration occurs when the equilibrium is reached between the hydrated system and the environment[20]. The amount of efflorescence depends upon the quantity of salts in solution, and the porosity of the masonry, or the masonry components (mortar, bricks or stones). These soluble salts may come from the soil or from the masonry components themselves.

Efflorescence is not dangerous as far as the aesthetical aspects of the masonry are concerned, but it can generate mechanical damage causing cyclical stresses similar to freeze–thaw action if the salts crystallise out beneath the faces of the masonry and the bricks are underfired.

The salts which most frequently feature in construction are the alkaline and alkaline-earth sulphates. Table 8.1 shows the frequency of the detected efflorescence in Italian bricks[21].

Sodium sulphate for instance, crystallises from a water solution in different shapes and volumes according to the thermodynamic cycles of the environment. Between −1.42 and 32.4°C it crystallises deca-hydrate as monoclinic prisms (Fig. 8.17). Above 32.4°C it crystallises in an anhydrous phase. Under certain conditions the anhydrous sulphate can crystallise below 24.4°C as $Na_2SO_4.7H_2O$. This is an unstable phase for the sulphate which, in a humid environment is transformed into $Na_2SO_4.10H_2O$.

The vapour pressure p corresponding to the equilibrium[22, 23]:

$$Na_2SO_4.10H_2O \Leftrightarrow Na_2SO_4 + 10H_2O + \text{vapour}$$

Table 8.1 Index of salt presence in efflorescences found on some Italian bricks

Nature of the salt	On dried	On fired	On the masonry
Na_2SO_4	+++	+++	+++
K_2SO_4	++	++	++
$CaSO_4$	+++	+++	+++
$MgSO_4$	+	++	++
$NaAl(SO_4)$	–	+	–
$Kal(SO_4)_3$	–	+	–
Na_2CO_3	–	+	+
K_2CO_3	–	+	+
$CaCO_3$	–	+	+
NaCl	+	+	++
KCl	+	+	++
$NaNO_3$	+	–	++
KNO_3	+	–	++
$Fe(SO_4)_3$	–	+	+
$FeSO_4$	–	+	+

+++ always present
++ often present
+ rarely
– absent

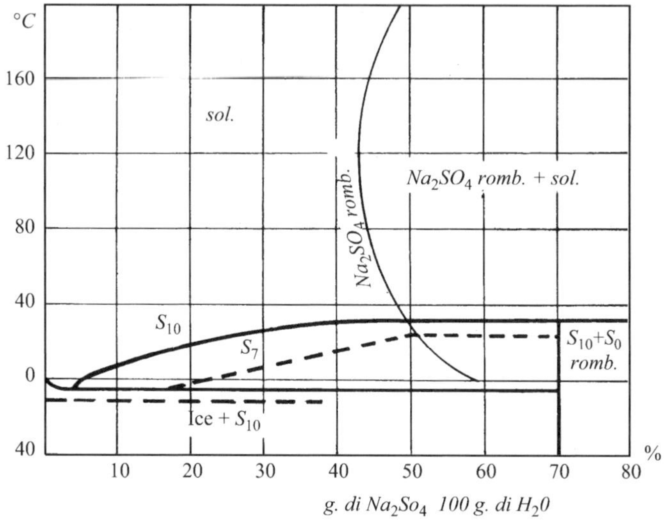

Figure 8.17 *Stability fields for Na_2SO_4 phases.*

assumes the following values as a function of the temperature:

Temp°C	0	5	10	15	20	25	29
p (mmHg)	3.8	5.2	7.0	9.7	13.9	19.0	24.0

When the relative humidity is near 100% the compound, $Na_2SO_4.10H_2O$ does not de-hydrate. In order for de-hydration to occur, it is necessary that the atmosphere vapour pressure is less than the vapour pressure of the hydrated salt. In these conditions, Na_2SO_4 can again absorb water in a liquid or gaseous state and become $Na_2SO_4.10H_2O$.

These changes from one crystalline phase to another following temperature and relative humidity variations, may cause fatigue phenomena in the brick, mortar or stone, which bring it to failure[24]. The time necessary to reach failure depends strongly on the material's pore structure and on the frequency of the anhydrous and hydrated cycles.

Another important factor that causes physical damage is the freeze–thaw action, which, in some climates, such as high mountains and cold countries, can cause considerable damage to exposed parts of buildings. A material is defined as vulnerable to frost action when, after being subjected to water saturation to a specified number of freeze–thaw cycles, it loses its integrity. In fact, when the water passes from a liquid phase to a solid phase, it increases in volume by about 9%. This transformation also occurs at atmospheric pressure when the temperature reaches 0°C. It can also occur at much lower temperatures under high pressure. The water, changing to ice, starts at the external surface of the material and proceeds to the internal part. The pores with the greatest diameter are the first to be affected, then the ice develops in the smaller pores as the temperature reduces. Due to the increased volume of the ice, the remaining water finds its way through the smaller pores; interstitial pressures develop which increase as the pore diameter decreases[25].

According to recent research, deterioration can occur in porous materials even when saturated with organic liquids which decrease their volume, when subjected to freeze–thaw cycles[26]. Therefore the volume increase of the water during the change of state is not the only cause of the increased occurrence of fatigue in the materials.

Whatever the mechanism of fatigue failure produced by ice formation, all the research agrees that materials with a high percentage of small pores are less resistant to frost damage than material with large pores. In fact, the ones with larger pores resist saturation by water. Nevertheless, no agreement has been yet reached on the values of the critical diameters and on their critical percentage. The frost resistance of bricks has not been investigated fully because contradictory results have been obtained.

Many parameters can influence a material's resistance to frost:

- the mineralogical nature of the clays;
- the grain size distribution of the clays; and
- the brick firing temperature and duration of heat input (Fig. 8.18).

A high firing temperature and low porosity produce more resistant bricks.

8.3.2 Stone alteration

Polluting agents

Once any type of stone has been taken from the quarry and exposed to the environment, it will oxidise and then decay. This decay can also occur in the quarry itself. With industrialisation, and the consequent increase in air pollution, stone degradation has rapidly increased[27] (Fig. 8.19).

Polluting agents can be defined as particles which produce a variation in the atmospheric composition (Fig. 8.20). Pollutants can be subdivided into primary and secondary as shown in Table 8.2.

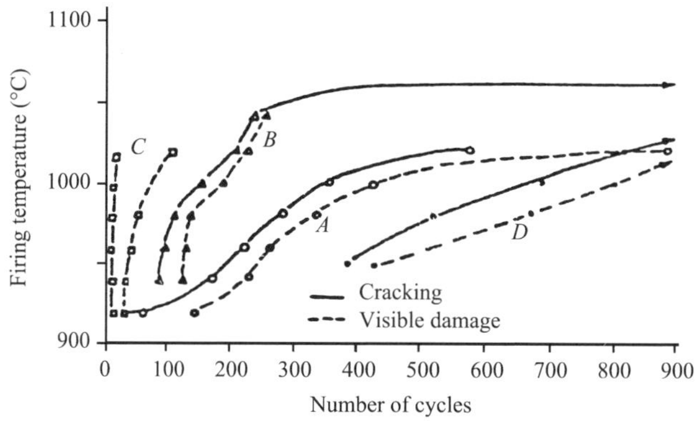

Figure 8.18 *Influence of the firing temperature on frost resistance of bricks prepared with four different clays [21].*

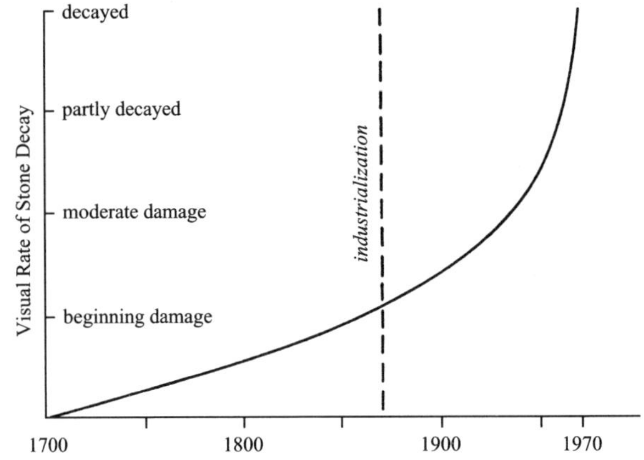

Figure 8.19 *Approximate rate of decay over time for a statue in the Castle of Herten di Westfalia (Germany)* [27].

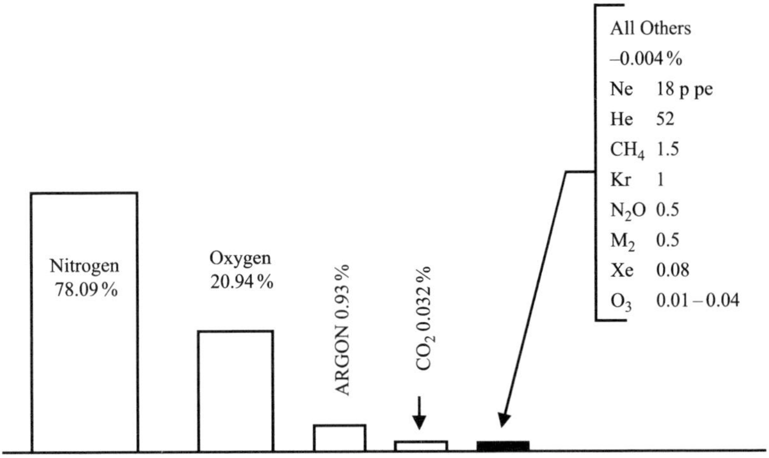

Figure 8.20 *The composition of clean, dry air at the surface level (Germany) by P. Urone.*

Effects on calcareous stones

The most important agents are given in Tables 8.2 and 8.3.

Effects of CO_2

The weak acid solution obtained by the dissolution of CO_2 in rain, dissolves calcium and magnesium carbonates in calcareous stones, dolomitic marbles, mortars and renderings. Soluble calcium and magnesium bicarbonates are obtained according to the following reactions:

$$CaCO_3 + CO_2 + H_2O \Rightarrow Ca(HCO_3)_2$$
$$MgCO_3 + CO_2 + H_2O \Rightarrow Mg(HCO_3)_2$$

The quantity of calcium and magnesium bicarbonate obtained is dependent upon water temperature, the proportion of dissolved CO_2, the solubility of the different minerals, etc. (Fig. 8.21).

Table 8.2 Classification of pollutants. Courtesy of C.E. Junge

Source	State of aggregations of matter	Size of particles	Physical state	Air pollution situations

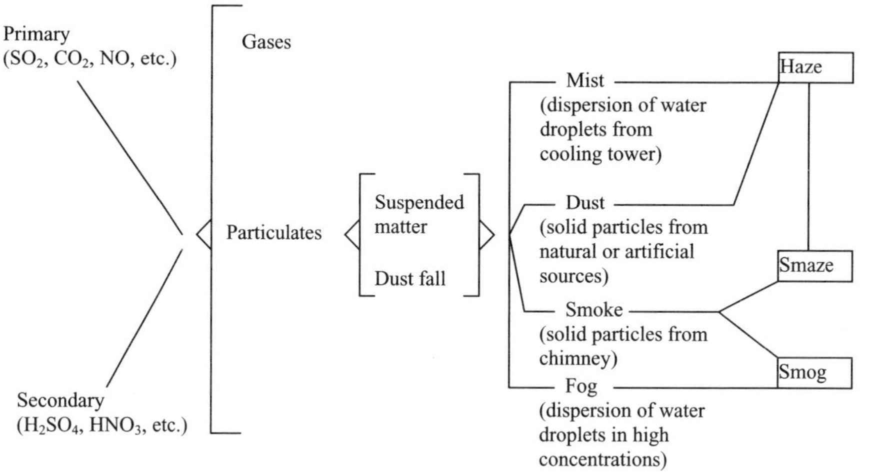

The rate of solubilisation is higher in the suburban and rural areas (Fig. 8.22)[27]. It also depends on the intensity of the rain, the surface roughness of the stone, the angle of the exposed surface and the direction of the rain[32]. The solubilisation of the carbonates due to acid rain can cause the formation of efflorescence which occurs according to the following reaction:

$$Ca(HCO_3)_2 \Rightarrow CaCO_3 + CO_2 + H_2O$$

Efflorescences, or cryptoefflorescences, can be formed, depending on the evaporation rate.

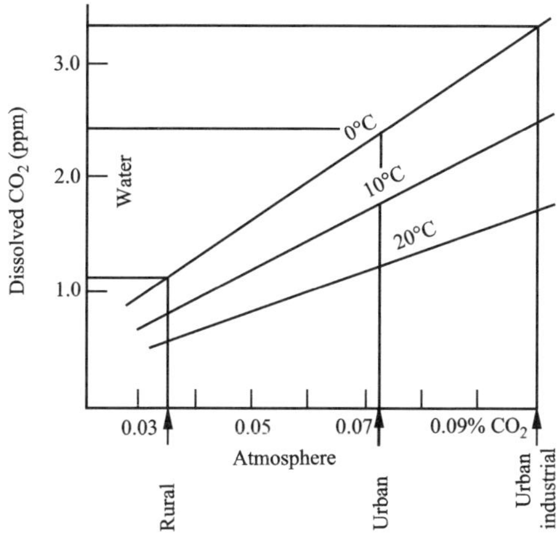

Figure 8.21 *Saturation equilibrium of carbon dioxide in the atmosphere with water, as dissolved CO_2 (after Winkler 1966).*

Table 8.3 Summary of sources, concentrations, and major reactions of atmospheric trace gases [79]

Contaminant	Major pollution sources	Natural sources	Estimated emissions (ton)		Atmospheric background concentrations	Calculated atmospheric residence time	Reactions removal and sinks	Remarks
			Pollution	Natural				
O_2	Combustion of coal and oil	Volcanoes	146×10^6	No estimate	0.2 ppb	4 days	Oxidation to sulphate by ozone or after absorption, by solid and liquid aerosols	Photochemical oxidation with NO_2 and NO may be the process needed to give rapid transformation of $SO_2 \rightarrow SO_4$
H_2S	Chemical processes sewage treatment	Volcanoes biological action in swamp areas	3×10^6	100×10^6	0.2 ppb	2 days	Oxidation to SO_2	Only one set of background concentrations available
CO	Auto exhaust and other combustion	Forest fires, oceans, terpene reactions	304×10^6	33×10^6	0.1 ppm	< 3 years	Probably soil organisms	Ocean contribution to natural source probably low
NO/NO_2	Combustion	Bacterial action in soil	53×10^6	NO:430×10^6 NO_2:658×10^6	NO: 0.2–2 ppb NO_2: 0.5–4 ppb	2 days	Oxidation to nitrate after sorption by solid and liquid aerosols, hydrocarbon photochemical reactions	Very little work done on natural processes
NH_3	Waste treatment	Biological decay	4×10^6	1160×10^6	6 ppb to 20 ppb	7 days	Reaction with SO_2 to form $(NH_4)_2SO_4$ oxidation to nitrate	Formation of ammonium salts in major NH_4 sink
N_2O	None	Biological action in soil	None	590×10^6	0.25 ppm	4 years	Photodissociation in stratosphere, biological action in soil	No information on proposed absorption of N_2O by vegetation
Hydrocarbons	Combustion exhaust	Biological processes	88×10^6	CH_4: 1.6×10^9	CH_4: 1.5 ppm Non CH_4:<1 ppb	4 years (CH_4)	Photochemical reaction with NO/NO_2, O_2	'Reactive' hydrocarbon emissions from
	Chemical processes			Terpenes: 200×10^6			Large sink necessary for CH_4	Pollution = 27×10^4 tons
CO_2	Combustion	Biological decay, release from oceans	1.4×10^{10}	10^{11}	320 ppm	2–4 years	Biological adsorption and photosynthesis, absorption in oceans	Atmospheric concentrations increasing by 0.7 ppm/year

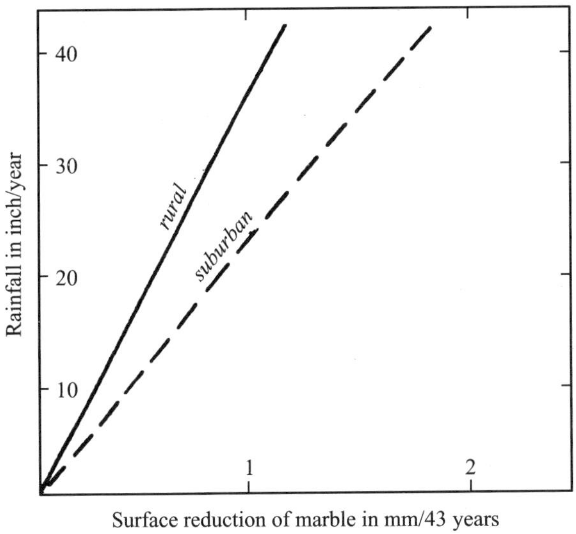

Figure 8.22 *Surface reduction of marble for 43 years of exposure against annual rainfall. Courtesy of E.M. Winkler* [27].

Effects of CO on calcareous stone

CO is present in the atmosphere in rather high quantities when compared to other substances. It is produced by combustion processes during which the quantity of oxygen is lower than the stoichiometric one. CO, as such, is not harmful to the stones unless it is transformed into CO_2.

Effects of NO_x on the calcareous stones

Around 70–80% of NO_x is produced by antropic pollution in industrialised areas[30]. The role of nitrates in the deterioration of stone is not yet completely understood. It seems that the action correlates to the action of soluble salts into walls. According to Kaufmann[28] the mineralogical alteration of calcareous stones is based on nitrification according to the following reaction:

$$2CaCO_3 + (NH_4)_2SO_4 + 4O_2 \Leftrightarrow Ca(NO_3) + CaSO_4 + 2CO_2 + 4H_2O$$

This reaction is also obtained in gypsum. Nevertheless, the presence of gypsum on the surface of the calcareous stones is due mainly to the attack of sulphuric acid in the air.

When ozone is present, nitrates can oxidate according to the following mechanisms:

$$NO_2 + O_3 \Rightarrow NO_3 + O_2$$
$$NO_3 + NO_2 \Rightarrow N_2O_5$$
$$N_2O_5 + H_2O \Rightarrow 2HNO_3$$

Nitric acid can react with the calcareous stones as follows:

$$CaCO_3 + 2HNO_3 \Rightarrow Ca(NO_3) + H_2O + CO_2$$

Fluorides and hydrofluoric acid: effects on calcareous and siliceous stones

Fluorhydric acid is a gaseous pollutant which attacks both calcareous and siliceous stones. Fluorhydric acid's reaction with calcareous stones brings about the formation of CaF_2. The attack on siliceous stones is of two types:

- nucleation and formation of new crystalline phases;
- formation of micro- and macro-cracks and orifices.

Chlorides and chloric acid

Chlorides are rarely found in atmospheric pollution. However, chloric acid is a common pollutant, coming from industrial emissions and from the combustion of carbon, paper and organic plastic materials. Chloric acid renders the rain slightly acid. The acidified water dissolves and solubilises the limestone. The soluble salts formed can cause efflorescence followed by delamination and spalling of the material.

Sulphur oxides on limestone

Sulphur oxides are gaseous compounds present in the atmosphere as SO_2 and SO_3. They are produced by the combustion of carbon compounds, oils, etc. Even if combustion occurs with a lack of oxygen, SO_2 is formed at a high percentage. The quantity of SO_3 depends upon temperature and is only 1% to 10% of the total sulphur oxides (SO_x). SO_2 exists in the atmosphere only if the concentration of water vapour is very low. Normally, water vapour combines with SO_3 forming sulphuric acid:

$$SO_3 + H_2O \Rightarrow H_2SO_4$$

The mechanism and rate of oxidation of SO_2 cannot be predicted. This is due to the complicated nature of this reaction, and to the influence of many factors such as: humidity, solar light, temperature and to the presence of catalysts[28]. The presence of sulphuric acid can lower the pH of rain and dissolve limestone with the formation of soluble salts (calcium sulphate). These salts can cause efflorescence on the stone surface. The quantity of SO_2 is particularly high in winter (Fig. 8.23).

Aerosols

Aerosols are liquid and solid particles dispersed in the atmosphere. Their formation and distribution are determined by natural, anthropic, meteorological and topographic causes. On the basis of particle diameter they can be classified as:

- particles with a radius $< 0.1\,\mu m$;
- particles with a radius between 0.1 and $1.0\,\mu m$;
- particles with a radius $> 1.0\,\mu m$.

Aerosols are usually composed of the following substances:

- sulphuric acid (in particularly polluted areas);
- sulphates;
- letovicite $(NH_4)_3{\cdot}H(SO_4)_2$
- ammonium sulphate $(NH_4)_2{\cdot}SO_4$

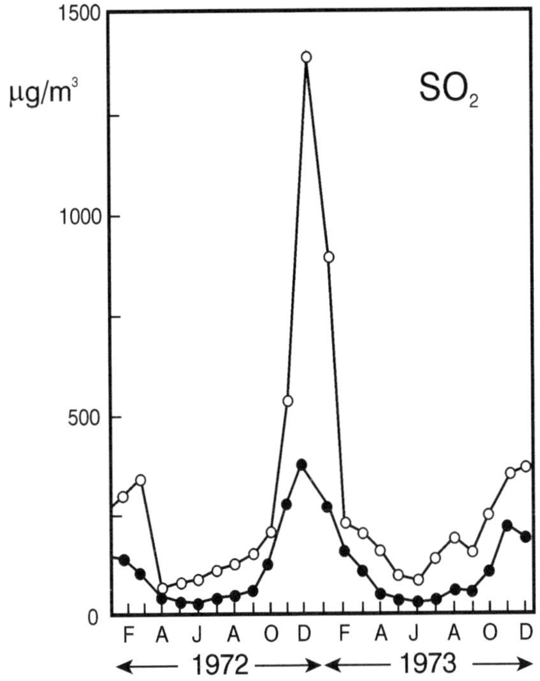

Figure 8.23 *Monthly average concentration (•) and maximum daily values for the month (o) measured in Venice during 1972 and 1973* [28].

- ammonium bisulphate $(NH_4)\cdot HSO_4$
- magnesium sulphate $MgSO_4$
- calcium sulphate $CaSO_4$ (with rather low solubility),
- sodium sulphate Na_2SO_4,
- bisulphate ion $H^+SO_4^-$
- sulphate ion SO_4^{2-}; and
- metal ions complexes.

Aerosols play an important role in the decay of stones and marbles. By deposition on external surfaces, aerosols form black crusts which can cause serious damage.

Effects on siliceous stones

Siliceous stones are much more resistant to aggressive agents than calcareous stones. The atmospheric agent most harmful to them is hydrofluoric acid. Nevertheless, these stones, once attacked, become more porous and are then more at risk to the cyclic freeze–thaw action and to the action of salt crystallisation.

8.3.3 Mortar decay

Definitions

Mortar is a mix of organic and inorganic binders, with fine aggregates, water and organic and inorganic additives, mixed in order to give the fresh mortar good workability and the hardened mortar, plus an adequate physical (porosity, vapour permeability, etc.), and mechanical (strength, deformability, adhesion, etc.) behaviour as well as good appearance and durability.

The function of the sand (or aggregate) is to increase the volume of the mix, to facilitate the penetration of CO_2 (necessary for the mortar hardening), and to reduce the shrinkage and hence the formation of cracks. The binder in ancient mortars could be air lime, hydraulic lime and lime-pozzolana or lime-pozzolanic materials (brick dust, etc.). It is useful to remember that it was only in the second half of the 19th century, that Portland cement was produced (e.g. in Italy only after 1876 was cement produced industrially). Therefore it is impossible to find cement mortars before that time.

Mortar in masonry

Before the stone-age, humans did not use mortar in shelters but during the stone-age clay was used as a mortar. The first use of air lime is not known; nevertheless it seems that it was already frequently in use in the bronze-age. The first historically proved use of hydraulic mortar goes back to King Solomon's times (10th century BC). In fact, in order to obtain waterproof cisterns, Phoenician artisans used to fix them with a rendering made of air lime and crushed bricks. At about the same time the Phoenicians also used pozzolanic mortars made with air lime mixed with pozzolana, a material from Santorini in Greece.

Expert in both techniques, Roman masons also found a great amount of pozzolana in the Italian Southern regions such as Lazio, Campania and in Etruria.

Air lime mortar

Air lime mortar is obtained by mixing sand, slaked lime and water in different proportions according to the mix requirements. This mortar can be used for the jointing of stone and brick masonry and as a rendering. These mortars have been the commonest used from ancient times until the 1950s.

The limitation of hydrated lime mortars are: low mechanical strength, slow hardening, high vulnerability to freeze–thaw actions, and the possibility of the formation of CaO pebbles which can hydrate and increase their volume in contact with air moisture both during and after the construction. These mortars can be affected by the same decay which takes place in limestone. Both CO_2 and SO_3 can cause the deterioration. CO_2 transforms calcium carbonate into soluble calcium bicarbonate according to the reaction:

$$CaCO_3 + CO_2 + H_2O \Rightarrow Ca(HCO_3)_2$$

SO_3 transforms calcium carbonate into gypsum:

$$CaCO_3 + SO_3 + 3H_2O \Rightarrow CaSO_4.2H_2O + CO_2 + H_2O$$

Other types of decay are given in Section 3.2 on stones.

Pozzolanic mortars

These mortars are obtained by mixing slaked lime with pozzolana. They have a good resistance to water dissolution and are in fact used in humid environments and where the masonry wall foundations are in water. The pozzolanes have a slightly acid behaviour and so can fix the calcium hydrate and make it insoluble.

Hydraulic mortars

These mortars are obtained by using hydraulic lime as a binder. Hydraulic limes are produced by burning argillaceous or siliceous limestones and reduced to powder by slaking – with or without grinding. All hydraulic limes can set and harden under water. Atmospheric carbon dioxide contributes to the hardening process.

Composite mortars

These mortars can be obtained by using two or more binders (lime-cement or others) with the aim of combining the advantages of the two binders. The most common mortars are lime-cement or cement-hydraulic lime mortars. Their decay depends upon the binders used.

8.3.4 Masonry decay

The decay of masonry is highly influenced by its components and by the environment. (Fig. 8.24). It is practically impossible to predict the durability of masonry from the behaviour of the single components. Therefore the study of masonry durability can only be carried out by experimental investigation. Of course, the difficulties can be overcome if the characteristics of the macro- and micro-environments are known, and if the presence, and movements, of water inside the masonry can be detected. The types of damage that affect masonry are well described in[31]. An expert system to help architects and engineers understand and interpret the mechanisms of decay in masonry, and to find reliable solutions for its repair and preservation, has been developed as through an EC contract: 'Expert System for Evaluation of Deterioration of Ancient Brick Masonry Structure'[32].

8.4 Moisture movements in masonry

As made clear in the previous sections, decay phenomena are always connected with the presence of water in its various forms. Hilsdorf uses a very interesting diagram to

Figure 8.24 *Decay of the stone models.*

describe the presence and distribution of moisture in masonry buildings (Fig. 8.25)[33]. Massari in[34] gives a comprehensive and practical explanation of the phenomena described in Fig. 8.25, with qualitative and quantitative data, collected during his long personal experience of various cases of water penetration. His conclusions are briefly summarised here.

Capillary rise moisture

The quantity of water, and its distribution inside masonry, depends upon several parameters such as the nature of the soil, the quantity of the soil's moisture content, the external and internal relative humidity and temperature, and, critically, on the material characteristics especially the size distribution of pores (and their combination in the case of composite materials as masonry). In Fig. 8.26, the different influences are presented for bricks which are more porous than mortars (and vice versa). Fig. 8.27 shows the case for stones which are, usually, less porous than mortars. Ref. [34] also shows the different height of capillary rise as a function of wall dimensions and of the type of exposure (Fig. 8.28).

Condensation moisture

A discontinuous phenomenon connected to the differences in temperature between the external environment and the internal environment in the building. Condensation can be present in the winter as well as in the summer (Fig. 8.29).

Moisture due to driving rain

Rain penetrates a wall as a result of capillary action. Nevertheless, according to Ref. [34] rain very seldom penetrates through the whole section in old walls. When soluble salts

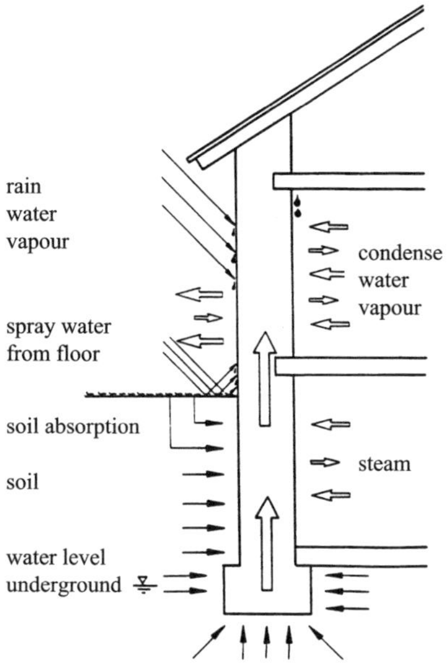

Figure 8.25 *Examples of different possibilities for penetration by water. Courtesy of H. Hilsdorf.*

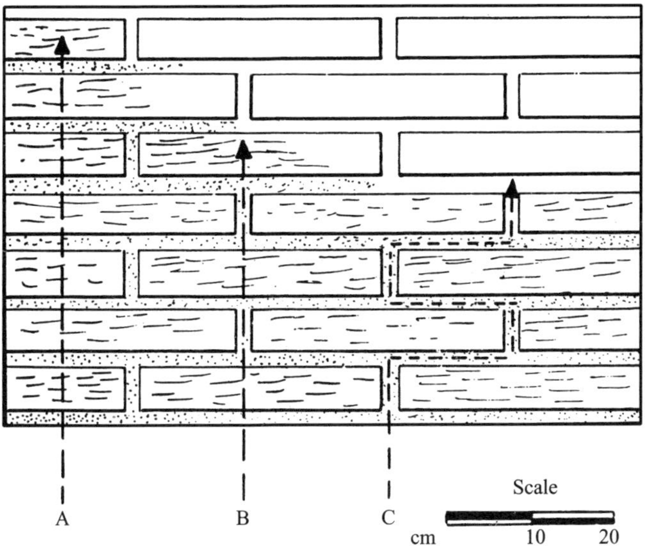

Figure 8.26 *The capillary follows the shortest way (A) or (B) when elements absorb more than the mortar. The longest route (C) is followed when the material has very low absorption* [34].

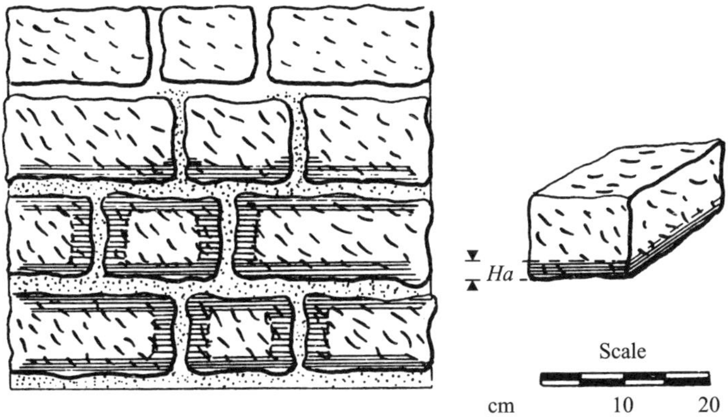

Figure 8.27 *When the stone height is greater than H_a, the maximum level of absorption, the moisture is going through joints and edges but the central core remains dry* [34].

are contained in water, in the soil or in the material itself, they are transported by evaporation towards the external surface of the wall. The salts tend to crystallise on the external surface as efflorescence or underneath the surface as cryptoefflorescence. This last crystallisation process causes the most damage by delamination.

Fig. 8.30 shows the typical type of damage caused by the different types of salts and the position where it occurs in a masonry wall as in Ref. [35]. Their presence influences the rate of capillary rise and the alteration determined by the crystallisation.

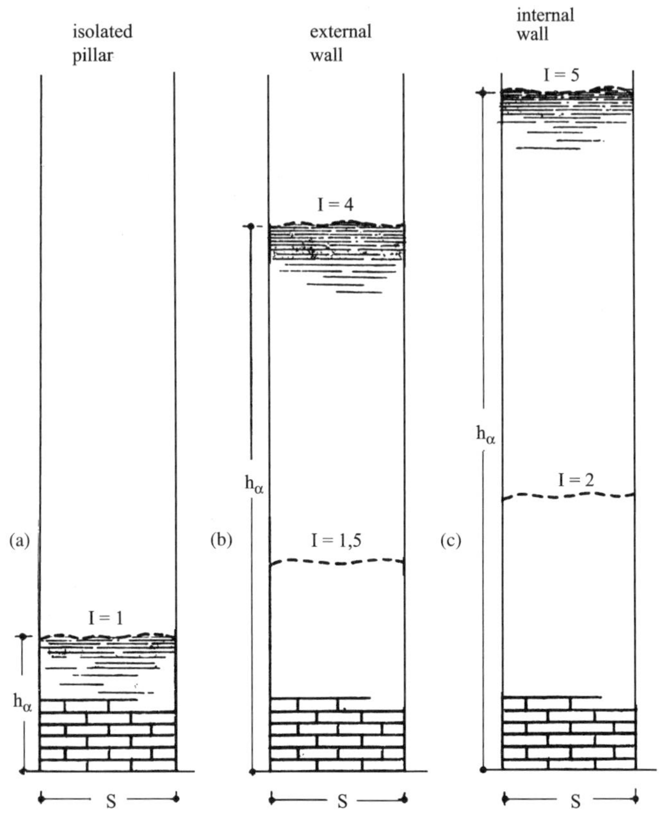

Figure 8.28 *A rise index ($I = h_d/S$) can be calculated; it depends on the wall thickness and on the external conditions* [34]*:*

- *a) in isolated piers I =1;*
- *b) in external walls I =1.5 ÷ 4S;*
- *c) in central walls I = 2 ÷ 5.*

The negative effects of water and salts can also be caused by bad design, misuse of the building and lack of maintenance.

Fig. 8.31 (a, b, c, d) shows some of the causes of moisture presence as reported by Massari[34].

8.5 Research on the durability of masonry and of their components: a brief 'State of the Art'

According to the conclusions reached by the previously mentioned joint CIB/RILEM TC, there is, in general a need to develop and/or enhance methods that will generate data on the lifetime of materials and buildings, in order to:

- have reliable data on effective on-site behaviour;
- increase knowledge on the mechanisms of failure;

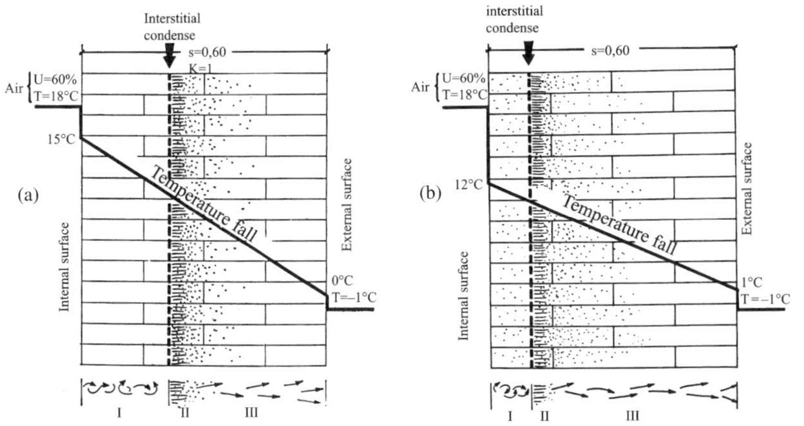

Figure 8.29 *(a) Scheme of the condensation water through the wall during the winter: I) water vapour diffusion, II) condensation area (liquid water), III) capillary movements from wet to dry; (b) presence of moisture from the soil (the condensation area can reach the internal surface)* [34].

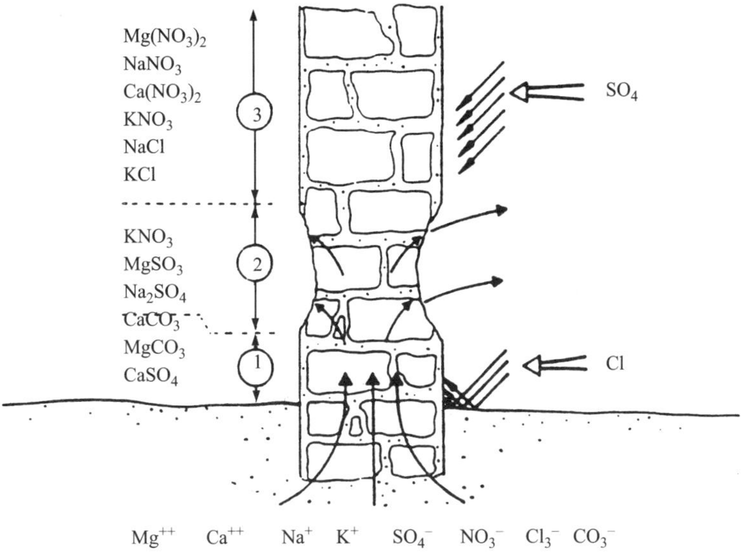

Figure 8.30 *Position of salts under capillary rise and major damage (Na_2SO_4) by Hilsdorf.*

- develop methods and procedures for the measurement of decay;
- develop knowledge on aggressive agents, on their quantity and influence;
- implement methods which are able to simulate or take into account the synergetic effects of aggressive agents; and
- implement deterministic or probabilistic mathematical models to describe the behaviour of materials in a specific aggressive environment.

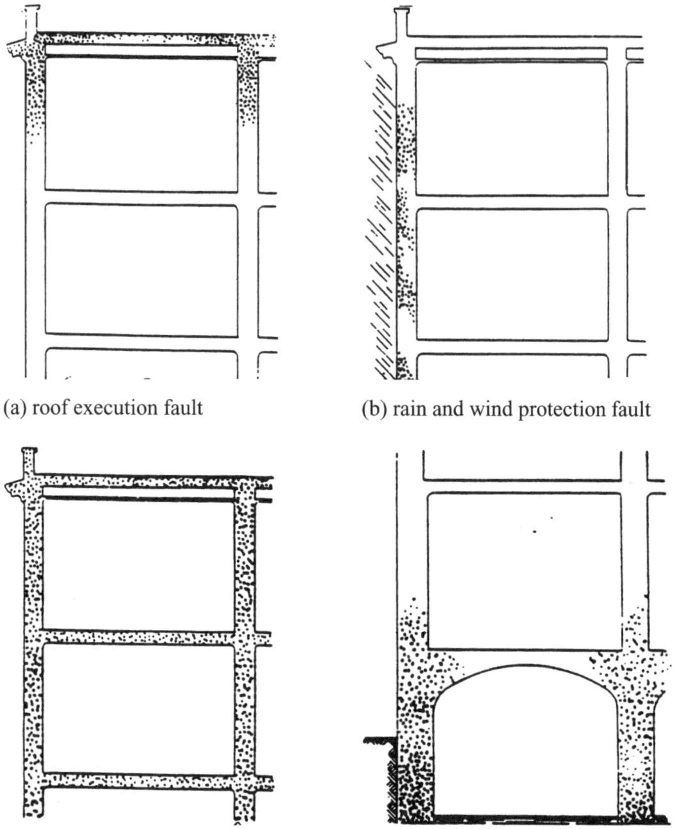

(a) roof execution fault (b) rain and wind protection fault

(c) early use of new construction (d) water dispersed underground

Figure 8.31 *Cases of lack of maintenance or of bad use of the building.*

Up to now, the best methodologies have been developed to study metallic materials and concrete. The study of masonry still has a long way to go. At the International Conferences on Durability of Building Materials (DBMC) very few papers were presented on masonry decay. The same was true of the RILEM International Conference on Durability[36]. All the papers dealt with a few points suggested by the CIB/ RILEM TC.

In Refs [1] and [7], the research on aggressive agents and on their effects on building materials is described.

In Ref. [37] the decay of stones is described, with special attention to the transformation of calcium carbonate into calcium sulphate. Nevertheless, Ref. [37] is mainly descriptive and the research concerns behaviour in a natural, non-aggressive environment.

In Ref. [38] a description of the physical, chemical and mechanical causes of decay and of their effects on masonry is presented, together with a consideration of synergisms.

The problem of durability has been addressed in specific conferences and workshops on the alteration of stones and bricks. International conferences on the alteration of stones take place every three years (Athens, Bologna, Venice, Lausanne, Torun, Lisbon, Berlin) as do a number of other international and national conferences. Many papers have been published on the subject.

The research developed is comprehensive and detailed on the study of:

- aggressive agents[28, 30, 39];
- the physical and chemical properties of materials[28, 39, 40];
- the on-site behaviour of components[41, 42, 43, 44, 45];
- investigation methods on microstructure[28, 46, 47, 48];
- mechanisms of decay[28, 49, 50]; and
- waterproofing and consolidation.

Little research has been carried out on:

- the measurement of long-term decay by quantifying reliable parameters;
- the implementation of mathematical, deterministic, models to interpret and reproduce the phenomena or probabilistic models for the prevention of decay; and
- the study of the interaction of the components in masonry and/or of the effects of synergetic agents.

The same can be said for bricks which were for a long time considered as inert materials with a long lifetime, lasting centuries. Little research on bricks concerned the chemical attack by sulphuric, fluoric, chloric and nitric acids[51, 52, 53]. The effects of salt crystallisation were studied in relation to sulphate and chlorides and their connection with the porous structure of the bricks[54, 55, 56, 57, 58].

Interesting work was published in the ASCE *Durability of Building Materials* journal which unfortunately ceased publication after many years of interesting activity. Interesting work was presented at a conference, held in Rome, on 'Air Pollution and Conservation – Safeguarding our Architectural Heritage', organised by the Swedish Institute of Classical Studies[59]. Nevertheless, even in this case, the research was more concerned with stones and mortars rather than masonry. The lack of a systematic study on masonry as a composite, was pointed out during the Dahlem Conference which took place in Berlin in 1996[60].

8.6 A systematic approach to the study of masonry durability

The idea of initiating a systematic approach, dealing with masonry as a composite material came to the minds of Binda and Baronio after considerable experience of research on masonry components. In Mediterranean countries even if freeze–thaw action is an important factor in decay, the most influential factor in masonry deterioration is without doubt, salt crystallisation. Efflorescence and cryptoefflorescence are always present on the decayed or decaying Italian, Greek and Spanish masonry, and they are distributed according to the combination of masonry components and to the techniques of construction (e.g. dry masonry, one-leaf, multiple-leaf masonry, brick or stone masonry, rendered masonry, etc.).

While a great deal of research has been carried out on components, very few papers have been dedicated to masonry as a composite material at the time when the authors decided to dedicate their efforts to masonry both in the laboratory (working on wallettes), and outdoors, starting with an extensive photographic investigation of Italian historic buildings.

8.6.1 Choice of a reference environment

The authors' research started with laboratory-accelerated ageing tests on single components (mortars and bricks). In order to define temperature and relative humidity conditions which could produce the mechanisms of decay found in natural environments, a reference environment was studied. Data collected for the 10 years, 1970 to 1979, in Milan from the Brera Observatory was taken into account. The temperature and relative humidity data collected was compared with the equilibrium curves of different soluble salts (sodium and magnesium sulphates, the most diffused in the Milan efflorescences, and sodium chloride). It was observed that the conditions for the formation of thenardite and mirabilite, the two more stable phases of sodium sulphate, can be found in Milan every month of the year, and several times a month, with an average duration of 48 hours[61].

8.6.2 On-site investigation: decay measured through photographic survey

Since 1981, the authors have carried out a systematic photographic investigation of several buildings in Milan, and in other nearby cities (Figs 8.32 and 8.33). This investigation was very useful for the study of the mechanisms and rates of decay in real situations. It was also possible to make an attempt to measure approximately the rate of decay *in situ*. Pictures of the details of facades were taken approximately four times a year and the damage was calculated by referral to an original undamaged area as follows:

$$d_n = \frac{S - S_r}{S}$$

Figure 8.32 *Efflorescence on the southern side of a garden wall.*

Figure 8.33 *Detail of Figure 8.32.*

where S is the total reference area, and S_r is the area remaining at subsequent measurements.

An example is shown in Fig. 8.34. The decay follows similar paths in all cases. After an initial high rate of deterioration, a steady state is reached. Subsequently, a quick increase in the rate of decay follows, then another steady state is achieved, and so on. The steady states represent the time that cryptoefflorescence needs to find its way through the material.

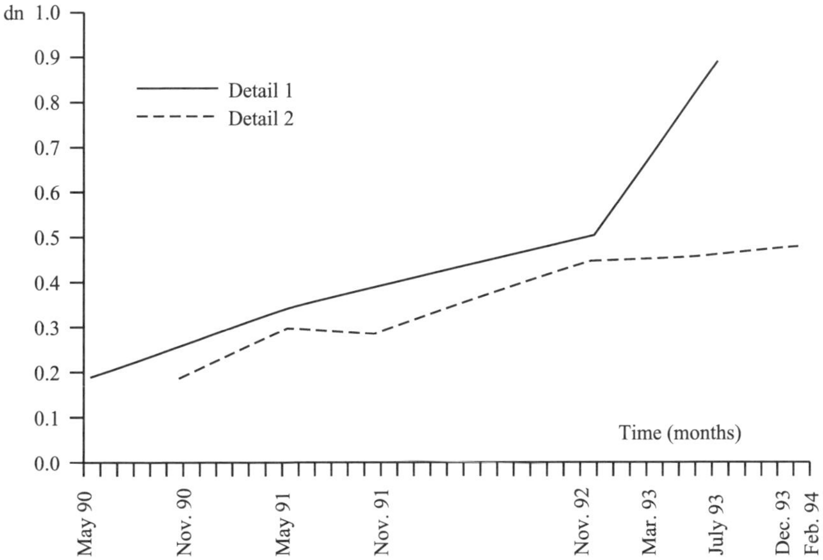

Figure 8.34 *Damage measured on two facades.*

8.6.3 Interaction of mortar with bricks/stones

The mechanisms of decay in combined materials are influenced by their interactive combination in the composite material. Masonry prisms were prepared by the authors using the same type of brick with three different types of mortars. The binders were as follows: hydrated lime-cement, cement, hydrated lime-pozzolana[62]. The influence of the mortar properties on the behaviour of the masonry is shown in Figs 8.35 (a, b, c). When the porosity of the mortar is similar to that of the brick, then deterioration occurs in both materials [Fig. 8.35 (a)]. If mortar porosity is very low, deterioration occurs to the bricks [Fig. 8.35 (b)]. If the porosity of the mortar is higher than that of the brick, then the decay occurs to the mortar [Fig. 8.35 (c)].

It is also known that the same material can decay in different ways according to its position in the building[25].

8.6.4 Laboratory crystallisation tests

Two types of laboratory tests were carried out on bricks, mortars and stones: a total immersion cyclic test (useful for the study of small elements such as thin columns or decorations), and a test based on the capillary rise of a salt solution (more useful to the study of thick structural elements and masonry where the decay must not compromise

(a) (b) (c)

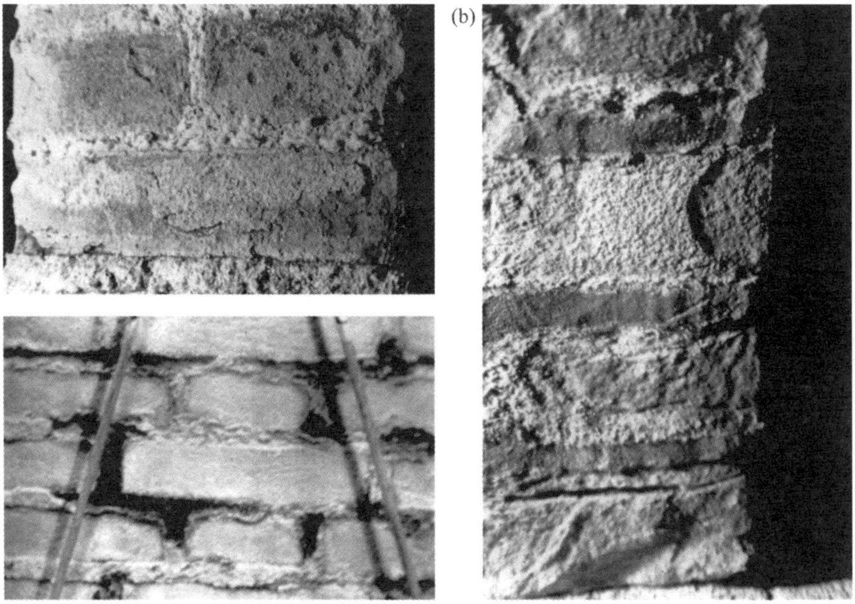

Figure 8.35 *Interaction of different mortars with brick:*

a) hydrated lime-cement mortar;
b) cement mortar;
c) hydrated lime-pozzolana mortar.

the core of the elements). Each test was carried out under controlled thermodynamic conditions as follows:

- total immersion, or partial immersion by capillary rise, of the specimen in a salt solution at a defined concentration for a time t_1 (a saturated solution is the most destructive, of course, but also gives the highest acceleration);
- exposure at a chosen temperature T and under chosen relative humidity conditions for time t_2 or more times t_2, t_3 according to the type of salt chosen.

The cycle was repeated until failure of the specimen occurred.

In Table 8.4, the times and duration adopted for three different types of salts are shown. The damage was measured as loss of surface area. Fig. 8.36 shows the damage compared to the duration of the cycles, in the case of sodium sulphate.

Different types of soluble salts were considered: NaCl, $MgSO_4$, Na_2SO_4[63]. In Fig. 8.37, the damage caused by the different salts under the different conditions

Table 8.4 Different temperatures, R.H. and cycle duration chosen for the three salts

Salts	Type of cycle	
NaCl	A. t_1 =	2 h
MgSO	t_2=	46 h (50% RH)
Na_2SO_4	B. t_1 =	2 h
	t_2=	94 h (50% RH)
	C. t_1 =	2 h
	t_2=	166 h (50% RH)
	D. t_1 =	2 h
	t_2 =	46 h (50% RH)
	t_3 =	120 h (60°C)
	D. t_1 =	2 h
	t_2 =	46 h (50% RH)
	t_3 =	120 h (100°C)

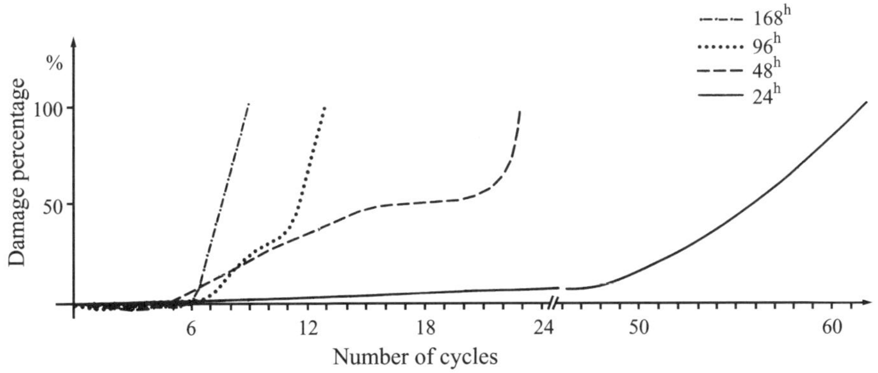

Figure 8.36 *Damage curves for different drying times (cycles type 1).*

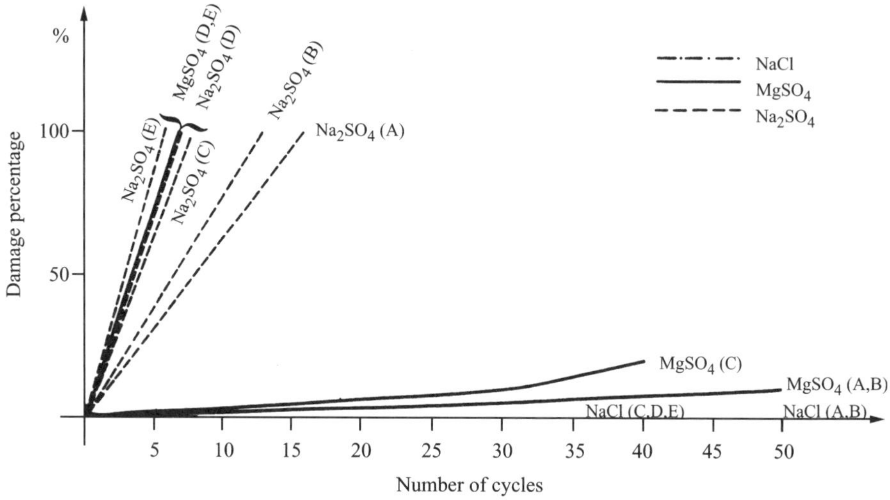

Figure 8.37 *Total amount of damage induced by different salts for every type of cycle (see Table 8.4).*

shown in Table 8.4 is compared. Na_2SO_4 and $MgSO_4$, are the most destructive salts; Na_2SO_4 was therefore chosen as the aggressive agent for the tests, both on site and in the laboratory, being so common in nature and offering the possibility of demonstrating its crystallisation.

8.6.5 Mechanisms of failure, and penetration of decay, due to salt crystallisation

Salts can be found in masonry for many reasons. Salts can be present in the original materials, come from the soil or be deposited on the masonry surface. If water is present in the masonry, salt can easily be solubilised and migrate inside the masonry. During evaporation salts can be transported towards the external surface of the material and can crystallise. More water feeding into the masonry will cause a new solubilisation of the salts, a new movement toward the surface and again cause crystallisation. This behaviour can continue for many thermic–hygrometric cycles causing deterioration of the material. As declared in Ref. [64], this decay presents as continuous delamination from the external surface of the material (Fig. 8.38). This phenomenon is due to the fatigue effects caused by wet–dry cycles especially when soluble sulphates, for any reason, are present inside the material (Fig. 8.39)[64]. The thickness of the delaminated layer depends upon: the temperature and relative humidity conditions of the micro-environment, the porosity and the mechanical strength of the material, and the duration of wet–dry cycles[64, 65].

Nevertheless, whatever the thickness of the layer, the material properties are only modified in a narrow millimetric area underneath the detached layer. This has been observed several times by the authors and can be explained according to the mechanism of decay[66, 67]. Figs 8.40 and 8.41 show the penetration of the decay in the case of two specimens, brick and stone, respectively.

Figure 8.38 *Delamination of a brick specimen after the crystallisation test.*

Figure 8.39 *Fatigue effect of wet–dry cycling under the external surface.*

The two specimens were cut through to the deteriorated surface after being subjected to a severe crystallisation test which badly destroyed the upper part of the specimens. Figs 8.42 and 8.43 show the same cut surface seen through a stereomicroscope; the depth of the decay is no more than 1.0–1.5 mm.

The material below this layer is practically untouched by deterioration. A similar phenomenon can be seen in Fig. 8.44, which shows the deteriorated layer below the delaminated and deeply excavated surface of a brick (sampled with its mortar joint) from a badly decayed wall (Fig. 8.6). The thickness of this layer is no more than 1.5 mm. The same phenomenon is extended to the mortar joint. The surface of the mortar, apparently composed of loose material, appears to be in very bad condition but, after cutting, the decayed area is clearly only a few millimetres deep.

Figure 8.40 *Cut portion of a badly damaged brick (top surface).*

Figure 8.41 *Cut portion of a badly damaged stone (top surface).*

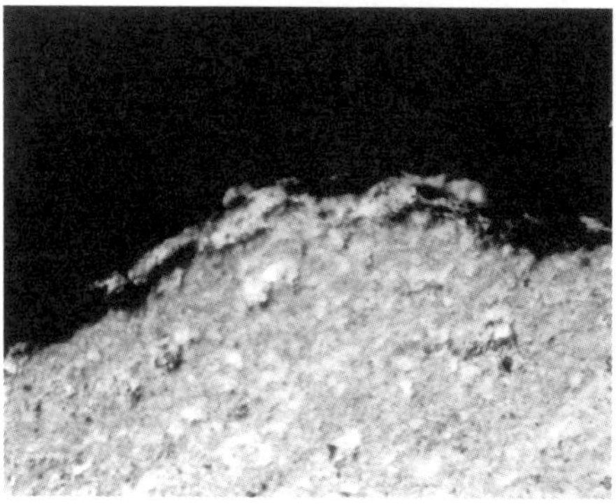

Figure 8.42 *Detail of the damaged area of Fig. 8.40 under the stereomicroscope.*

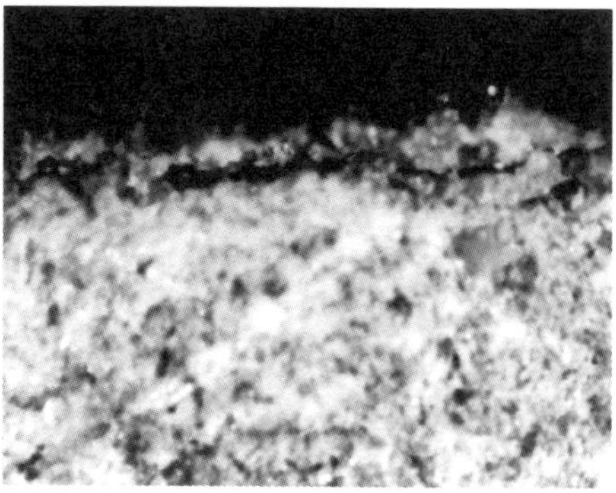

Figure 8.43 *Detail of the damaged area of Fig. 8.41 under the stereomicroscope.*

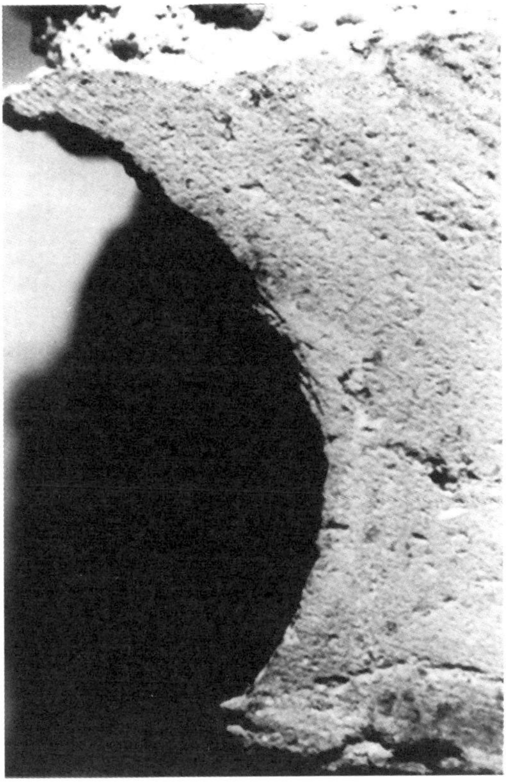

Figure 8.44 *Penetration of the decay in the system brick mortar joint (brick deeply damaged).*

8.6.6 Descriptions of full-scale models

In order to study more realistic situations than the ones created in the laboratory, some full-scale physical models were built in stone and brick masonry, and exposed to the open air.

Geometry, materials and details of construction

The models were built intentionally in a polluted area of Milan. This was made possible due to an offer of the land from ESEM (Milan Province School of Masons), with the construction being supported financially by ICITE-CNR (Milan). Three models were built. The models were one-floor constructions with the principal facades divided into modular orthogonal panels exposed to the south and to the west (Fig. 8.45).

Two models, one with a sandstone façade, one with a soft-mud facing brick façade, each had five pairs of orthogonal panels. The third, of mixed stone and brick, had only four pairs of orthogonal panels. The construction of the models ended in September 1990.

The buildings, instead of having independent walls, were constructed in order to have a thermic–hygrometric gradient inside the walls, as in a normal residential building. Therefore, the models were heated indoors during the winter.

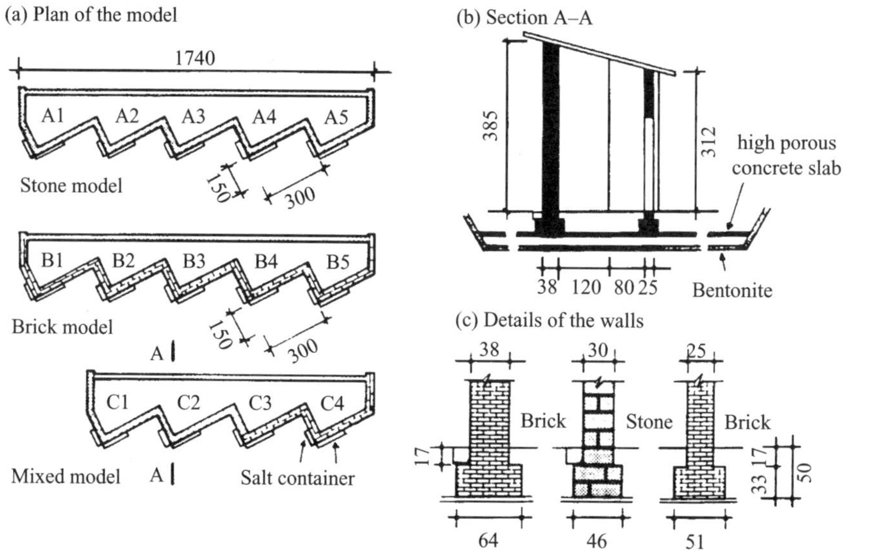

Figure 8.45 *Full-scale models in Milan's plan, section and details of walls and foundations.*

In order to study the effects of salt crystallisation, artificial decay was caused in some areas of the walls by introducing a salt solution (Na_2SO_4) in small containers placed at the bottom of the walls [Fig. 8.45 (a)]. The subsoil of the construction was excavated down to a set level and coated with a layer of bentonite [Fig. 8.45 (b)]. This operation was carried out in order to ensure the capillary rise of water into the masonry. The water was fed naturally by rain or fed artificially. The presence of water in the subsoil was controlled by five piezometers. The thickness of the walls is given in Fig. 8.45 (c).

Environmental data

The temperature and relative humidity of the outside air and inside air of the models was continuously monitored from November 1990. An attempt to correlate the conditions of the micro-environment to those measured in the centre of Milan, gave the following results[68]: whilst a correlation was found between the temperatures measured in the models and in the city data, no correlation was found with regard to relative humidity. The temperatures in the models were found to be approximately 1.5°C lower than those measured in the city.

The monitoring of the local environmental parameters will continue. The aim of the data collection is to define the environmental stresses acting on masonry surfaces and to compare their effects to those obtained in the laboratory through accelerated testing, and, eventually to the implementation of probabilistic models.

As an example, Figs 8.46 and 8.47 elaborate on the data collected in terms of freeze–thaw and crystallisation cycles for 1991 and 1992. The total number of freeze–thaw cycles within the two years was approximately 60, while a much higher number (more than 400) of possible crystallisation cycles took place in the same period. The decay is obviously influenced by factors other than the environmental conditions, the

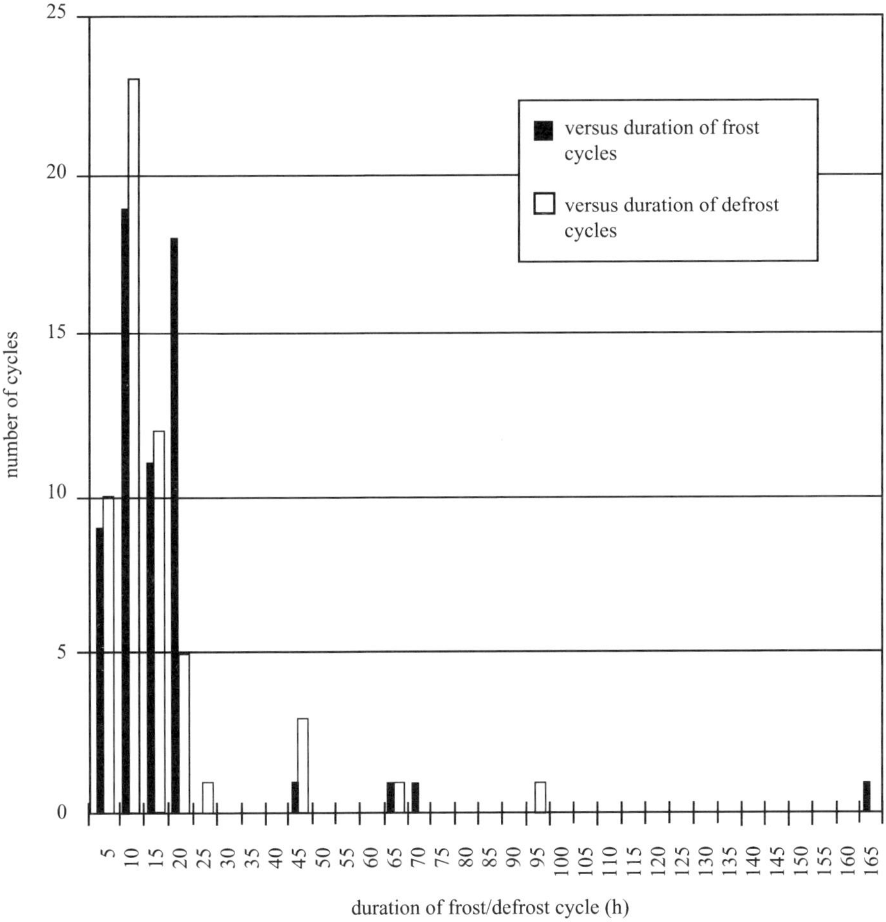

Figure 8.46 *Freeze–thaw cycles plotted against the duration of the freeze–thaw cycle time.*

surface conditions and the gradients of the internal and external surfaces. To gather additional data, a series of thermocouples was applied at different strategic points of models A and B[69].

The differences in temperature detected between the open air and the surfaces of the walls, and between the external surfaces and the internal surfaces were recorded, together with the minimum and maximum temperatures for freeze–thaw and the rate of temperature variation. All these data are important in explaining the phenomena of decay.

Moisture and salts movement

The capillary rise was surveyed visually and controlled by gravimetric measurement and radar detection[70]. Since February 1991, the capillary rise of water was controlled by measuring the apparent height of the water level on the facades at set points, while looking for a better procedure of monitoring moisture content[69].

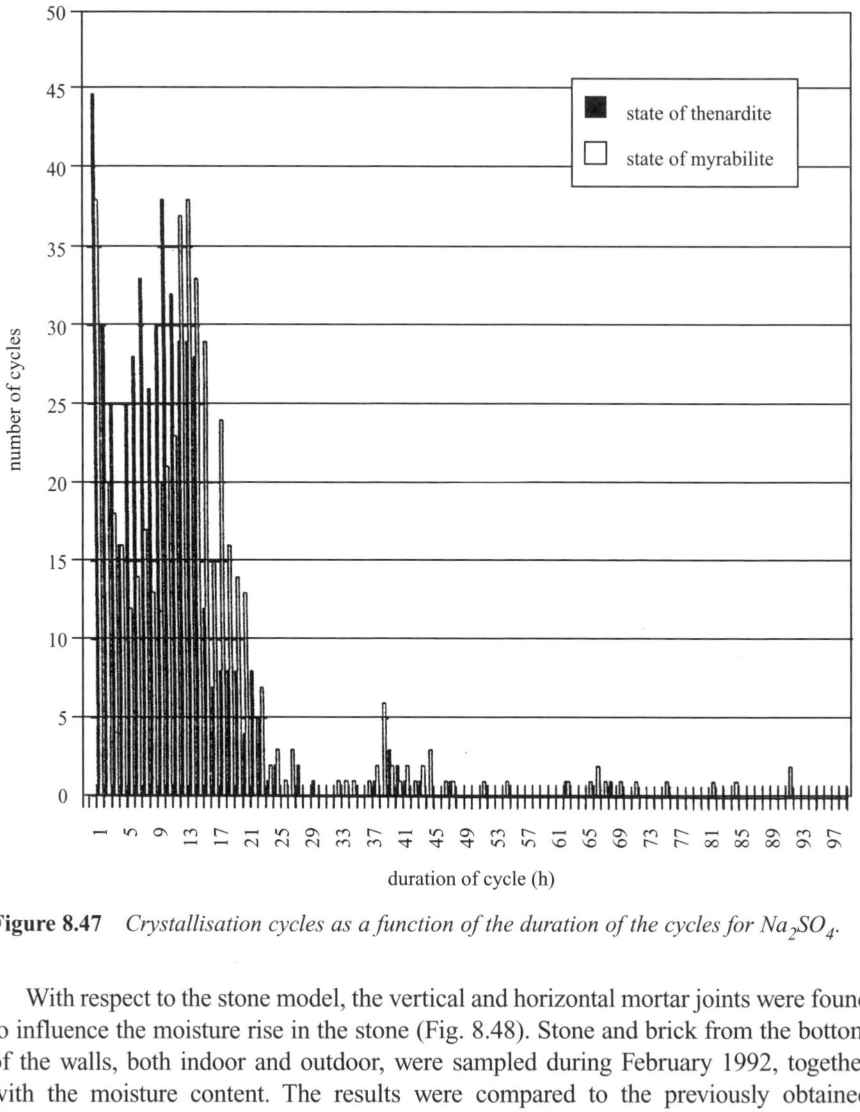

Figure 8.47 *Crystallisation cycles as a function of the duration of the cycles for* Na_2SO_4.

With respect to the stone model, the vertical and horizontal mortar joints were found to influence the moisture rise in the stone (Fig. 8.48). Stone and brick from the bottom of the walls, both indoor and outdoor, were sampled during February 1992, together with the moisture content. The results were compared to the previously obtained absorption data. The data showed that the stone and brick were probably saturated. By comparison with the data, it is possible to see that capillary rise is higher in the brick panels than in the stone panels. The presence of Na_2SO_4 accelerates the rise of the water and salt solution in the brick wall. The difference (approximately 30 cm) is, in comparison to the masonry where Na_2SO_4 is absent, maintained after 37 months. In the stone masonry, on the contrary, after a higher rate of rise of the salt solution in panel A4, the level is almost the same as in the model where Na_2SO_4 is absent.

Measure of the damage as a function of time

The same mechanism of failure has taken place both in the laboratory and on site in the cases presented above. This is clearly a long-term mechanism occurring under

Figure 8.48 *Water and salt distribution in a stone wall.*

repeated cycles, a deterioration-under-fatigue phenomenon. The external surface is gradually delaminated as the deposition of crystallising salts takes place underneath the masonry surface, as can be seen from Fig. 8.49. The depth of deterioration corresponds to the thickness of the detached layer. From a structural point of view, the surface decay can be represented as a decrease of the thickness of an element (load-bearing wall, column, pier, arch, etc.), leading to a lower load-carrying capacity in the element itself (Fig. 8.30).

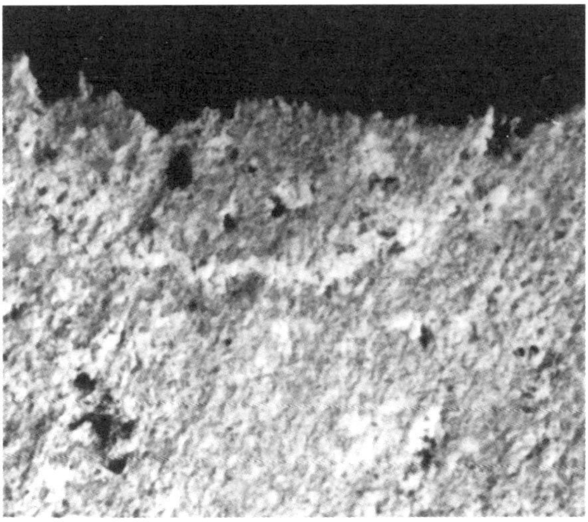

Figure 8.49 *Formation of cryptoefflorescences underneath the external surface of a brick.*

This type of deterioration cannot easily be modelled due to the complicated laws which govern the physical phenomenon, as can be deduced from Ref. [71]. More successful models have been proposed to study the probability of salt crystallisation decay as shown in Ref. [72]. For probabilistic models, more information is required on the environmental conditions that cause crystallisation (temperature and relative humidity), and on the rate of decay of the material.

Degradation of surface properties cannot be represented by mechanical tests which are destructive and misleading when the pores of the material contain soluble salts. The dimensional variation, more precisely the reduction of the cross-section of the structural element, has been considered as a parameter to measure deterioration[73]. This parameter can also describe the surface roughness of the element. The measurement procedure proposed by the authors has been applied previously during laboratory ageing tests. Specimens of brick and stone were subjected to a cyclic crystallisation test. In order to measure the degradation at every cycle, a simple device was used consisting of a dial gauge[74]. The measurement was repeated and the data recorded at every cycle as a decrease in the height of the specimen was plotted against the number of cycles. Fig. 8.50 represents some curves obtained for a specimen of sandstone. Local height losses represent the thicknesses of the layers which were detached in the area around the measurement point. Several peaks appear clearly in the diagram. The peaks have a precise meaning: every apparent swelling of the surface represents the formation of a new layer which will be detached in subsequent cycles. In this way the onset of the decay can be recorded.

The dial gauge measurement has also been applied to study the resistance of treated surfaces to the crystallisation of salts. The specimens were treated, on the external surface,with different commercial treatments and subjected to the crystallisation test.

The loss of height was then measured at every cycle. As an example, the damage curves of limestone, both with and without treatments are presented in Fig. 8.51 (NT refers to the untreated specimens). Specimens were taken from a historic building in Milan. These were treated with ten different types of products named A, B, C, D, E,

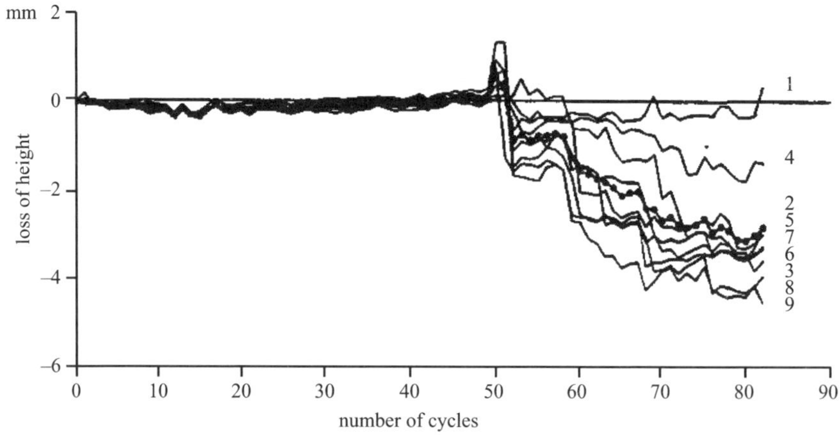

Figure 8.50 *Data recorded in 9 points in a stone specimen during the ageing test.*

F, G, H, I and NT (water repellents and consolidants), and subjected to crystallisation cycles. Fig. 8.51 shows the average curves for the treated/untreated specimens; it is clear that only three treatments D, E and G were successful.

Use of a laser sensor as a profile recording device

The usual techniques adopted to detect the profile and roughness of material surfaces in rock mechanics are based on punctual measures performed using mechanical or electrical displacement transducers, that are in direct contact with the surface. However, such techniques have drawbacks. They are time consuming both in measuring and processing, and there is a risk of surface damage, especially in soft and very deteriorated materials. The laser sensor, is a non-contact means to obtain reliable information concerning the surface characteristics of materials. Its use has been reported on rock joints to evaluate the joint roughness coefficient (JRC) and on stone aggregates in order to ascertain the average roughness, profile and undulation depth[73, 75].

The laser sensor works via a semiconductor laser driver circuit which emits a laser beam, directed perpendicularly to the surface to be examined[73]. The reflected light is converged by a receiver lens and cast onto the optical position detector element as a small spot (Fig. 8.52). If the distance from the surface changes, the light spot moves accordingly. The optical element encodes the amount of movement into an electric signal. An arithmetic circuit in the laser unit controller, translates the electric signal into an indicator of the surface displacement. Some measurement of profiles have been carried out, using a laser sensor with a spot of 1.0 × 2.0 mm, a resolution of 40 microns, and a linearity of 0.8% for a measurement range of 80 mm. The standard distance of the sensor from the surface is 100 mm with a measure range of ± 40 mm. The sensor is connected to a vertical steel frame through a dragging trolley, moving on a linear support beam by means of a step-by-step electric motor (Fig. 8.53).

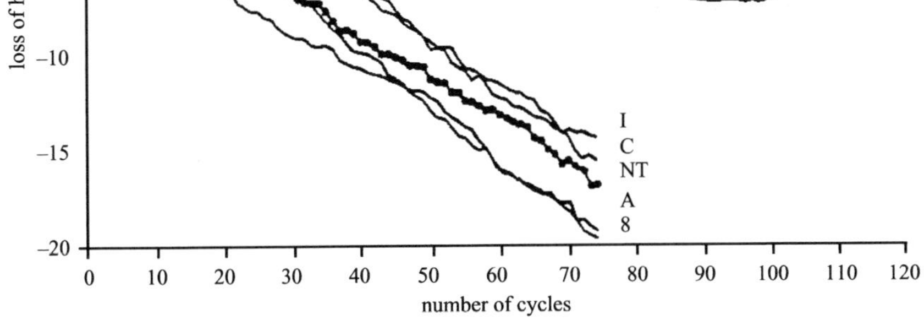

Figure 8.51 *Average damage curves for stone specimens both untreated and treated with 7 different products.*

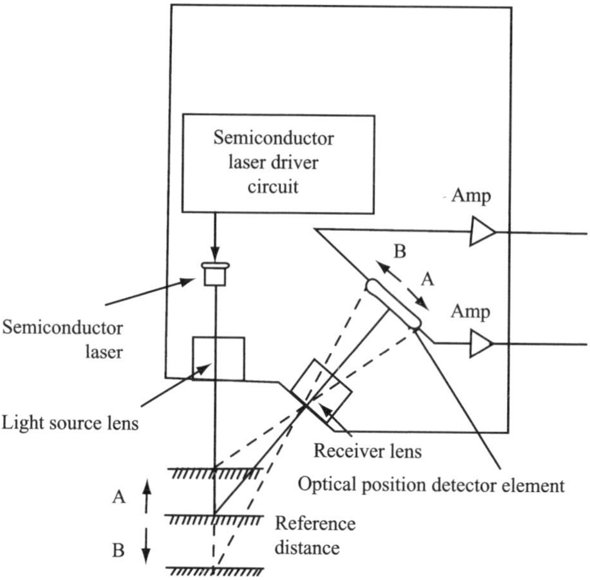

Figure 8.52 *Opening layout of the laser sensor.*

Figure 8.53 *Profile recording laser device used in the laboratory.*

Preliminary trials carried out on the frame with a centesimal comparator, led to evaluation of the mechanical oscillations of the trolley in a range of $\pm$ 0.01 mm. Profiles are automatically measured, plotted and recorded through a suitable data acquisition system. The layout is shown in Fig. 8.54. Based on acquired experience, it has been ascertained that the environmental light variation does not affect the measurements.

Some problems arise when surfaces have crystals with reflecting cleavage planes, that can deviate the emitted ray away from the sensor receiver lens. Black spots on the surfaces

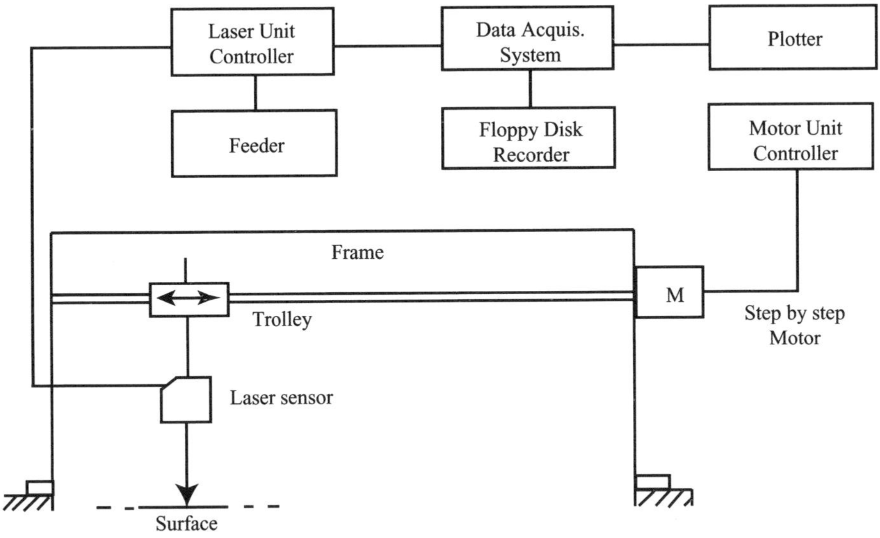

Figure 8.54 *Layout of the instrumentation system used for measurement.*

can also cause measurement variations. These problems can, however, be solved by applying thin, removable opaque coatings to the surface. Two frames of different sizes, are used for measurements on specimens in laboratory and on walls *in situ*, respectively.

Experimental results and discussion

The application of the procedures described have the main purpose of recording the long-term behaviour of masonry and masonry materials, under the effect of soluble salts crystallisation. The values of the chosen parameter (reduction of the thickness of the specimen or element) measured over time, can constitute the input data for a mathematical model to study the behaviour of materials[76].

The aims of this study are as follows:

- to compare the durability of different masonry materials;
- to measure the mutual influence of mortars, brick, or stone on the durability of masonry;
- to compare the durability of surface consolidants and/or water repellents;
- to provide a data base for modelling the phenomenon.

The following results were obtained from laboratory and in situ experiments and show that the laser sensor is able to fulfil much of the scope of the research.

Laboratory tests

In the laboratory, the maximum measurement of the detectable profile was 300 mm. The laser sensor was positioned at a distance of about 70 mm above the surface of the specimen and the velocity of the trolley was 2 mm/s. The use of the sensor was successful for different applications as is shown in the following.

Small masonry panels, K1.1 (Fig. 8.55) were subjected to a crystallisation test set up by TNO (Toegepast Natuurwetenschappelijk Onderzoek), Delft, Netherlands. The research was carried out within the EC Contract 'Evaluation of the Performance of Surface Treatments for the conservation of brick masonry', number EV-CT94-0515 under which TNO and the Politecnico of Milan were partners together with KIK (Koninklijk Instituut voor Het Kunst- patrimonium), of Belgium. The aim of the contract was to study the durability of water-repellent and consolidant treatments for brick masonry. The tests were carried out in order to simulate the decay which can take place in the case of treated walls containing salt solutions. For each specimen, four profiles were measured at defined intervals of time as shown in Fig. 8.55. Clearly the mortar joints seem to be the weakest part of the specimen.

Untreated prisms

Two untreated specimens made with different types of bricks and mortars, T2.5 and K1.5, and were compared. Fig. 8.56 (a, b) shows the first and the last profile measured

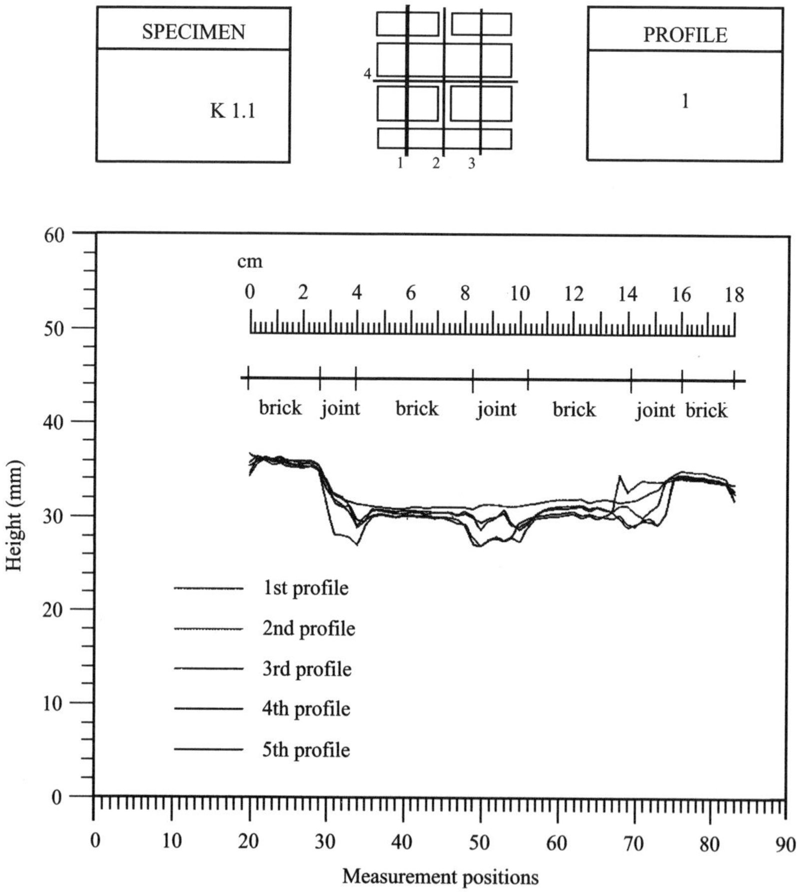

Figure 8.55 *Plan of the prism K1.1, measurement positions and profiles recorded during the ageing test.*

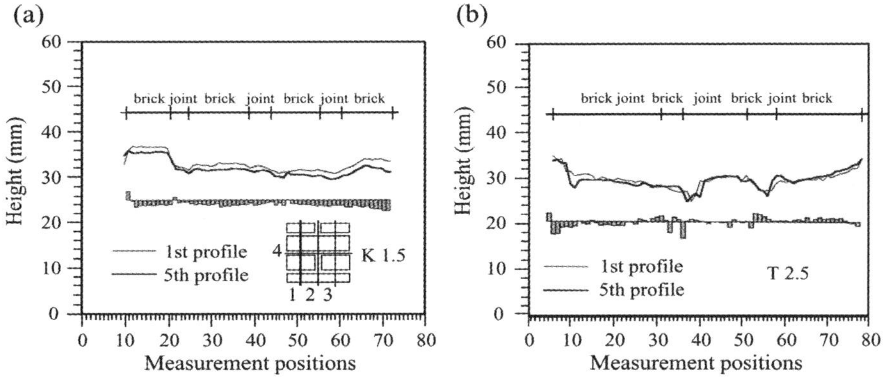

Figure 8.56 *First and last profile recorded for K1.5 and T2.5.*

respectively on the two specimens. The loss of material can be deduced from the bar chart in the figures. The results show that while K1.5 is behaving uniformly, T2.5 is partially losing material, partially bulging. So, even if the loss is greater for the first specimen, T2.5 could lose more material in the future. The influence of the mortar can also be seen from Fig. 8.55 and Fig. 8.56 (b).

Treated prisms

The specimens were treated with 4 types of water-repellent from different commercial productions (W1, W2, W3, W4). Two cases are presented in Figs 8.57 (a, b) and Figs 8.58 (a, b). In the first case [Fig. 8.57 (a, b)], the behaviour of two different masonries treated with the same water-repellent (W2) is shown. The results are represented in three-dimensional graphs where the profiles are plotted in their position on the specimens. T2.2 demonstrates the greatest loss of material. This is certainly due to the porosity of the bricks and mortars, which had been measured previously.

In the second case, two types of water-repellent (W1, W2) were used for the same masonry (K2.1 and K2.2). While K2.1 shows most decay on the mortar joints, confirming the low durability of mortar MK1, K2.2 shows damage to the bricks as well. As the brick and the mortar are the same for both specimens, the difference in

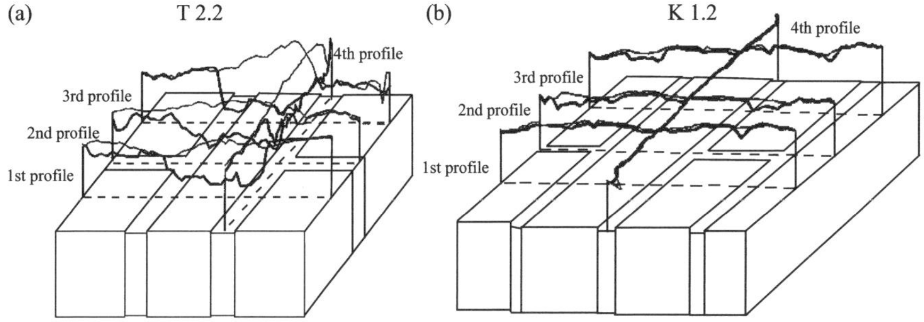

Figure 8.57 *First and last profile recorded for T2.2 and K1.2.*

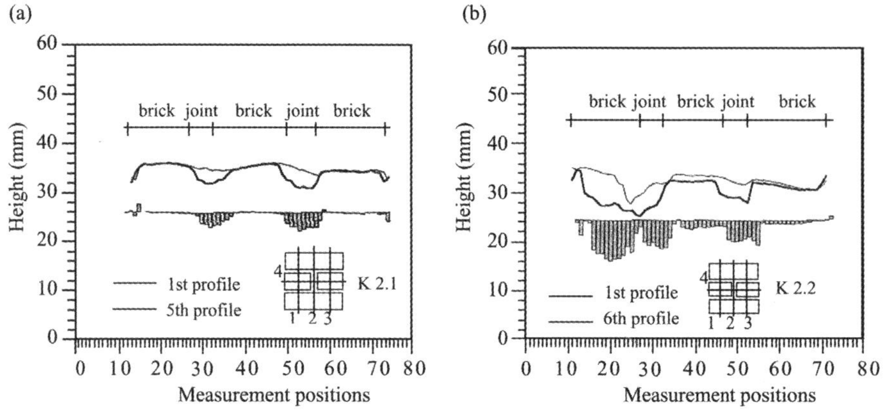

Figure 8.58 *First and last profile recorded for K2.1 and K2.2.*

behaviour is given by the treatment. The results indicated that any treatment must be tested carefully before use.

In situ measurements

The use of the laser sensor to measure decay was carried out on site on full-scale physical models. A special frame was attached to the walls in order to apply the measurement device (Fig. 8.59). For the *in situ* measurements, the vertical frame was fixed initially onto two steel bars, each having a series of 36 holes in order to be able to move the frame in steps of 20 mm and to detect parallel profiles.

The frame and the bars were in turn fixed on the walls through steel supports, suitably inserted in the walls themselves. This allowed accurate repetition of the profile measurements of the deteriorated surfaces at defined times. The maximum length of detectable profile was 620 mm. The laser sensor was positioned at a distance

Figure 8.59 *In-situ application of the laser device.*

of about 100 mm above the surface of the walls and the velocity of the trolley was 4 mm/s. Three and four different measurements were carried out on the brick and the stone walls, respectively. Figs 8.60 and 8.61 show an example of different vertical profiles, one for the stone model A1 and one for the brick model B1. From the profiles in the two figures, the different behaviour of the two masonries can easily be checked. B1 is far less deteriorated than A1. In the same figures, the bar charts corresponding to the differences between the first and the fourth profiles are reported. From the bar charts, it can easily be seen that besides the loss of material in some positions of the walls, there is also in other points a tendency for the wall to bulge. As said above, the bulging has to be interpreted as predictable decay due to cryptoefflorescences.

The recorded profiles enable an understanding of the type and rate of decay. They can be used successfully on site to control the effectiveness of surface treatments. The laser sensor, besides being a ND technique useful for the study of time-dependent

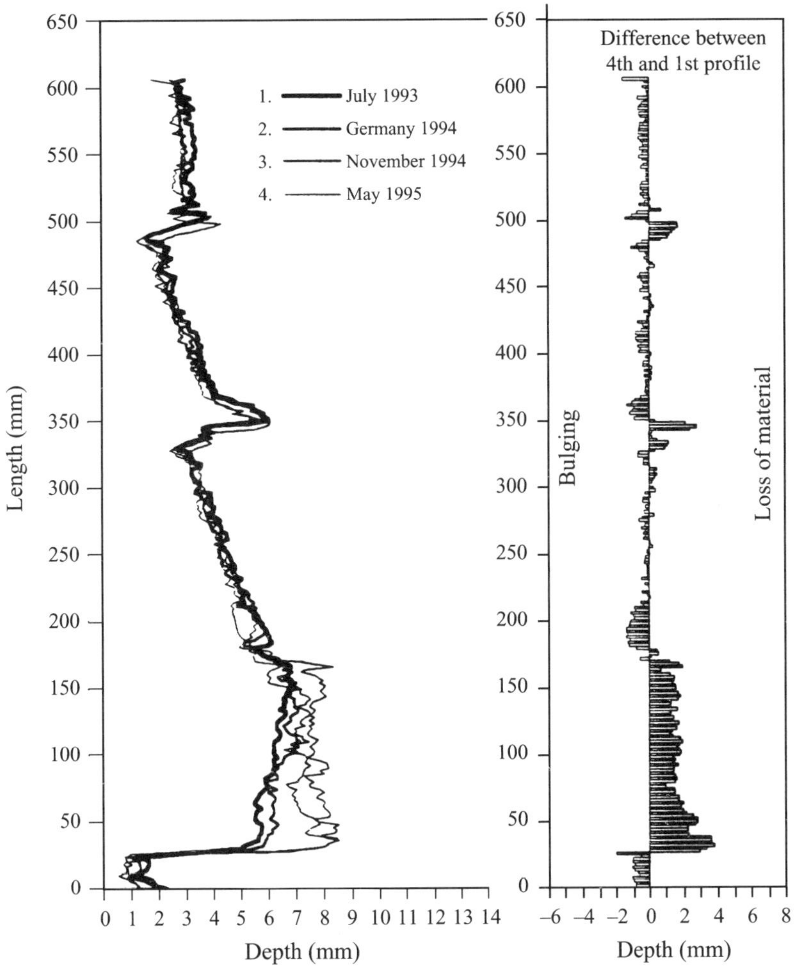

Figure 8.60 *Decay measurement of stone walls: A1 profile 4 vertical.*

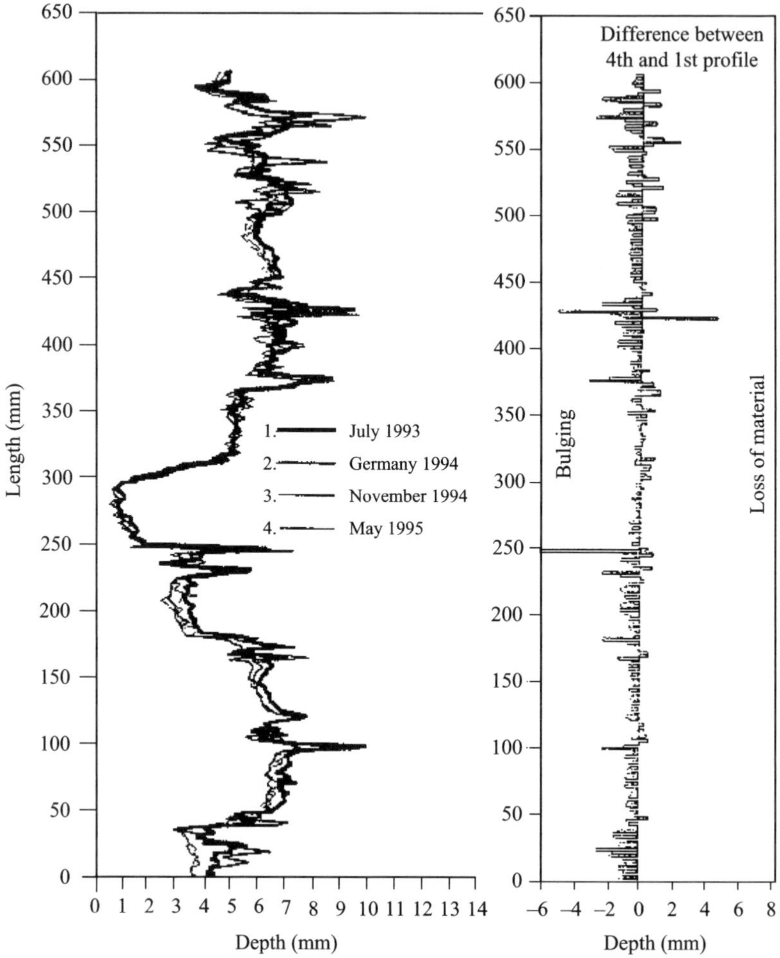

Figure 8.61 *Decay measurement of brick walls: B1 profile 11 vertical.*

phenomena, also has the advantage of producing quantitative results which can constitute input data for deterministic or probabilistic models[72].

By summation over the areas of loss of the section and over the areas of bulging, the plots of Figs 8.62 and 8.63 were obtained. In the figures, the calculated areas are plotted against the dates when the profiles were measured. The remarks given above appear to be confirmed: panel A1 has reached the highest decay compared to panel B. In the meantime, the bulging area is greater for panel B1 than for panel A1. This can indicate that cryptoefflorescences are acting below the brick masonry surfaces more then below the stone masonry surfaces.

8.6.7 Recent laboratory crystallisation tests

Taking into account that some percentage of soluble salts is always present in masonry, the aim of another research carried out within the EC Contract N. EV-CT98-0710 *Salt*

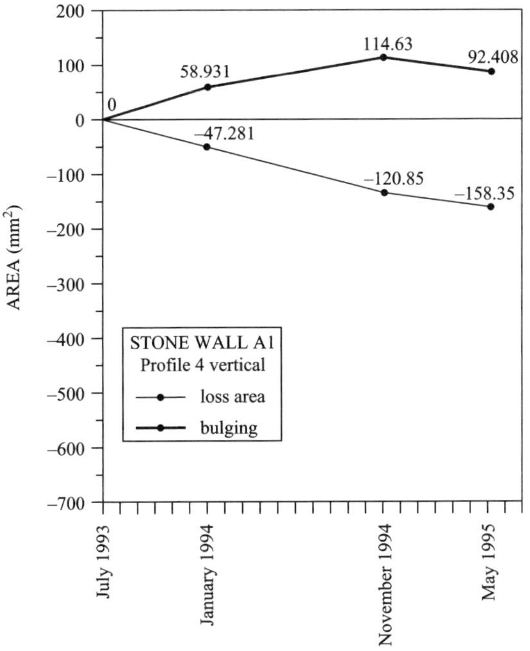

Figure 8.62 *Lost and bulging area in profile 4 of A1.*

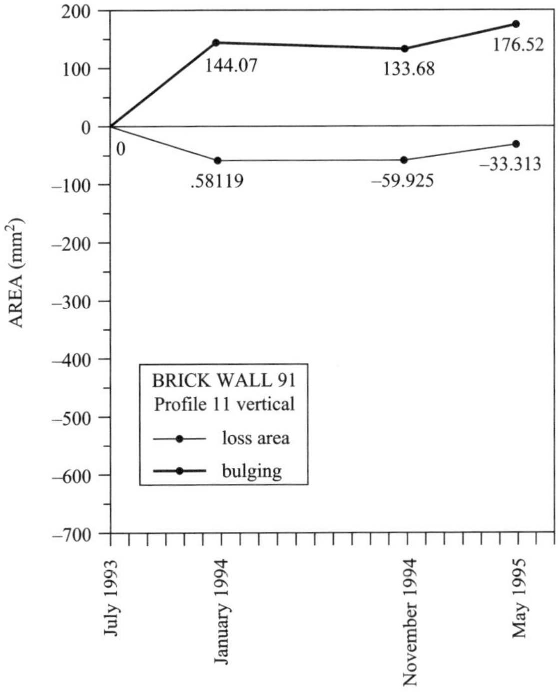

Figure 8.63 *Lost and bulging area in profile 11 of B1.*

compatibility of surface treatments, was to establish the salt content below which the surface treatment does not fail, in the presence of water. Crystallisation tests, following RILEM TC 127 MS Recommendation, were carried out, on single components (bricks and stones) and on wallettes treated with a water repellent and a consolidant. Different types of salt solutions and different salt concentrations were used. The used salts were Na_2SO_4, NaCl and $MgSO_4$.

Crystallisation tests were carried out on one type of softmud contemporary brick used for restoration and on three different natural building stones: (a) Noto limestone, (b) Serena sandstone, (c) Savonnière limestone from France.

The test was carried out first on brick and stone units in order to control whether the threshold value of salt concentration found in a previous part of the research for the single materials could be applied also to masonry walls[80]. The results of the salt crystallisation tests suggested that, for the treated single substrates, the lowest concentrations chosen (1% and 2.5% of the CapMC – capillary moisture content absorbed in 48 hours) can be considered as threshold value of the salt compatibility. Nevertheless in the equivalent untreated single substrates, after 7 months, the lowest salt concentrations were also able to produce small efflorescences.

All wallettes were then built with the above mentioned single substrates and with bedding joints, 15 mm high, based on putty lime, except for the Savonnière stone wallettes, where a mix of hydrated and hydraulic lime was used. In line with the results achieved with the crystallisation test on single masonry units[80], damage was always observed for the highest concentrations tested (5% or 7.5% of CapMC) and in many cases for lower concentrations (2.5% or 5% of CapMC), but never for the minimum (1% of CapMC), which could be considered as a threshold value[81]. Subsequent laser-surveyed profiles show how the surface is changing

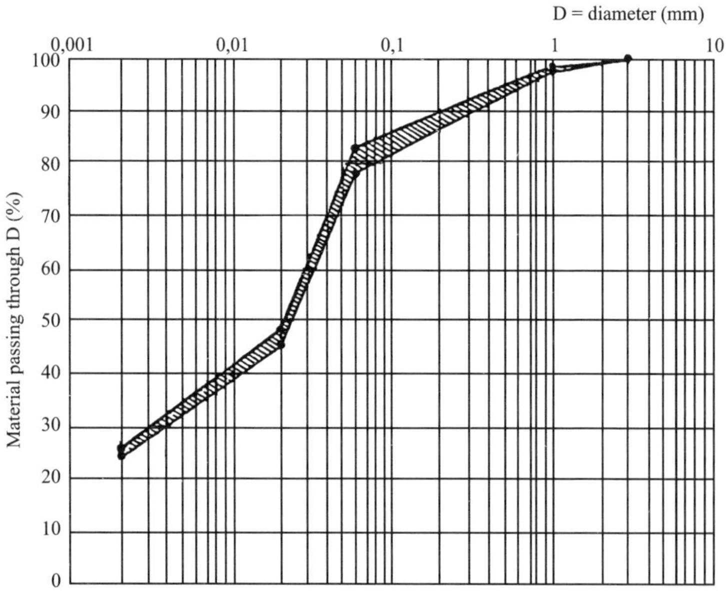

Figure 8.64 *Grain size distribution of clay for bricks frost resistance (Ravaglioli 1977).*

over time due to the progress of the decay. The loss of material can be measured and through a simple model the experimental measurements can easily be converted in new deterioration diagrams where the bulging is eliminated, but considered as a starting point of decay. By calculating (at each survey) the percentage of the area loss of the specimen section, normal to the profile, area loss against time can be represented.

The example presented (Fig. 8.65)[81] concerns the crystallisation test with Na_2SO_4 carried out on soft mud, brick wallettes with the following results: *a)* for *untreated wallettes,* while the damage for the lowest concentration (1%) started uniformly only after three months both in the bricks and in the mortar joints, for the highest concentration (2.5%) the damage soon became serious due to cryptoefflorescences; *b)* for *water repellent treatment* after three months damage was visible in large parts of the mortar joints only, as the only vehicle to water evaporation; *c)* for *consolidant treatment,* the damage started from the brick and then diffused to the mortar. The treated bricks presented some scaling of surface layers: from a thickness of 2–3 mm after 6 weeks to one of 1.2 mm after 8 months [Fig. 8.65 (a)].

As the material loss in untreated wallettes after the same period of time is equal to those treated with consolidant, this treatment applied on this type of bricks seems to be unnecessary to prevent salt crystallisation decay. Up to now no damage is visible on bricks treated with water repellent also with the highest salt concentration [Fig. 8.65 (b)], but past experiences suggest to continue the test. In fact, after cutting of the specimen treated with water repellent at 2.5% of salt concentration, a small line of salt was clearly visible under the treated layer [Fig. 8.65 (c)].

The deterioration process could be interpreted as a stochastic process, function of time and damage. In this way, for different damage levels it is possible to build fragility curves[82]. A fragility curve describes the probability of reaching or exceeding a given damage over time. By using this approach the magnitude of the expected damage over time and the occurrence time of it can be predicted.

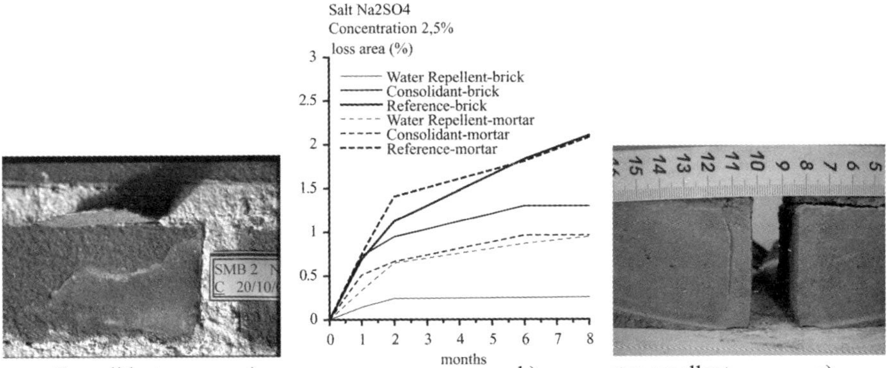

Consalidant a) b) water repellent c)

Figure 8.65 *a) Soft mud brick wallettes with Na_2SO_4 with consolidant b) Deterioration vs. time damage diagram for brick-wallettes with 2.5% salt concentration c) Specimen with water repellent at 2.5% salt concentration after cutting.*

8.7 Conclusions

This chapter reports on definitions of service life and durability as agreed by the International Technical Committee, and describes the causes and processes of decay in masonry and masonry components. It briefly describes the state of the art of research on the durability of masonry structures, and points out the research still required. Parts of the chapter explore a systematic approach to research on the durability of masonry as carried out by the authors. Some concluding remarks are made on the effectiveness of the proposed procedure:

1. When the decay of a masonry surface appears as delamination or crumbling and the material properties underneath the deteriorated surface are not radically changed, the dimensional variation of the masonry section (loss of thickness) versus time can be used as a suitable parameter to define the damage.
2. The measurement of this parameter can be carried out in a non-destructive way; this creates the opportunity of following the performance of the material over time.
3. The profile-recording laser device is a successful tool for collecting these measurements both on site and in the laboratory. Complete and reliable profiles can be obtained in a short time without any contact with the deteriorated surface.
4. The loss of thickness of the section appears to be a good parameter with which to compare the durability of different treatments and different materials.
5. The limits of the device can be defined as follows:

    - difficulty in the definition of the reading speed;
    - impossibility of application in the case of decorations or built-in details which require a larger distance than 10 mm from the laser sensor;
    - impossibility of being used in the present configuration (i.e. need for a frame attached to the wall) in the case of historic buildings;
    - necessity of a support not influenced by the site temperature variations.

Further research will be carried out to solve the above mentioned problems.

Acknowledgements

The authors wish to thank G. Cavallini, P. Franceschi, B. Fusi, E. Gaini, and B. Lubelli for data collection and elaboration. This research has been supported by the Italian Ministry of Cultural Properties and by European Community (Contract EV-CT94-0515).

References

[1] Brimblecombe, P., Rodhe, H., 'Air pollution – Historical trends', in *Air Pollution and Conservation, Durability of Building Materials*, Vol. 5, Amsterdam: Elsevier, 1988.

[2] Feilden, B.M., 'A short introduction to conservation of cultural property in Europe', in *Durability of Building Materials,* Vol. 5, Amsterdam: Elsevier, 1988.

[3] Binda, L., Anti, L., *Prestazione e tempo di vita utile dei materiali porosi,* Ricerca sperimentale e modelli previsionali sulla durabilità delle muratura, su: *Costruire in Laterizio,* n.13, pp. 38–46, 1990.

[4] Masters, L.W., 'Service life prediction: a state of the art', *Fourth International Conference on Durability of Building Materials & Components, Singapore*, 1987.

[5] *Durability of Building Materials and Components, Proceedings, First International Conference*, Sereda, Livtan (Eds), ASTM STP 691, 1980.
[6] *Durability of Building Materials and Components, Proceedings, Ninth International Conference, Vancouver, Canada*, 1998.
[7] Frohnsdorff, G., Masters, L.W, 'The meaning of durability and durability prediction', in *Proceedings, Second International Conference on Durability of Building Materials and Components*, National Bureau of Standards, Gaithersburg, Maryland, USA, pp. 17–29, 1981.
[8] Sentler, L., *Service Life Prediction of Concrete Structures*, Dept. of Structural Engineering. Lund Institute of Technology, Rep. TVBK – 3'18, Lund, Sweden, 1983.
[9] Kreijger, P.C., 'Environmental factors affecting building durability', *Proc. Seminaire sur la Durabilité des Bétons et des Pierres, Collège Int. des Sciences de la Construction, Saint Rémy-les-Chevreuse, France*, pp. 37–80, 1981.
[10] Sijstrom, C., 'Overview of methodologies of prediction of service life', Atti del Workshop *Problems in Service Life Prediction of Building Construction Materials*, L.W. Masters (Ed), NATO-ASI Series, Martinus Nijhoff Publ., pp. 3–20, 1985.
[11] Eurin, P., 'Topic 3, Action of time environment: durability of construction materials', *Materials and Structures*, 21, pp. 131–136, 1988.
[12] Saunders, S.C., 'Topic 3, Prediction and control of the effect of time and environment on the reliability of materials', *Materials and Structures*, 21, pp. 90–96, 1988.
[13] 'Prediction of service life of building materials and components', Final Report CIBW80/RILEM 71-PSL, *Materials and Structures*, 20, n.115, pp. 55–77, 1987.
[14] Baronio, G., and Binda, L., *Studio sull'alterazione delle murature in laterizio*, in *L'Edilizia e l'Industrializzazione,* n.6 pp. 323–320, 1989 (1^{a} parte); n. 7/8, pp. 366–373, 1989 (2^{a} parte).
[15] Vincenzini, P., and Fiori, C., *Caratteristiche naturali di argille italiane e correlazione con le proprietà tecnologiche dei prodotti da esse ottenibili*, Caramurgia, VII 3., pp. 119–134, 1977.
[16] Ravaglioli, A., Il problema della gelività nei cotti ceramici: parametri chimico-fisici connessi col fenomeno e metodi di misura, Ceramica Informazione, 135, pp. 543–557, 1977.
[17] Maggi, E., *Appunti e cifre sperimentali nell'industria dei laterizi,* R.D.B., Piacenza (Ed), 1979.
[18] Lewin, S.Z., Charola, A.F., 'The physical chemistry of deteriorated brick and its impregnation technique', Atti Conv. Il Mattone di Venezia, Ottobre, pp. 189–214, 1979.
[19] Stambolog, T., 'The corrosive action of salts', *Lithoclastia*, 3–8, 1976.
[20] Glasstone, S., *Trattato di Chimica Fisica,* Carlo Manfredi (Ed.), Milano, 1956.
[21] Fiori, C., 'Efflorescenze su laterizi, parte II: composizione e origine', *Ceramurgia*, V, 3, pp. 155–157, 1975.
[22] *Gmelins Handbuch der Anorganischen Chenie, Natrium*, n. 21, Verlag Chemie, GmbH, Weinheim, Bergstr., Berlin, 1966.
[23] Pascal, P., *Nouveau Traitè de Chimie Minerale*, Tome II, Prem. Fasc., Masson et C. (Ed.), Paris, 1966.
[24] Herrero, J.I., *Alteration des calcaires et des gres utilises dans la construction*, Eyrolles (Ed.), Paris, 1967.
[25] Mamillan, M., *Alteration et Durabilité des Pierres. Seminaire Alteration et Durabilité des Bétons ed des Pierres*, 17–19/11/1981, St. Remy-Les-Chevreuse, 1981.
[26] Litvan, G.G., *Frost Action in Porous Systems, Seminaire Alteration et Durabilité des Bétons ed des Pierres*, 17–19/11/1981, St. Remy-Les-Chevreuse, 1981.
[27] Winkler, E.M., 'Durability index for stone', *Fifth International Congress on Deterioration and Conservation of Stone*, Lausanne, Suisse, pp. 151–156, 1985.

[28] Amoroso, G.G., and Fassina, V., 'Stone decay and conservation', *Atmospheric Pollution, Materials Science Monographs*, 11, New York, USA, 1983.

[29] Guidobaldi, F., and Macchi, A.M., 'Corrosion of marble by rain. The influence of surface roughness, rain intensity and additional washing', *Fifth International Congress on Deterioration and Conservation of Stone*, Lausanne, Suisse, pp. 467–474, 1985.

[30] Beilke, S., and Gravenhorst, G., 'Cycles of pollutants in the atmosphere', in B. Versin and H. Otto (Eds) *Proceedings, First European Symposium on Physico-Chemical Behaviour of Atmospheric Pollutants*, Ispra, 1979, Commission of the European Communities, pp. 331–353, 1979–1980.

[31] Franke, L., Schumann, I., van Hees, R.P.J., van der Klugt, L.J.A.R., Naldini, S., Binda, L., Baronio, G., van Balen, K., and Mateus, J.M., *Damage Atlas – Classification and Analysis of Damage Patterns Found in Brick Masonry*, Stuttgart: Fraunhofer IRB Verlag, 1998 (ISBN 3-8167-4702-7).

[32] Van Balen, K., Mateus, J., Binda, L., and Baronio, G., 'Scientific background of the Damage Atlas and the Masonry Damage Diagnostic System, expert system for the evaluation of the deterioration of ancient brick structures', *Research Report 6*, 1997.

[33] Haardt, P., and Hilsdorf, H.K., 'Tests on eccentrically loaded reinforced masonry pillars', *Fifth North American Masonry Conference,* Urbana, Vol. III, pp. 1017–1028, 1990.

[34] Massari, G., and Massari, I., *Risanamento igienico dei locali umidi*, Hoepli (Ed.), Milano, 1987.

[35] Hilsdorf Garrecht, H., Hilsdorf, H.K., and Kropp, J., 'Hygroscopic salts – influence on the moisture behaviour of structural elements', *Durability of Building Materials and Components,* J.M. Baker, P.J. Nixon, A.J. Majumdar and H. Davies (Eds), pp. 313–324, 1990.

[36] *Proceedings, First International RILEM Congress,* Vol. 1, 2, 3, Versailles, France, 1987.

[37] Mamillan, M., *Alteration et Durabilité des Pierres,* Collège Int. des Sciences de la Construction, St. Remy-Les-Chevreuse, pp. 233–242, 1981.

[38] Tassios, T.P., 'Altération et durabilité des maconneries et des joints', *Proc. Seminaire sur la Durabilité des Bétons et des Pierres, Collège Int. des Sciences de la Construction, St. Remy-Les-Chevreuse*, pp. 175–196, 1981.

[39] Girardet, F., and Furlan, V., 'Measure de la vitesse dì'accumulation des composés en atmosphère rurale et urbain', *Proceedings, Fourth International Congress on the Deterioration and Preservation of Stone Objects, Louisville, Kentucky*, pp. 159–168, 1982.

[40] Robertson, W.D., 'A three dimensional description of the pore structure in limestone', *Fifth International Congress on the Deterioration and Conservation of Stone*, Lausanne, Suisse, pp. 87–92, 1985.

[41] Butlin, R.N., Cooke, R.U., Jaynes, S.M., and Sharp, A.S., 'Research on limestone decay in the United Kingdom', *Fifth International Congress on Deterioration and Conservation of Stone*, Lausanne, Suisse, pp. 537–546, 1985.

[42] Charola, A.E., and Lewin, S.Z., 'Example of stone decay due to salt efflorescence', *Third International Congress on Deterioration and Preservation of Stones*, Venezia, pp. 153–163, 1979.

[43] Cortes, V., and Martin, A., 'The effects of atmospheric pollution by fluorides on stone alteration', I Siliceous Materials, *Third International Congress on Deterioration and Preservation of Stones*, pp. 1–12, 1979.

Cortes, V., and Martin, A., 'The effects of atmospheric pollution by fluorides on stone alteration', II Calcareous Materials, *Third International Congress on Deterioration and Preservation of Stones*, pp. 13–21, 1979.

Cortes, V., and Martin, A., 'The effects of atmospheric pollution by fluorides on stone alteration', III The enhancing of the deterioration actions by water and other natural agents, *Third International Congress on Deterioration and Preservation of Stones*, pp. 23–24, 1979.

[44] Efes, Y, and Luckat, S., 'Relation between corrosion of sandstones and uptake rates of air pollutants at the Cologne Cathedral', *Second International Symposium on the Deterioration of Building Stones,* Athens, pp. 193–200, 1976.

[45] Fassina, V., Lazzarini, L., and Biscontin, G., 'Effects of atmospheric pollutants on the composition of black crusts deposited on venetian marbles and stones', *Second International Symposium on the Deterioration of Building Stones,* Athens, pp. 201–211, 1976.

[46] Fassina, V., *Influenza dell'inquinamento atmosferico sui procesi di degrado dei materiali lapidei*, Bollettino d'Arte – Serie Speciale su Studi relativi ai problemi di degrado e alle tecniche di intervento sui materiali lapidei, 1986.

[47] Lazzarini, L., and Fassina, V., *Studio scientifico sullo stato di conservazione delle pietre e dei marmi della Prota della Carta di Venezia*, *Third International Congress on Deterioration and Preservation of Stones*, pp. 646–660, 1979.

[48] Lewin, S.Z., and Charola, A.E., *Aspect of crystal growth and recrystallization mechanisms as revealed by scanning electron microscopy*, SEM Inc. AMF d'Hare, Chicago, USA, pp. 551–558.

[49] Chatterji, S., Christensen, P., and Overgrad, G., 'Mechanisms of breakdown of natural stones caused by sodium salts', *Third International Congress on Deterioration and Preservation of Stones*, pp. 131–134, 1979.

[50] Niesel, K., and Schimmelwitz, P., 'Deux méthodes pour caractériser des processurs lors de l'altération de pierre de construction', *Second International Symposium on Deterioration of Building Stones*, *Athens*, pp. 127–162, 1979.

[51] Charola, A.E., and Koesler, R.J., 'The action of salt-water solutions in the deterioration of the silico-aluminate matrix of bricks', *Concorso di idee su patologie, diagnosi e terapie del Mattone di Venezia*, Ateneo Veneto: Venezia, pp. 67–76, 1982.

[52] Lewin, S.Z., and Charola, A.E., 'The physical chemistry of deteriorated brick and its impregnation technique', *Conv. Il Mattone di Venezia – Stato delle Conoscenze Tecnico-Scientifiche,* pp. 189–211, 1979.

[53] Nriagu, J.O., 'Deterioration effects of sulfur pollution on materials', *Sulfur in the Environment Part II: ecological impact*, J.O. Nriagu (Ed), Burlington, Canada: Wiley, 1978.

[54] Bardin, G., 'I fenomeni di efflorescenza e le alterazioni delle murature e dei lori rivestimenti', *Costruire* n. 1, I parte, pp. 51–53, 1972.
Bardin, G., 'I fenomeni di efflorescenza e le alterazioni delle murature e dei lori rivestimenti', *Costruire* n. 2, pp. 111–113, 1972.
Bardin, G., 'I fenomeni di efflorescenza e le alterazioni delle murature e dei lori rivestimenti', *Costruire* n. 3, pp. 249–253, 1972.

[55] Erfurth, U., and Tenge, H.W., 'Decay of masonry due to crystallization of soluble salt and effective countermeasures', *Conv. Il Mattone di Venezia – Stato delle Conoscenze Tecnico-Scientifiche,* pp. 379–394, 1979.

[56] Robinson, G.C., Holman, J.R., and Edwards, J.F., 'Relation between physical properties and durability of commercially marked brick', *American Ceramics Society* n. 12, pp. 1071–1079, 1977.

[57] Robinson, G.C., 'The relation between pore structure and durability of brick', *Ceramic Bull.*, Vol. 63, n.2, pp. 295–300, 1984.

[58] Torraca, G., 'Physico-chemical deterioration of porous rigid building materials', Notes for a general model, *Conv. Il Mattone di Venezia – Stato delle Conoscenze Tecnico-Scientifiche,* pp. 95–144, 1979.

[59] 'Air Pollution and Conservation', *Durability of Building Materials*, Vol. 5, Amsterdam: Elsevier, 1988.

[60] *Saving Our Architectural Heritage: The Conservation of Historic Stone Structures*, N.S. Baer and R. Snethlage (Eds), Dahlem Workshop Reports, ISBN 047196526X, 1996.

[61] Baronio, G., Binda, L., and Scirocco, F., 'Accelerated ageing and models for the progressive chemico-physical deterioration of bricks in the diagnosis of masonry structures', *Proceedings, International Conference on Material Sciences and Restoration, Esslingen, Germany*, pp. 391–395, 1983.

[62] Binda, L., and Baronio, G., 'Alteration of the mechanical properties of masonry prisms due to ageing', *Seventh IBMaC*, Melbourne, Australia, Vol. 1, pp. 605–616, 1985.

[63] Binda L., Charola, A.E., and Baronio, G., 'Deterioration of porous materials due to salt crystallization under different thermohygrometric conditions, I Bricks', *Fifth International Conference on Deterioration and Conservation of Stone,* Lausanne, Suisse, pp. 279–288, 1985.

[64] Binda, L., and Baronio, G., 'Mechanisms of masonry decay due to salt crystallization', *J. Durability of Buildings Materials*, N.4, Amsterdam: Elsevier, pp. 227–240, 1987.

[65] Lewin, S.Z., 'The mechanism of masonry decay through crystallization', *Conservation of Historic Stone Buildings and Monuments*, Washington Nat. Academy of Sciences, Washington, D.C., 1981.

[66] Larsen, E.S., and Nielsen, C.B., 'Decay of bricks due to salt', *Materials and Structures*, 23, pp. 16–25, 1990.

[67] Binda, L., Baronio, G., and Squarcina, T., 'Evaluation of the durability of bricks and stones and of preservation treatments', *Seventh International Congress on Deterioration and Conservation of Stone,* Lisbon, pp. 753–761, 1992.

[68] Binda, L., and Molina, C., 'Building materials durability: Semimarkov approach', *J. of Materials in Civil Engineering, ASCE*, Vol.12, n.4, pp. 223–239, 1990.

[69] Baronio, G., Binda, L., Cantoni, F., and Rocca, P., 'Durability of stone and brickwork protectives and consolidants: experimental research on full-scale models', *Sixth International Conference on Durability of Building, Materials and Components,* Omiya, Vol. 2, pp. 824–833, 1993.

[70] Forde, M.C., McCavitt, N., Binda, L., and Colla, C., 'Identification of moisture capillarity in masonry using digital impulse radar', *Proceedings, Fifth International Conference on Structural Faults and Repair, Edinburgh*, pp. 397–403, 1993.

[71] Binda, L., and Molina, C., 'Building materials durability: Semi-markov approach', *J. of Mat. in Civil Engineering.*, ASCE, Vol. 12, n.4, pp. 223–239, 1990.

[72] Molina, C., Garavaglia, E., Bekker, P.F.C., and Binda, L., 'A service life prediction model for masonry based on accelerated testing and Weibull interpreted testing results', *7DBMC, Stoccolma*, Vol. 1, pp. 75–84, 1996.

[73] 'Profilografo a laser per il rilievo della rugosità dei giunti in roccia', ENEL/CRIS Internal Report n. 3788, 1989.

[74] *Proceedings, Third Conference on Non-destructive Evaluation of Civil Structures and Materials*, M.P. Schuller and D.B. Woodham (Co-eds), Boulder, Colorado, 1996.

[75] Peroni, G., Fraschetti, F.M., and Lanucara, R., *Studio sugli inerti impiegati nella manutenzione delle pavimentazioni*, Indagine eseguita con l'uso di rugosimetro a laser, Autostrade n. 3, pp. 66–73, 1991.

[76] Berra, M., Binda, L., Squarcina, T., and Fatticcioni, A., 'Laboratory and in-situ measurement procedure of the decay of masonry surfaces', *Sixth International Conference on Durability of Building Materials and Components*, Japan, Vol. 2, pp. 834–843, 1993.

[77] Urone, P., 'The primary air pollutants – gaseous, the occurrence, sources and effects', in A.C. Stern (Ed.), *Air Pollution*, Third edition., Vol. I, New York: Academic Press, 1976, Ch. 2, pp. 23–75.

[78] Junge, C.E., *Air Chemistry and Radioactivity,* New York: Academic Press, 1963.
[79] Robinson, E., and Robbins, R.C., 'Air Pollution Control', W. Strauss (Ed.), Part II, New York: Wiley (Interscience), 1972, pp. 1–93.
[80] Garavaglia, E., Cardani, G., and Binda, L., 'A probabilistic model to predict the durability of surface treatments', *Int. Journal for Restoration of Buildings and Monuments (IJR)*, Vol. 8, n.2/3, pp. 233–254, 2002.
[81] Cardani, G., Tedeschi, C., Binda L., and Baronio G., 'Crystallisation test on treated brick/stone masonry specimens for damage evaluation', *Proceedings, Ninth International Conference on Durability of Building Materials, 17–20 March 2002, Brisbane, Australia*, CD-ROM, paper 039, 2002.
[82] Garavaglia, E., Lubelli, B., and Binda, L., 'Two different stochastic approaches modelling the decay process of masonry surfaces over time', *Materials and Structures,* RILEM Pub. s.a.r.l., Vol. 35, n.248, ISSN1359-5997, pp. 246–256, 2002.

9 The Durability of Masonry, Mortar, Stone and Ancillary Components

Dr RC de Vekey

9.1 Introduction

When built in the right place with the appropriate materials and the correct detailing and design for the climate, masonry is a fantastically durable material, often lasting thousands of years. A typical example is St Albans Abbey (Fig. 9.1), built around 900 years ago partly with bricks from the nearby Roman city of Verulamium which are now around 1800 years old. Masonry is inherently poor in tension and is normally built so that the structure is largely maintained in compression. In cases where masonry is required to withstand limited tension, or requires a tension connection to other materials such as timber and metal, ancillary components were developed. However, masonry components are made of rock or similar materials and, as is common knowledge, mountains formed from rocks rise up due to volcanic and tectonic activity but are also worn away again by the action of the weather. This process is driven mainly by interaction with water and thus mountains and masonry will survive for exceptionally long periods in dry desert conditions because they are refractory materials, i.e. not affected by climatic temperature cycles in the absence of water.

So how does water wreak such havoc with such materials including the common rocks, fired clay, concrete, calcium silicate, cement mortar, lime and metals? Most masonry materials are porous. This means that water does not just flow over them or around them but often through them, albeit slowly. Although most masonry materials are nearly insoluble in pure water they are slightly soluble and so will gradually lose material by dissolution over long periods of time. Of course water is usually impure because it picks up carbon dioxide, sulphur dioxide and other industrial and natural pollutant gases from the atmosphere, and can pick up acid effluents from peat moors and industrial processes. Many materials are much more vulnerable to acidified water than to pure water. Additionally, water can carry abrasive particles which can cause surface wearing of masonry materials. Water has an almost unique characteristic that it expands by around 8% when converting from liquid to solid ice at around 0°C. This behaviour is the reason for the destructive action of freeze–thaw cycles on water-soaked rigid porous materials. The water in the pore structure expands causing tensile stresses that sometimes lead to failure near the surface of the material.

Figure 9.1 *View of St Albans Abbey tower.*

Water is also a very good solvent of electrolytes, so ionic compounds may be moved into masonry from exterior sources. Indigenous compounds can also be transported to the surface and between the component materials, e.g. from brick to mortar and vice versa. The physical action of crystals forming and growing just below the surface of materials (termed cryptoefflorescence) has an effect similar to the freezing of water. Also, many compounds engage in deleterious chemical reactions, especially sulphates with Portland cement mortars, electrolytes with metals generally, and chlorides with ferrous fixings.

Lastly, water supports the growth of algae, lichens and mosses, which stain masonry and can generate acids which also attack masonry materials, while higher plants, particularly trees, can cause considerable damage by the splitting action of growing root systems. Also, plant life frequently blocks drainage systems, impeding the free flow of water and leading to the saturation of walls. Thus nearly all types of masonry are susceptible to the side effects of plant growth but there are few specific actions. Masonry is generally unaffected by fungal growth itself but dry-rot tendrils can penetrate walls and infect buried timber such as wall plates and joist ends, often causing structural problems. Masonry with soft mortar may be eroded locally by ant colonies and masonry bees but this is very rarely a major problem.

Processes which cause movement are another physico–chemical mechanism that can result in damage and disfigurement of masonry. Movements imposed by external factors such as ground subsidence, creep induced by load, and thermal expansion/contraction are not strictly inherent durability problems but they can generate cracking which allows other forms of deterioration. Moisture expansion of fired clay units might be regarded as an inherent durability problem and can cause build-up of stress with consequent crushing in some cases and cracking in others. Cement-based materials such as mortar and some units shrink after manufacture which can lead to tension cracking. Stones may have a small induced movement but it is not normally sufficient to cause problems in correctly designed structures.

Table 9.1 is an attempt to summarise the processes and the susceptibility of the commoner materials. It is stressed that the ticks in this table indicate that the mechanism has been observed for that material but is not necessarily common. The 'P' indicates that such materials are potentially susceptible but the mechanism is not, or has not, been observed or confirmed in practice. For example, Portland cement concrete is potentially susceptible to attack by sulphates but concrete bricks are resistant to all commonly observed levels of sulphates in the ground, and concrete blocks are affected only rarely. Nearly all cement-based materials will be attacked by strong acids in run-off from peat moors or industrial pollution. In the 'Material movement' column a plus sign indicates that an irreversible expansion usually occurs and a minus sign that there will be an irreversible contraction.

9.2 The failure mechanisms

9.2.1 Sulphate attack (and thaumasite formation)

Sulphates occur widely in nature as salts of sodium, potassium, magnesium and calcium. They are found, naturally occurring, in certain types of ground and may be generated by biological activity and from acid emissions in flue gases. Sulphates are also a common constituent of fired-clay bricks (de Vekey 1999) and may be introduced into flooring by the re-use of brick rubble, gypsum plaster debris and certain types of shale fill. They cause problems, mainly in cement mortars, due to the expansive reaction between sulphate ions in water solution, calcium hydroxide and the tricalcium aluminate component of set Portland cement to form gypsum and ettringite. Eglington in Lea (1998) suggests that the reaction between sodium sulphate and calcium hydroxide in uncarbonated mortar to give gypsum gives a doubling of volume:

$$Na_2SO_4 + Ca(OH)_2 + 2H_2O \rightarrow CaSO_4 \cdot 2H_2O + 2NaOH$$

The gypsum can then react with calcium aluminate to form ettringite with a further doubling of volume:

$$\begin{aligned} &4\,CaO \cdot Al_2O_3 \cdot 19H_2O + 3CaSO_4 \cdot 2H_2O + 16\,H_2O \\ &\rightarrow 3CaO \cdot Al_2O_3 \cdot 3CaSO_4 \cdot 31H_2O + Ca(OH)_2 \end{aligned}$$

Where gypsum is already present the second reaction can take place independently.

Thaumasite (DETR 1999) can also form in brickwork as the product of the reaction between dicalcium and tricalcium silicate, sulphate, carbonate and water. This, fairly rare, process can turn mortar to a mush. Although no known failures have occurred

Table 9.1 Materials problems table

Materials	Sulphate attack	Thaumasite attack	Salt crystallisation	SO_2	Acid rain (CO_2)	Acid run-off	Freeze-thaw	Wind scour	Water scour	Oxidation	Electrolytic corrosion	Chloride	Material Movement
Clay brick			✓				✓	✓	✓				✓+
Concrete brick	P					✓	P	✓	✓				✓ –
Calcil brick						✓	✓	✓	✓			✓	✓ –
Concrete block	P	P				✓	✓	✓	✓				✓ –
AAC block						✓	✓	✓	✓				✓ –
Cement lime mortar	✓	P	✓		✓	✓	✓	✓	✓				✓ –
Aerated mortar	✓	P	P			✓	P	✓	✓				✓ –
Lime mortar		✓	✓	✓	✓	✓	✓	✓	✓				✓ –
Sandstone			✓			✓	✓	✓	✓				
Limestone	✓		✓	✓	✓	✓	✓	✓	✓				
Granite													
Slate							P						
Wrought iron					✓	✓				✓	✓		
Mild steel					✓	✓				✓	✓	✓	
Galvanised steel				✓	✓	✓				✓	✓	✓	
Bronzes									✓		✓	✓	
Stainless steel												✓	
Mo-Ni-Cr stainless steel													

where sulphate-resisting cement has been used, this cement may not protect mortar against thaumasite formation.

These processes only occur in situations where there is a consistent supply of moisture and thus are almost nonexistent in properly designed, detailed and specified masonry walls of buildings. In buildings it is most likely to occur in the most exposed parts of clay brick masonry, such as parapet walls exposed on two sides, and may be exacerbated by faulty rendering or impervious finishes such as roof upstands which prevent evaporation of water leaking through faulty cappings or copings. It can also occur in retaining walls where there are sulphates in the soil especially if recommendations such as the sealing of the back face, provision of a free-draining backfill and drains or weep holes at the foot of the wall have not been observed. It may affect concrete in contact with the ground and guidance on specification is contained in *BRE Special Digest 1* (2001 & 2003).

Recognition

Sulphation causes an expansion which can be as much as several percent in persistently moist conditions. In milder cases it shows as a crack running along the centre of the mortar joint where the internal mortar has expanded more than the drier surface layer. In severe cases it will cause observable expansive movement of the masonry. If chimneys or parapet walls are affected they may 'curl' due to there being more expansion on the weather face relative to that on the lee face. Thaumasite reduces the mortar to a soft mushy consistency and has been observed mainly in masonry-lined tunnels.

Avoidance and treatment

The British standard for clay bricks (BSI 1974) specifies two levels of soluble salts content: Low (L) and Normal (N). In very exposed masonry and masonry in contact with salts from the ground the L-class bricks should always be used if possible, or as an alternative, the mortar should be formulated with sulphate-resisting Portland cement and a well graded, washed sand to give the maximum durability. Concrete (BSI 1981, de Vekey 2001) and calcium silicate (BSI 1978) bricks do not contain soluble sulphates.

BRE has developed a mortar durability test (Harrison 1986, 1990) which evaluates the resistance towards either freeze–thaw cycles, sulphate attack or the combination of both. This test was made into an international standard (RILEM, 1998b). By using this test to evaluate mortars, BRE has proposed, in Digest 362 (1991), a 'universal' mortar which gives good all-round resistance and comprises sulphate-resisting cement:lime:sand (1:1:5.5) with plasticiser to give air entrainment.

Mild cases may be arrested by preventing ingress of water, e.g. by installing a damp-proof membrane (DPM) under a faulty coping or by replacing a brick capping with a proper overhung coping with drip channels. Treatment of the most exposed face of a wall with a colourless 'waterproofer' may be successful provided the wall is left with a breathing face but sealing both faces may exacerbate the problem if some moisture still gets in. Severe cases may require demolition and replacement by more appropriate materials.

Figure 9.2 *Severe sulphate expansion causing splitting and disruption of a brick pier.*

9.2.2 Salt crystallisation

This is a deterioration process caused by the crystallisation of soluble salts carried in water to a surface at which evaporation occurs. If the crystals grow right at the surface it is termed efflorescence and is harmless causing temporary disfigurement, but under some humidity conditions the crystals grow just beneath the surface. The subsurface growth of the crystals, termed subflorescence or cryptoefflorescence, exerts a force on the surface layer of the masonry material and often results in spalling or scaling of thin layers of the unit. Fig. 9.3 is a typical example from a Victorian industrial building. If the process remains untreated then the successive scaling can eventually erode quite thick layers of the masonry. It is most common on natural stone and clay brick units and may occasionally affect mortar as well, but most concrete units have too open a texture. It is most likely in situations where rain water is entering via a solid exterior wall and is evaporating at a sheltered or interior face at a higher temperature or lower humidity. It can also occur as a result of moisture rising up walls (rising damp) where there is no DPC or where the DPC is breached or bridged by rendering or soil. It is thus, more likely in older buildings predating the cavity wall and the wide use of DPCs. Further information is available from Binda et al. (1985), and test methods to measure susceptibility of materials are given by RILEM (1998a).

Figure 9.3 *Salt crystallisation eroding a brick column.*

Recognition

Erosion of units and mortar associated with crystalline deposits are usually white in colour but often stained by other impurities. The erosion is usually associated with a flaky layer of debris and crystals of the salts. The most common salt is sodium sulphate which crystallises as the anhydrous Na_2SO_4. Thenardite or hydrated forms such as Mirabilite $Na_2SO_4 \cdot 12H_2O$. Magnesium and potassium sulphates may also occur. Moisture from rising damp may contain chlorides especially in marine locations and nitrates from animal excreta and fertilisers in rural areas. The crystals will normally dissolve easily in water and surface deposits are usually easy to brush off.

Avoidance and treatment

Modern cavity walls and the use of damp-proof membranes to prevent moisture ingress up from floors, foundations and retained ground in basements or partial basements, normally prevent its occurrence. The process may be arrested by cutting off the supply of water, e.g. by installing DPMs or a waterproofer at the point of entry of the water. The process is also stopped by preventing evaporation at the affected surface, e.g. by applying a waterproofed Portland cement render, but the trapped moisture may cause other problems. Severe cases may require cutting out and replacement of affected units.

9.2.3 Sulphur dioxide (SO_2)

This is a byproduct of the burning of fossil fuels and of some industrial processes such as brickmaking. The SO_2 forms sulphurous acid in the presence of moisture and this is oxidised to sulphuric acid by oxygen and ozone. It is a special case of acid rain. This then reacts with calcium carbonate near the surface of lime mortar and limestone to form gypsum (hydrated calcium sulphate) crust:

$$CaCO_3 + SO_2 + \tfrac{1}{2}O_2 + 2H_2O \rightarrow CO_2 + CaSO_4 \cdot 2H_2O$$

This is an expansive process which damages masonry and erodes the surface. Such acids will also increase the rate of corrosion of metals and it can be shown that metals, such as zinc galvanising will be eroded at a faster rate in industrial atmospheres (Fig. 9.9. Useful references are Schaffer (1932), Livingston (1983), Amoroso and Fassina (1983), Camuffo et al. (1983), Sabbioni et al. (1992), Del Monte et al. (1984).

Recognition

A crust forms on the surface of the stone or mortar which will usually be discoloured by particulates in the atmosphere to a grey to black deposit. The associated expansion will eventually lead to spalling and erosion of the surface.

Figure 9.4 *Sulphate crust on limestone and associated erosion.*

Avoidance

It is difficult to avoid the process since the designer has no control over the atmosphere. The problem can be avoided at the design stage by the selection of inherently resistant stones and mortar formulations. Consolidants, e.g. silanes, which strengthen the surface layers and reduce moisture ingress may slow down the process but they can sometimes make the situation worse and should ideally be trialled in non-critical areas before wholesale use.

9.2.4 Acid rain

Normal acid rain is a solution of carbon dioxide in water which forms carbonic acid H_2CO_3. This reacts with limestone $CaCO_3$ to form calcium hydrogen carbonate (calcium bicarbonate) $CaHCO_3$ which is partially soluble in water and is carried away by the rain. Thus porous limestone masonry, lime-bonded sandstones, open-textured concrete with limestone aggregate, carbonated lime mortar and weak Portland cement mortars are susceptible to leaching in this way and will be weakened by becoming progressively more porous. They are particularly susceptible to continuous leaching by groundwater in structures such as bridges, tunnels, retaining walls, and wet foundations. Several examples of failure have been reported in concrete block foundations in soil with highly mobile groundwater containing carbon dioxide. Galvanised steel will gradually lose its protection, and unprotected mild steel will be corroded rapidly. Austenitic stainless steel is not normally affected.

Recognition

Stones and concretes become progressively more porous and weakened. Lime mortar is either softened to a mush or completely leached out of the masonry and this allows units to fall out of tunnel crowns.

Figure 9.5 *Loss of mortar from arch rings of a bridge by wash-out by acid rain.*

Avoidance

The problem can be prevented at the design stage by using dense, low-porosity masonry units and hydraulic binders in the mortar but prevention of further damage is difficult except where the amount of water can be reduced, e.g. by adding a waterproof deck to a bridge.

9.2.5 Acid run-off

This is a special problem caused, usually, by the run-off from peat moors which can be quite strongly acidic but it can also result from effluents from industrial processes. Such material may attack most concrete products (except the very dense, well-compacted materials) and any porous limestone, lime-bonded sandstones or mortar. Clay bricks and acid rocks, such as basalts, will normally be resistant but special waterproof mortars are required to build acid-resistant masonry. Metals will behave as in acid rain but corrosion/erosion will be more rapid.

Recognition

Erosion or swelling and crumbling of cement-based materials and rapid corrosion of many metals although the austenitic stainless steels are usually resistant.

Avoidance and treatment

Choose inherently resistant materials. There is very little that can be done to prevent attack of susceptible materials except to install liners in drains and gulleys. Metals can be covered with protective coatings.

9.2.6 Freeze–thaw

This is one of the processes that can reduce mountains to plains in the long term and it can, in principle, affect any porous material. The mechanism is due to an 8% increase in volume of water as it turns from the liquid form to thc solid ice form. If a porous material is saturated with water and subjected to temperatures of below 0°C then the first zone to freeze will be a surface layer. At this level the increase in volume can be accommodated by exuding ice from surface pores. Once this layer has formed the remaining internal pore water is driven through the pore system against the capillary pore pressure. Successive frost cycles can trap meltwater between ice layers and as this freezes and expands it exerts a splitting force on the surface layer. Fig. 9.6 illustrates one of the most destructive mechanisms caused by successive freezing and partial melting due to diurnal temperature cycles. The materials most affected are those with medium porosity, mainly fine pores and inherently low tensile strength. This includes many clay brick types and rocks. Materials with a mixture of pore sizes or mainly very large pores, such as gap-graded concretes and many mud-process bricks, are very unlikely to fully saturate so there is usually somewhere harmless for the expansion to occur. Materials with very low porosity, around 1% or less, usually have a very high tensile strength and are able to tolerate the limited expansion without failure. Green (unhardened) mortar is very susceptible to frost attack although aerated mortar is less vulnerable than lime-plasticised mixes. Metal products are unaffected.

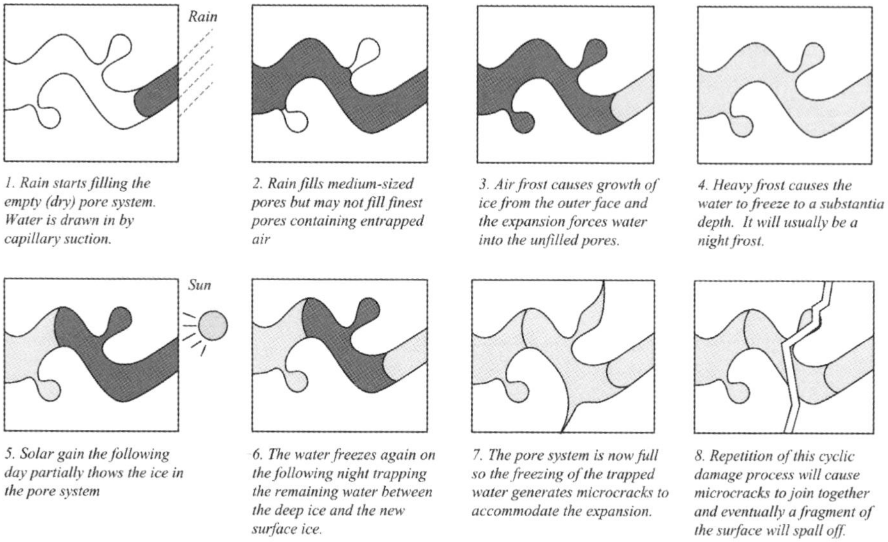

Figure 9.6 *Diagram showing a common mechanism of frost failure.*

Recognition

Scaling or sometimes more substantial loss of the surface of units or mortars in successive layers. Often either the mortar or the units will be more resistant and be left standing proud of the more vulnerable material. Products will often change colour due to either erosion of weathered surfaces or loss of special surface finishes that have different colours to the body. Mortar affected by frost attack when green will usually be very friable and easily scraped away. Fig. 9.7 shows typical frost scaling of M-class clay bricks used wrongly as a capping of a freestanding wall.

Avoidance

Only use the most frost-resistant materials in the most exposed position, e.g. copings, cappings, freestanding walls, earth-retaining walls, paving, gullies and roof tiles. If materials are not fully frost resistant then detail the structure such that they never get saturated with water under normal service conditions. Clay bricks will normally be designated in one of three durability categories:

- frost resistant – F
- not frost resistant – O; and
- intermediate – M

West et al. (1984), at the BCRL, developed a simulation brick panel test to measure frost resistance, on the basis of data about frost cycling from Beardmore and Ford (1987). This is now being developed as an EN test (2000). Further development of the cycling regime was suggested by Stupart (1996). Generally, concrete and calcium silicate bricks are frost resistant but low strength products can, in some circumstances, be eroded.

Figure 9.7 *Frost failure of M-class clay bricks used wrongly as a capping of a freestanding wall.*

Choose air-entrained mortar in cold weather conditions but never build masonry if a frost has been forecast and if surprised then cover with bubble plastic or other equivalent material. The EN1015-14 test can also be used to evaluate frost resistance of mortars. Affected masonry should be checked and dismantled if necessary.

9.2.7 Wind scour and water scour

These are mechanical processes caused by fluid flow over the masonry surface. Both wind and water can carry abrasive particles that simply wear away the surface but they can also generate suction forces by the same process that occurs on ship's propellers which leads to surface loss by cavitation (Fig. 9.8).

Recognition

Sharp details and arises will be lost. Mortar will be lost from the brickwork of bridges, locks, quays, exposed features of buildings etc.

Avoidance and treatment

Choose inherently resistant materials. There is very little that can be done to prevent the attack of susceptible materials except to install liners in water systems. Re-pointing with a well-graded or waterproofed mortar can be quite effective provided the mortar is 'ironed' within an hour of initial placement to eliminate micro-cracks formed due to plastic shrinkage and to develop a consolidated surface film.

9.2.8 Oxidation, electrolytic corrosion and chlorides

All metals oxidise, to some extent, on exposure to the atmosphere. Mild steel embedded in an alkaline material, e.g. concrete or fresh mortar will develop a thin oxide film which will then stabilise and prevent further corrosion. This process is termed passivation.

Figure 9.8 *Typical loss of masonry at the water line by scour from boat-wash.*

In mildly acid conditions, however, mild steel loses the passivation layer and oxidises rapidly in moist air to form hydrated ferric oxide according to the reaction:

$$2Fe + 2H_2O + O_2 \rightarrow Fe_2O_3,H_2O + H_2$$

This is why carbonation of concrete or mortar cover is a problem because it lowers the pH to the point where the passivation fails. Zinc coatings are lost more rapidly in acid (polluted) conditions but may last for many years depending on thickness. Based on data in BS 5493, Fig. 9.9 shows the effect of various conditions on the corrosion rate of zinc in zinc galvanised products exposed to the atmosphere. Galvanised ties embedded in carbonated mortar lose zinc at between 1 and 3 μm/year (7–21 $g/m^2/yr$). Digest 401 (1996) gives some further data on the rate of loss of zinc coatings when buried in mortar. Because zinc is easier to oxidise than steel (coming lower in the electrochemical series) it is corroded in preference to steel and thus while some zinc is still present in a coating it will protect the steel even if the steel is exposed in places. This mechanism is termed sacrificial protection. Metals higher in the series, e.g. copper, are not suitable as thin protective layers since if they are breached the steel will be corroded preferentially. Stainless steels develop this oxidised layer under a wider range of conditions and retain their passivation down to quite low pH levels.

Electrolytic corrosion is caused where two dissimilar metals form an electric cell if they are electrically connected and immersed in an electrolyte–commonly water containing soluble salts. This can occur if ties of one metal are fixed to masonry using fixings of a different metal and the masonry gets wet. A classic case is the usc of shot-fired high-carbon steel nails to fix stainless steel ties and straps. It can also occur where alloys are not perfectly homogeneous and so auto-form cells between different areas of a single metal component and lastly it can occur at points where a surface film becomes damaged. When electrolytic corrosion occurs there is oxidation at the anode and reduction at the cathode of the cell.

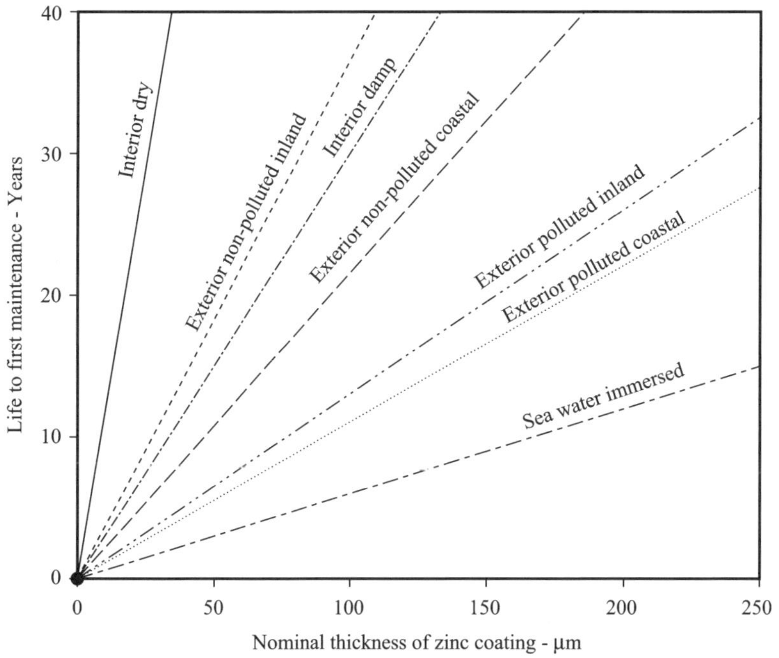

Figure 9.9 *Life of zinc-galvanised steel sheet products subject to various exposure conditions.*

Figure 9.10 *Corroded galvanised steel tie showing white and red rust.*

Chlorides are a problem because, in their presence, passification layers fail even in alkaline conditions. They can also promote rapid failure in fine crevices even of ordinary 18/8 austenitic stainless steel but molybdenum–chrome–nickel steels are normally resistant.

Recognition

Steels will develop a layer of red rust wherever oxidation takes place. Zinc will form a white layer of hydrated oxide 'white rust' under some conditions but can form black dcposits in other circumstances. Copper and bronzes will form green hydrated copper oxide.

The formation of rust on steel is expansive due to a four-fold increase in volume for the formation of hydrated ferric oxide from steel. This means that rusting steel ancillary components, e.g. fixings ties, straps, cramps and reinforcement of any kind may cause cracking of the masonry in which it is buried (Figs 9.10 and 9.11).

Figure 9.11 *Corroded beam cracking wall.*

Avoidance

Unprotected steel will be stable almost indefinitely in chloride-free, uncarbonated concrete or mortar and in dry, warm internal conditions. However it is unwise to rely on mortar cover for externally exposed masonry since all but the strongest mortars formulated with very well-graded sand tend to be carbonated through in a period of around 10 years.

In any other situation the most reliable option is to use austenitic stainless steel (containing at least 18% nickel and 8% chromium) which is stable under most inland exposure conditions. For applications that are subject to salt spray, such as walls adjacent to roads or exposed to marine conditions, the molybdenum–nickel–chrome alloys should be used. Steel protected with zinc galvanising or paint systems or resin films can be used for externally exposed masonry but will have a finite life and should only be used for short-life structures or those which are easily accessible to allow future replacement.

9.2.9 Growth of plants, algae, etc

The direct effect is usually fairly minor except that masonry is often disfigured due to the contrasting colour as shown in Fig. 9.12. If left unattended the chemical byproducts

Figure 9.12 *Green algae stains on a reconstituted stone wall due to splash-up from rain run-off.*

of plant growth may reduce masonry life. Another problem is mechanical damage caused by the growth of root systems of larger plants/trees which can disrupt masonry (Fig. 9.13). This is especially common on transport structures such as canal lock-sides and bridges and railway retaining walls. A wide range of trees and shrubs can take root and a particular offender is the buddleia which is now very widespread in the wild.

Recognition

Algae deposits are usually roughly circular and may have a colour ranging from almost white through greens, oranges and browns to almost black. Most other plant life is usually green.

Avoidance

Plants are able to survive only if there is a supply of water for all or part of the year. Design to prevent splash-up of water falling from the edges of roofs onto hard paving and leakage of gutter/down-pipe systems.

9.2.10 Material movements

Unrestrained masonry expands when its temperature and/or its moisture content increases and shrinks when they fall, so subjecting externally exposed masonry to cyclic movement over its lifetime. Additionally, it suffers non-cyclic, irreversible, movement due to the following mechanisms:

- permanent creep due to dead or imposed loads;
- permanent shrinkage due to drying/carbonation of mortar, concrete and calcium silicate materials; and
- permanent expansion of fired clay products due to the slow absorption of water into the pore system and a reactive adsorption.

Figure 9.13 *Growth of larger plant species on a poorly maintained wall.*

While strictly not a durability problem in the conventional sense, the effect of movement is dependant on the properties of the material that is being subjected to variations of the thermal, moisture or chemical environment. In the case of masonry, cracking and disruption often result from movement and so it can be said to influence durability. Table 9.2 contains some typical ranges and design values for common masonry components as given in codes of practice, etc.

In simple loadbearing masonry elements, vertical movements are accommodated by the structure going up and down as required. Problems can arise, however, where materials with different movement characteristics are bonded together or connected together with stiff fixings. Horizontal movements are more difficult to design for and lightly stressed and unrestrained elements will need a slip plane or soft joint between the elements to avoid problems. Well-restrained elements, e.g. those with a significant vertical prestress will often be more tolerant of the movement. There is a potential for bowing where thick elements have differential temperature/moisture gradients through them. This often manifests itself as a progressive bow in poorly restrained masonry which requires the installation of restraints at floor lines (see BRE GBG29, (1997)).

Table 9.2 gives data on the coefficient of linear thermal expansion of masonry materials. It is, however, difficult to predict accurately the likely movement of a wall

Table 9.2 Movement data for concrete, masonry unit materials and mortar

Source of typical values based on test measurements	Material	Thermal expansion coefficient per °C × 10^{-6}	Reversible moisture movement (%)	Irreversible shrinkage or expansion (%)	Creep coefficient	E modulus of elasticity (kN/mm²)
BRE Digest 228	Mortar	10 to 13	0.02 to 0.06	–0.04 to –0.10	–	20 to 35
	Gravel aggregate concrete units	12 to 14	0.02 to 0.06	–0.03 to –0.08	–	15 to 36
	Crushed rock (not limestone) units	10 to 13	0.03 to 0.10	–0.03 to –0.08	–	15 to 36
	Crushed limestone units	7 to 8	0.02 to 0.03	–0.03 to –0.04	–	20 to 36
	LWA Medium units	8 to 12	0.03 to 0.06	–0.03 to –0.09	–	8
	Fired clay bricks	5 to 8	0.02 to 0.04	+0.02 to +0.15	–	–
Draft of BS5628:	Mortar	–	0.03 to 0.06	+0.04 to +0.10	–	–
Part 3:1999	Concrete units	7 to 14	–	–0.02 to –0.06	–	–
BS DD ENV1996-1-1 Table 3.8	Dense aggregate concrete masonry & manufactured stone	6 to 12	–	–0.06 to –0.01	1 to 2	–
	LWA concrete masonry	8 to 12	–	–0.1 to –0.02	1 to 3	–
	Fired clay bricks	4 to 8	–	–0.02 to +0.1	0.5 to 1.5	–
	Calcium silicate bricks	7 to 11	–	–0.04 to –0.01	1 to 2	–
	Natural stone	3 to 12	–	–0.04 to +0.07	Very low	–

Figure 9.14 *Shrinkage crack in calcium silicate masonry.*

because of restraints due to internal friction and interaction with other components. Further, because the two surfaces of a wall are often heated unequally, there may be a tendency to differential movement within the wall thickness. Although thermal movement is reversible, this is not wholly true of masonry in practice because of friction and other effects which cause ratcheting (see BRE Digest 441 (1999)).

In a building, most concrete and calcium silicate masonry will shrink slightly as the manufactured moisture content dries to the final equilibrium level. Products with an open texture, that allow diffusion of carbon dioxide, can shrink further due to carbonation. None of the measuring methods can distinguish between the two shrinkage mechanisms so overall values are quoted in Table 9.2. If the masonry is in a situation where the moisture content can fluctuate over time it will also exhibit a reversible movement.

Recognition

A net shrinkage will put a wall spanning horizontally between end restraints into tension. If this tensile stress exceeds the tensile resistance of the wall there will be a tendency for vertical cracking to occur. Such cracking often originates at openings, because the opening reduces the local tensile resistance of the wall. Fig. 9.14 shows typical shrinkage cracking in the concrete brick wall of a domestic dwelling. Modest expansion is usually beneficial as it puts walls into compression and reduces any tendency to crack but expansion of masonry which is supported eccentrically can cause bulging and spalling of the unsupported area and severe cracking can occur where horizontal runs of masonry are interrupted by a short return (see Fig. 9.15).

Prevention

To prevent the occurrence of cracking, the masonry should be broken into lengths by movement joints. The recommended length in the current code versions is 6–9 m

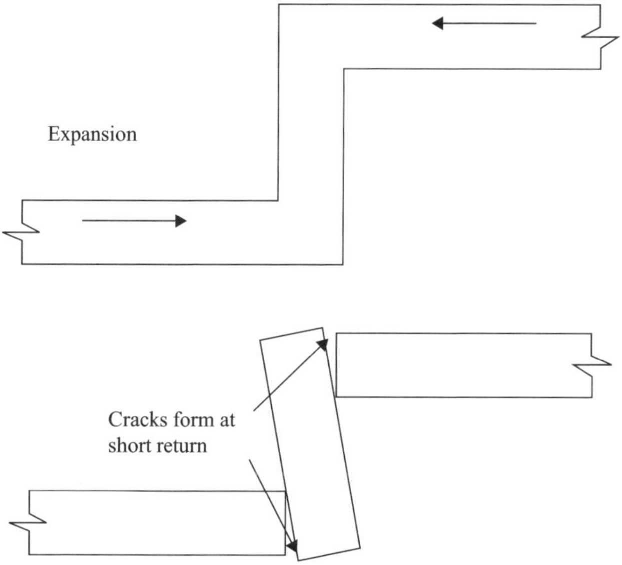

Figure 9.15 *Damage mechanism at short returns in long masonry walls.*

maximum for shrinkable materials. It is also advisable to reinforce over and under openings with bed-joint reinforcement to reduce any tendency to crack from the corners – especially for externally exposed walls. Lengths may be increased for some units or where systematic bed-joint reinforcement is used. Expansive masonry can usually tolerate lengths of up to 15 m without damage provided it has some restraint (e.g. dead load) and short returns are avoided. Lightly restrained expansive masonry, such as parapet walls, may need more frequent movement joints.

Bibliography

Bomley, A.V. and Pettifer, K. (1997) 'Sulfide-related degradation of concrete in Southwest England – the Mundic Problem' *BRE Report BR325,* Watford: Building Research Establishment.

BRE CP24/70 (1970) 'Some results of exposure tests on durability of calcium silicate bricks'.

BRE CP23/77 (1977) 'Chemical resistance of concrete', *Concrete,* 11, No. 5, 35-7.

BRE (1984) 'Performance specifications for wall ties', BRE report.

BRE (1991) 'Repairing brickwork', *BRE Digest 359.*

BRE (1991) 'Why do buildings crack?', *BRE Digest 361.*

British Cement Association (1954) 'The effects of sulphates on Portland cement concretes and other products', Technical report TRA/145.

Concrete (1986) 'The concrete engineering quality brick', *Concrete,* April.

Concrete Block Association (1998) *Aggregate Concrete Blocks for Use in Sulphate Soil Conditions,* Guidance Sheet, June 1998.

de Vekey, R.C. (1990a), 'Corrosion of steel wall ties: background, history of occurrence and treatment', *BRE Information Paper IP12/90*, Watford: Building Research Establishment.

de Vekey, R.C. (1990b), 'Corrosion of steel wall ties: recognition and assessment', *BRE Information Paper IP13/90.*

de Vekey, R.C. (1998) 'The effect of weathering on the flexural strength of masonry' (Updated to include 1995 measurements), 362 Building mortar, *Proceedings, Fifth International Masonry Conference* (BHS Proc. 8) pp. 251–256.
de Vekey, R.C. (2001) 'Corrosion of Metal Components in Walls', *BRE Digest 461.*
de Vekey, R.C. (2002a) 'Autoclaved Aerated Concrete "Aircrete" Blocks and Masonry: Part 1 – Performance Requirements, *BRE Digest 468*, Part 1.
de Vekey, R.C. (2002b) 'Autoclaved Aerated Concrete "Aircrete" Blocks and Masonry: Part 2 – Appearance and Environmental Aspects', *BRE Digest 468*, Part 2.
Harding, J.R. and Smith, R.A. (1983) 'The performance of calcium silicate brickwork in high sulphate environments', BDA Design Note 7, *Brickwork Durability,* Technical Note 368.
Harrison, W.H. (1987) 'Durability of concrete in acidic soils and groundwaters', *Concrete,* 21, No. 2.
Royal Institute of Chartered Surveyors, (1997), *The 'Mundic' Problem – A Guidance Note*, Second edition.
SP56 (1980) 'Model specification for clay and calcium silicate structural brickwork', Supplement No.1 to SP56, *Glossary of terms relating to the interaction of bricks and brickwork with water.*

References

Amoroso, G. and Fassina, V. (1983) *Stone Decay and Conservation*, Amsterdam: Elsevier.
Beardmore, C. and Ford, R.W. (1987) 'Winter weather records relating to potential frost failure of brickwork (1)', *Brit. Ceram. Trans. J.*, Vol. 86, p.7.
Binda, L., Baronio, G. and Charola, A., (1985), 'Deterioration of porous materials due to salt crystallization under different thermohygrometric conditions, 1: Brick', in *Proceedings, Fifth Congress on the Deterioration and Conservation of Stone, Lausanne, September 1985, (Vol. II) Presses Polytechniques Romandes, Lausanne*, pp. 279–288.
Binda, L., and Baronio, G. (1987) 'Mechanism of masonry decay due to salt crystallization', *Durability of Building Materials*, n.4, Amsterdam: Elsevier, pp. 227–240.
Bowler, G.K., Harrison, W.H., Gaze, M.E., and Russell, A.D., (1995) 'Mortar durability: an update', *Masonry International Journal*, Vol. 8, No.3, pp. 85–90.
Building Research Establishment (1991) 'Building Mortar', *BRE Digest 362*, BRE, Watford.
Building Research Establishment, 'Concrete in aggressive ground : Assessing the aggressive chemical environment' (inc. amendment of Mar. 2003), *BRE Special Digest 1*, Part 1.
Building Research Establishment, 'Concrete in aggressive ground : Specifying concrete and additional protective measures' (inc. amendment of Mar. 2003), *BRE Special Digest 1*, Part 2.
Building Research Establishment, 'Concrete in aggressive ground' (inc. amendment of Mar. 2003), *BRE Special Digest 1*, Part 3.
Building Research Establishment, 'Concrete in aggressive ground : Design guides for specific precast products' (inc. amendment of Mar. 2003), *BRE Special Digest 1*, Part 4.
(Note *BRE Special Digest 1* is due to be replaced by a single part version in 2005.)
BS 187:1978, *Specification for calcium silicate (sandlime and flintlime) bricks*, London: BSI.
BS 6073:1981, *Precast concrete masonry units, Parts 1 and 2*, London: BSI.
BS 3921:1985, *British Standard specification for clay bricks*, London: BSI.
Camuffo, D., Del Monte, M., and Sabbidni. C. (1983) 'Origin and growth mechanisms of the sulphated crusts on urban limestone' *Water Air Soil Pollution* Vol. 19, pp. 351–359.
C&CA (1954) 'An investigation of the erosive effect on concrete of soft water of low pH value' TRA/143 Technical report.

C&CA (1954), 'The effects of sulphates on Portland cement concretes and other products', TRA/145 Technical report.

de Vekey, R.C. (1997) 'Connecting walls and floors, Part 1 – A practical guide *and* Part 2 – Design and performance', *Good Building Guide* GBG29

de Vekey, R.C. (1999a) 'Clay bricks and clay brick masonry', *BRE Digest 441*, Part 1, November 1999 Watford: Building Research Establishment.

de Vekey, R.C. (1999b) 'Clay bricks and clay brick masonry', *BRE Digest 441*, Part 2, November 1999.

de Vekey, R.C. (2001) 'Bricks blocks and masonry made from aggregate concrete: Part 1: 2001 – Performance requirements; Part 2: 2001 – Appearance and environmental aspects' *BRE Digest 460.*

Del Monte, M., Sabbioni, C. and Vittori, O. (1984) 'Urban stone sulphation and oil-fired carbonaceous particles', *Sci. Total Environ.* Vol. 36, pp. 369–376.

Eglinton, M. (1998) 'Chapter 7, Resistance of concrete to destructive agencies', *Lea's Chemistry of Cement and Concrete*, Fourth edition, London: Arnold.

EN 1015-14 (draft), (1999), 'Methods of test for mortar for masonry – Determination of durability of hardened mortar'.

EN 772-22 (final voting draft Dec 2002) 'Methods of test for masonry units – Determination of the frost resistance of clay masonry units'.

Harrison, W.H. (1981) 'Conditions for sulphate attack on brickwork', *Chemistry and Industry*,

Harrison, W.H. (1986), 'Durability of concrete in acidic soils and groundwaters', *Concrete Journal.*

Harrison, W.H. (1986) 'Durability tests on building mortars – Effect of sand grading', *Magazine of Concrete Research*, Vol. 38, No.135.

Harrison, W.H. and Bowler, G.K. (1990) 'Aspects of mortar durability', *Brit. Ceram Trans. J.*, Vol. 89, pp. 93–101.

Livingston, R. and Baer, S. (1983) 'Mechanisms of air pollution-induced damage to stone', In *Proceedings of Sixth World Congress on Air Quality, Paris*, Vol. 3, pp. 33–40.

Metha, P.K. (1973), 'Mechanism of expansion associated with ettringite formation', *Cement Concrete Res.* Vol. 3 1–6.

RILEM recommendations of durability tests for masonry materials, (1998a) MS.A.1, 'Resistance of wallettes against sulphates and chlorides', *Materials and Structures*, Vol. 31, pp. 2–5.

RILEM recommendations of durability tests for masonry materials, (1998b) MS.A.4, 'Determination of the durability of mortar', *Materials and Structures*, Vol. 31, pp 12–16.

RILEM recommendations, MS.A.2 (1998) 'Uni-directional salt crystallization test for masonry units', *Materials and Structures*, Vol. 31, pp. 5–8.

Sabbioni, C. and Zappia, G. (1992) 'Decay of sandstones in urban areas correlated with atmospheric aerosol', *Water Air Soil Pollut*, Vol. 65, pp. 305–316.

Schaffer, R. (1932) *The Weathering of Natural Building Stones*, HMSO, London.

Stupart, A.W. (1996) 'Possible extensions to developing a frost index', *Masonry International*, 7 (1) 4–9.

DETR (1999) 'The thaumasite form of sulphate attack: risks, diagnosis, remedial works and guidance on new construction' Report of the Thaumasite Expert Group, Department of Environment, Transport and the Regions, London.

West, H.W.H., Ford, R.W. and Peake, F.A. (1984) 'A panel freezing test for brickwork', *Proc. Brit. Ceram. Soc.,* Vol. 83, 112.

10 The Durability of Brickwork and Blockwork

Dr John Morton

10.1 Introduction

For brickwork and blockwork (masonry), the major failure component of durability in the northern European climate is frost failure. Other durability factors, include:

- efflorescence;
- concrete mix strength;
- calcium silicate bricks;
- sulphate attack;
- vegetation, moss and lichen;
- mortars

although some of these are only loosely connected.

10.1.1 Frost failure

Frost failure requires the simultaneous combination of saturation of the masonry and repeated freeze–thaw cycles. It must be noted that saturation does not mean 'wet' or 'very wet' but that the surface layer of the masonry is actually saturated. All porous materials, not just masonry, which are subject to both freeze–thaw and saturation conditions acting together are potentially at risk due to frost failure.

The number of freeze–thaw cycles is also important. For example, it is often not fully appreciated that in those parts of the world where the winter temperature is many degrees below freezing, materials are not as subject to frost attack as they are in more temperate climes. In the harsher winter climates of northern America or central and eastern Europe, there may only be a small number of freeze–thaw cycles for a short period in autumn and in spring. The temperature in summer is always above freezing and the converse in winter. In the more temperate climates found in parts of northern Europe, including the UK, the number of freeze–thaw cycles is much greater: it is unusual to have prolonged periods when temperatures remain continuously below freezing.

The temperature effect is only one part of the two-part equation. The other is saturation. If masonry which is surface-saturated is frozen, ice forms first at the surface then works progressively deeper into the structure. The increase in volume (as the water turns to ice) will merely 'push' the water trapped behind it further into the body of the unit. The presence of repeated freeze–thaw cycles, however, can result

Figure 10.1 *Frost failure of bricks. Note the characteristic delamination of the first few millimetres of the brick surface.*

in water being trapped between ice lenses. When this condition exists, it is believed that the repeated application of such pressure cycles within the structure disrupts the masonry. Frost failure therefore involves typically the loss of part, or all, of the face of the masonry for a depth of only a few millimetres (Fig. 10.1).

Not all structures will fail under such frost action. Indeed by adopting a full and proper specification, the probability of frost failure can be virtually eliminated in Britain.

10.1.2 Clay bricks

The British Standard for clay bricks is BS 3921[1]. In Britain clay bricks, manufactured and sold to BS 3921, are classified by the manufacturer as being either:

- fully frost resistant (known as Designation-F bricks); or
- moderately frost resistant (known as Designation-M bricks).

There is an additional designation category, 'Designation O', for bricks which are not frost resistant. Designers specifying masonry would not generally call on Designation O.

Specifying bricks as Designation F will ensure frost resistance. Frost failure, however, is not quite as 'cut and dried' as that. For example, many bricks of Designation M will give a long and perfectly satisfactory service life in many parts of Britain in many situations where they are not liable to become saturated. Generous overhangs are often the key to ensuring that, while the brickwork may become wet, it is virtually impossible for it to become saturated. The degree of exposure of the masonry to driving rain is also an important factor.

The European CEN Standard for clay masonry units, BS EN 771-1[2], broadly follows the above approach:

- severe exposure (CEN Designation F2) relates to BS 3921 Designation F;
- moderate exposure (CEN Designation F1) relates to BS 3921 Designation M.

For bricks which are not frost resistant, there is an additional designation (CEN Designation F0) which is directly similar to BS 3921 Designation O. As mentioned above, designers specifying masonry would not generally call on this designation for work exposed to weathering: it would be acceptable, however, for internal work or work protected by impervious cladding.

10.1.3 Concrete blockwork and concrete brickwork

Concrete blocks are very durable masonry units if they are well made using a sufficiently strong concrete mix. Concrete blockwork and concrete brickwork, therefore, possess good frost resistance. Problems due to frost action would therefore not normally occur.

The Code BS 5628 Part 3[3] gives guidance as to the 'strength of mix' required. It does this by specifying a combination of three conditions:

- minimum block density;
- type of aggregate;
- minimum strength.

These recommendations are contained in Table 13 'Durability of Masonry in Finished Construction'. In most normal construction, a block with a density of at least 1500 kg/m^3, or with a minimum compressive strength of 7 N/mm^2, or made with dense aggregate complying with BS 882[4] or BS 1047[5] is satisfactory. The same principles apply to concrete brickwork. A concrete brick of 20 N/mm^2 or greater covers the vast majority of construction forms. For copings and sills 30 N/mm^2 is recommended while 40 N/mm^2 is the recommended strength for applications where the masonry is in contact with foul drainage fluids.

It should be noted that for many situations, concrete bricks of a lesser strength should provide adequate durability.

For the detailed design and specification of masonry in the UK, it is believed at the time of writing that the European (CEN) Standard BS EN 1996-2[6] will effectively refer designers to similar rules as those contained in BS 5628 Part 3, which implies no change from the current position.

10.1.4 Calcium silicate bricks

Calcium silicate bricks have always enjoyed a small percentage of the overall UK brick market. It is believed that this percentage of market share is falling. They are, however, an important material since they are used quite frequently in particular regions.

The frost resistance of a brick is directly related to its strength. Bricks of Class 3 strength (21 N/mm^2) or greater are recommended for all but the most onerous situations. For capping and copings the minimum strength recommendation is Class 4 (28 N/mm^2).

A peculiarity of calcium silicate bricks is a propensity to suffer deterioration if they are impregnated with strong salt solutions and then subject to freeze–thaw cycles. Since freeze–thaw cycles are an expected part of the UK climate, it follows that calcium silicate bricks should not be used when the masonry can be directly wetted by seawater, seawater spray or where road de-icing salts can contaminate it.

The CEN Standard, BS EN 771-2[7], includes provision for 'frost classification' to be declared by the manufacturer as being freeze–thaw resistant for structural use and visual aspects, or for structural use alone. Test methods are now in place in BS EN 772-18:2000[8] for measuring this and it is expected that manufacturers will classify their products using these methods depending on the intended use.

10.2 Efflorescence

Efflorescence is the manifestation of a white or whitish powder on the face of clay brickwork. It is normally present at the time when the construction has just been completed.

The white/whitish powder is normally soluble salts which have come from the body of the brick. The salts, which are usually highly soluble, will have been present in the clay when it was excavated from the ground. Generally speaking, during the life of the building the brickwork masonry is never as wet as when it is first laid. As it dries, by the water evaporating from the front surface of the brick, the moisture within the masonry migrates to the surface carrying the soluble salts in solution. As the moisture evaporates from the surface, the salts crystallise and are left on the surface of the brickwork as a white 'bloom' (see Fig. 10.2).

Figure 10.2 *Characteristic unsightly white efflorescence on a recently completed boundary wall.*

Figure 10.3 *The same wall some months later when virtually all of the white soluble salts have been dissolved and washed down into the soil by the action of rain.*

Efflorescence, in the strictest of terms, is not a true durability consideration. It does, however, cause many designers and clients concern if it appears and it is therefore covered here for completeness. It should be clearly understood that while the appearance of the brickwork is undoubtedly affected temporarily, efflorescence is not a serious problem, but occasionally, it can lead to the decay of under-fired bricks, if the salts crystallise beneath the surface. Because the salts are usually highly soluble, the efflorescence will dissolve in rain and be carried down the surface of the masonry in solution by the action of gravity. In this way, the salts will normally disappear after the first winter's rain (Fig. 10.3). Occasionally, depending on location and the amount of local rainfall, two winters may be required to completely remove all the white or whitish salts.

It should be noted, of course, that fair-faced internal brickwork will not be subject to winter rain. Should efflorescence appear it will require to be removed using several applications of a damp sponge. The corollary, of course, is that great care must be taken to protect any internal, fair-faced clay brick masonry from the elements during the construction period. This is standard good practice for all masonry but is doubly essential for the avoidance of efflorescence staining on internal masonry.

The amount of salts contained in clay bricks is covered by BS 3921[1]. The full details are given in the standard. Where the salts content is less than prescribed amounts, the bricks are referred to as Designation L ('L' for Low). Bricks which have a salts content greater than the prescribed limits are referred to as Designation N ('N' for normal). In the 1995 amendment, an upper limit was placed on Designation-N bricks. These designations have no significance when considering efflorescence. When considering sulphate attack, however, it becomes more important whether the brick is Designation L or N.

The CEN Standard, broadly follows the above approach. The low salts content CEN Designation S2 is broadly, but not exactly, equivalent to BS 3921 Designation L.

10.3 Salt content in clay bricks

CEN Designation S1 is broadly, but not exactly, equivalent to BS 3921 Designation N. In addition there is a CEN Designation S0 which signifies that the manufacturer makes no claim regarding limits of soluble salt content. Its inclusion acknowledges that in some European countries soluble salts content is not regarded as significant as their bricks are normally only used in dry conditions and therefore sulphate attack of the mortar is not encountered. There is no equivalent designation in BS 3921.

10.4 Sulphate attack in clay brickwork

Sulphate attack is the term given to the phenomenon where the mortar between the bricks degrades. The mechanism is the conversion, by soluble sulphate ions, of the tri-calcium aluminate in the mortar (i.e. from the cement) into tri-calcium alumino-sulphate. The latter is a gargantuan-sized crystal that is so large that, when it forms and grows, it effectively splits the mortar bed. Indeed, a characteristic of sulphate attack is the presence of regular horizontal cracks in the bed joints. These are usually horizontal and are usually present at the mid-height of the joint. Initially the cracks within the joints are hairline: the tendency is for them to grow with time thus causing the brickwork to expand slightly. Of course, once there are cracks within the mortar, then rain can gain access deeper into the masonry and depending upon the location of the affected masonry even more severe durability problems may result (Figs 10.4 and 10.5).

Fortunately, sulphate attack is quite rare. It is, however, usually a difficult problem to resolve. Generally it involves taking the brickwork down and rebuilding using an improved mortar specification.

Figure 10.4 *A free standing wall showing the characteristic signs of sulphate attack in some of the upper mortar bed joints.*

Figure 10.5 *Close up of bed joints showing the fine horizontal cracking at the mid height of the joint that is characteristic of the onset of sulphate attack.*

It requires not only the presence of sulphate salts but also copious quantities of water by which the salts can be carried by gravity in suspension down into the matrix of the mortar bed. This tends to suggest that sulphate attack is usually associated with either a faulty design detail, a construction fault or is the direct result of the masonry being in a position of extreme exposure.

Although sulphate attack is rare, because of its serious implications it is fully considered in BS 5628 Part 3[3]. Table 13, BS 5628 Part 3 gives specific guidance for the majority of cases likely to be met in practice. The approach used is a combination of detailing and careful specification. This may involve the type of cement (sulphate-resisting cement may be required), the strength of the mortar and the designation of the clay brick (Designation-L brick may be required).

10.5 Vegetation, moss and lichen

There are a large number of masonry structures, covering a wide range of usage, that are 100 years of age or older. They encompass normal buildings as well as canal and railway structures, sewers and road bridges. In many structures and civil engineering works there is often much masonry hidden from view. The various interceptor sewers and many of the old bridges in London are examples of this.

From experience, considerable damage can be caused by the intrusion into masonry of root systems of any sort. The most common form of root infestation is from standard perennial plants or trees. Who has not seen a self-seeded buddleia growing with vigour from a chimney stack or parapet of an old brickwork building? In addition many modern masonry buildings, particularly new homes, are purposely clad in foliage. Some self-clinging creepers such as the Virginia creeper (*Parthenocissus quinquefolia*), adhere to walling using 'sucker' pads. Other climbers may be supported by specially provided trellis-work. In many situations these may prove perfectly acceptable. Certainly, the vigorous nature of the growth and the general invasiveness of the plant/s should be noted: suitable pruning regimes can then be introduced.

Some creepers, however, use penetrating root systems: Ivy is perhaps the best known. This type of climber is best avoided on masonry walls since their roots do invade into the masonry and can cause progressively more distress with time.

Experience also suggests that moss and lichen growth may be a sign of maintenance problems. Blocked rainwater hoppers or broken or cracked rainwater pipes will result in some areas of masonry becoming permanently or nearly-permanently wet. This will usually result in moss and lichen growth which should be treated as a warning that repair work is required. If left unrepaired, this will result, in time, in severe staining of the masonry and, possibly, frost failure of those parts of the masonry facade in the immediate vicinity of the leak.

Not all moss and lichen growth however falls into this category. With the growing popularity, during the late 1970s and 1980s, of 'hard' architectural detailing, many more masonry walls exist which are not generously protected by overhanging eaves or other weathering details. It is not uncommon to find some moss growth on the top and upper portions of parapet walls. This is something which has been seen for centuries on free-standing walls and does not appear to have any deleterious effect.

Indeed, the presence of some moss and lichen growth is often accepted as giving an 'Olde English' appearance to the masonry.

It is a noticeable trend that when masonry is overhung by trees, the growth of mosses and lichens is markedly increased. This is doubtless due to the upper masonry being kept wet for longer periods. In such situations moss growth should be expected.

10.6 Mortars

As might be expected, the porosity of mortars is related to the mortar density and, hence, strength. The higher the strength, the denser and less porous the mortar: the denser the mortar, the more difficult it is to become saturated with water.

Stronger mortars are therefore more durable than weaker mortars. This is also true when considering sulphate attack where the greater density (and, hence, reduced porosity) of stronger mortars militates against water percolating through the mortar matrix to carry the sulphate ions to the tri-calcium aluminate. For this reason, where sulphate-resisting mortars are suggested in BS 5628, they are normally Designation (i) or (ii) mortars.

In positions of severe exposure, the 'ironing' or 'tooling' of the mortar joint provides an extremely hard front surface to the mortar. 'Bucket handle' or 'weather struck' joints are both very useful in increasing the frost resistance of the mortar joints. The advantages of such joints, as well as their aesthetic appeal, have led to their widespread use in modern practice.

References

British Standards publications:

[1] BS 3921:1985 (1995), *Specification for clay bricks* (Partially replaced by BS EN 771-1:2000), London: BSI.

[2] BS EN 771-1, *Specification for masonry units – Part 1: Clay masonry units*, London: BSI.

[3] BS 5628 Part 3:1985, *The use of masonry – Part 3: Materials and components, design and workmanship*, London: BSI.

[4] BS 882:1992, *Specification for aggregates from natural sources for concrete*, London: BSI.

[5] BS 1047:1983, *Specification for air-cooled blast furnace slag aggregate for use in construction*, London: BSI.

[6] BS EN 1996-2 Eurocode 6: *Design of masonry structures – Part 2: Design, selection of materials and execution of masonry*, London: BSI.

[7] BS EN 771-2, *Specification for masonry units – Part 2: Calcium silicate masonry units*, London: BSI.

[8] BS EN 772-18, *Methods of test for masonry units – Part 18: determination of freeze-thaw resistance of calcium silicate masonry units*, London: BSI.

Other British Standards publications

BS 187:1978, *Specification for calcium silicate (sandlime and flintlime) bricks*, London: BSI.

BS 6073 Part 1:1981, *Precast concrete masonry units – Part 1: Specification for precast concrete masonry units*, London: BSI.

BS 6073 Part 2:1981, *Precast concrete masonry units – Part 2: Method for specifying precast concrete masonry units*, London: BSI.

BS EN 771-3, *Specification for masonry units – Part 3: Aggregate concrete masonry units (Dense and light-weight aggregates)*;
BS EN 771-4, *Specification for masonry units – Part 4: Autoclaved aerated concrete masonry units*;
BS EN 771-5, *Specification for masonry units – Part 5: Manufactured stone masonry units*;
BS prEN 771-6, *Specification for masonry units – Part 6: Natural stone masonry units*;
BS EN 772-1, *Methods of test for masonry units – Part 1: Determination of compressive strength.*
BS EN 845-1, *Specification for ancillary components for masonry – Part 1: Ties, tension straps, hangers and brackets.*
BS EN 998-2, *Specification for mortar for masonry – Part 2: Masonry mortar.*

11 Durability of Timber in Construction

Dr WWL Chan

11.1 Introduction

Timber is a most environmental friendly material, particularly in the case of softwoods. Trees are a naturally renewable resource, absorbing carbon dioxide and producing oxygen during their lifetime. Timber waste can be processed into reconstituted products, or burnt with little or no emission of toxic gases.

Trees undergo a biological growth–decay sequence. In principle, exposure to decay commences when the tree is felled, but decay can be controlled by suitable processing, handling, storage and design detailing to avoid retention of excessive moisture which may lead to fungal or insect attack. By using naturally durable species in favourable service conditions, preservative treatment and maintenance, decay can be further controlled.

Timber, in favourable conditions of use, lasts almost indefinitely with little or no degradation of structural properties, as evidenced by the well preserved condition of timbers in many old buildings and structures throughout Europe and other continents. Modern developments in durable and strong glues have enabled the size, shape and type of timber and timber-based structural products to be extended beyond their tree size and shape. The use of glue laminated timber (glulam), plywood; and reconstituted products including laminated veneer lumber (LVL), oriented strand board (OSB), particleboard and fibreboard, are examples.

The use of most species of tropical hardwood has been increasingly discouraged because they have a slow rate of growth, making them almost a non-renewable resource, and because of the damaging effect of deforestation on tropical rain forests.

Timber is easy to work with both on- and off-site. Thus *in situ* replacement, strengthening, repair and the preservative treatment of structural components is relatively straightforward.

As there are not too many practising engineers who are familiar with timber construction, an overview is given on timber species and wood-based panel products, their basic properties, and the way they are processed, supplied, and designed. As timber is a biological material, the principal objective in maximising durability is to reduce the risk of fungal or insect attack, or to increase timber's resistance to degradation by the use of chemical treatment.

Information given in this chapter is based primarily on United Kingdom and European practice and environmental conditions. References are made to British and European Standards by way of examples and are not exhaustive. References prefixed with BS EN indicate that the European Standard EN is published by BSI and, in the

BS version, will include a British National Annex. The National Annex enables the CEN member state to specify Nationally Determined Parameters where these are allowed in the normative text of the Eurocode.

At the time of writing this chapter, product standards and structural codes are being harmonised between European member states and other European Free Trade Area (EFTA) countries, collectively the CEN member states, prepared by the technical committees of the European Committee for Standardisation (CEN). A suite of European structural codes, covering structural design in timber and other structural materials, is nearly all in place. In the next two years, the National Annexes will be available. After that, the Eurocodes will co-exist with national codes (BS 5268-2 for timber) and standards for 3 or more years, after which conflicting national standards will be withdrawn.

BS EN 1995-1-1:2004 Eurocode 5: *Design of timber structures – Part 1-1: General – Common rules and rules for building*s was published in December 2004, and its National Annexes are in course of preparation. Its normative references include many ENs on the durability of timber and wood-based board materials.

It should be mentioned that BS 5268-2 is (uniquely among other BS structural codes) written in permissible stress format, in which global safety factors are incorporated in the permissible stress values, whereas BS EN 1995-1-1 is written in limit state terms with characteristic strength values and the use of partial safety factors for actions and material strength. The difference in design output by using either code is small, but is considered to be within acceptable margins of adequate safety and serviceability.

It is intended that when the Eurocodes eventually replace national codes, they will contribute to the removal of barriers to free trade between CEN member states. In principle, all CEN member states would then design structures on the same basis. In detail, however, there will still be a limited number of overriding exceptions whereby participating countries have the right to modify certain design rules that are affected by national statutory regulations, mainly concerning safety: these will be reflected in the National Annexes. Thus a design may still require minor modification for use in different CEN member states.

European standardisation has been a courageous venture. Although progress has been slow because of the large representation and the need for reconciliation, it has brought together the best of collective wisdom and practice across (currently) 28 CEN member states. The number of CEN members will increase with the expected enlargement of the EU and so the impact of European standardisation will grow.

11.2 Materials

11.2.1 Timber

Softwoods are used in the majority of timber structures because of their abundant availability, sustainable renewal, low cost and easy workability. Tropical hardwoods are used structurally in exposed conditions such as bridges and particularly in environments exposed to attack by marine organisms.

Natural durability varies with species. Biological durability of a species refers to its ability to resist attack from wood-destroying fungi. Higher resistance is provided by the naturally-occurring chemicals (extractives) in the heartwood of the tree, whereas the resistance of sapwood is much lower. The biological durability of a particular species may also vary according to climatic conditions in which the tree is grown and its age, with faster growing trees being generally of lower density, weaker and less durable.

Sapwood is easier to penetrate with preservatives (more treatable) than heartwood, but treatability also varies with species. As nearly all rectangular timbers sections are flat sawn across the tree trunk as shown in Fig. 11.1, most sawn timber sections contain some sapwood and some heartwood.

Biologically durable timbers may also, but not necessarily, have a natural resistance to attack from other organisms such as insects and marine borers. Timber containing active insect infestation is not permitted to be used structurally (BS 4978).

To obtain an adequate service life, a naturally durable timber species or one that is amenable to preservative treatment should be selected, provided it is commercially available in suitable sizes. Natural durability against wood destroying fungi and treatability classifications are listed in Tables 11.1 and 11.2. Natural durability ratings against attack by certain insects are given in Table 11.3.

It is emphasised that the predicted mean lives are based on tests in extremely onerous conditions of exposure with small stakes of timber being half-buried in the ground, providing a basis for comparison rather than absolute durability. In practice, service conditions are usually less onerous and actual service lives are much longer provided that the conditions of use and design detailing are appropriate to the hazard class and the species adopted.

Timber species may be identified visually by an experienced timber technologist or by a laboratory investigation on samples of timber.

Table 11.1 Durability classifications used by BRE and European Standards

BRE system		**European Standards system**		
Grade	**Mean life (years)**	**Class**	**Description**	**Mean life (relative to reference species)**
Perishable	up to 5	5	Not durable	< 1 times
Non-durable	5 to 10	4	Slightly durable	1 to < 2 times
Moderately durable	10 to 15	3	Moderately durable	> 2 but < 3 times
Durable	15 to 25	2	Durable	> 3 but < 4 times
Very durable	> 25	1	Very durable	> 4 times

Source: BRE Digest 429 (reproduced by permission of BRE) and BS EN 350-1.

Table 11.2 Treatability classifications used by BRE and European Standards

BRE system		**European Standards system**	
Group 1	Permeable	Class 1	Easy to treat
Group 2	Moderately resistant	Class 2	Moderately easy to treat
Group 3	Resistant	Class 3	Difficult to treat
Group 4	Extremely resistant	Class 4	Extremely difficult to treat

Source: BRE Digest 429 (reproduced by permission of BRE), and BS EN 350-2.

Table 11.3 Natural durability ratings against insects

Rating	Definition
Insects:	
D	Durable
S	Susceptible
SH	Susceptible heartwood
Termites and marine borers:	
D	Durable
M	Moderately durable
S	Susceptible

The most widely available species in the UK for timber structures are those listed in BS 5268 Part 2, and for European species generally, draft EN13556. Their natural durability and treatability are given in Table 11.3. A more detailed list of over 160 species is given in BRE Digest 429, although most of these are not normally used for structures.

11.2.2 Softwood sizes

The availability of rectangular sawn sizes is governed by the cross-sectional size and shape of felled trees and by the mix of rectangular plank sizes which can be obtained economically from the round wood with minimum wastage. With increasing demand, smaller trees are being felled, and today sawn timber exceeding 75 mm × 225 mm in section and 5 m in length is more difficult to obtain and would attract a cost premium. However, larger sections are available by glue laminating a number of boards, or extended lengths by glued finger jointing. It is important to note that structural sawn timber is produced in a range of standard or preferred sizes. They are listed in BS EN 336 for Europe, but different ranges and sizes apply to USA and Canadian sawn and planed timbers (BS 5268 Part 2).

Timber swells or shrinks with increased or decreased moisture content, respectively, when the fibre saturation level is below 30%. Between 18% and 20% moisture content, softwoods swell or shrink by about 1% in cross-sectional dimensions across the grain for every 1% change in moisture content. Moisture movement longitudinally along the grain in solid timber is small enough to be ignored in design.

There is also a significant difference in shrinkage across the grain in the radial and the tangential directions (Fig. 11.1), tangential shrinkage being about twice the radial shrinkage. This explains the distortions that may occur due to the tendency for the growth rings to straighten out in tangentially-sawn sections (Figs 11.1 and 11.2). Splitting may occur, particularly in larger sections, when the internal stresses induced by anisotropic shrinkage are relieved. Such distortions and splitting can be reduced by kiln drying at a controlled rate.

British and European Standards for sawn timber specify maximum limits for bow, twist and spring over a defined length (L), and cup over a defined width.

Timber attains an equilibrium moisture according to the temperature and humidity of the service environment. The moisture content may be lower near the surfaces.

Table 11.4 Natural durability and treatability of certain timbers

Timber	Durability	Treatability	
	Heartwood EN class	Heartwood EN class	Sapwood EN class
UK imported:			
Redwood	4	2	1
Whitewood	4	3	2
Parana pine	4	2	1
Pitch pine	4	2	1
Western red cedar	2	3	3
Douglas fir-larch (Canada and USA)	4	3–4	1–3
Hem-fir (Canada and USA)	4	3	2
Spruce-pine-fir (Canada and USA)	4	3	2
Sitka spruce (Canada)	4	3	2
Western white woods (USA)	4	3	2
Southern pine (USA)	4	2	1
British grown:			
British larch	3	3	2
British pine	4	2	1
British spruce	4	3	2
Douglas fir	4	3	1
Tropical hardwoods:			
Balau	2	4	1
Ekki	1	4	2
Greenheart	1	4	2
Iroko	1	4	1
Jarrah	1	4	1
Kapur	1	4	1
Karri	1	4	1
Kempas	2	3	2
Keruing	3	3	2
Opepe	1	2	1
Teak	1	4	2

Source: BRE Digest 429 (abstracted by permission of BRE).

Timbers with the lesser lateral dimension of 100 mm or more are difficult to dry out and should be designed assuming hazard class 3 conditions (see Table 11.5).

11.2.3 Structural properties

The structural properties of timber (compression, tension, bending and shear strength, elastic and shear moduli) vary with species, density, the growth rate, permissible defects and moisture content. These characteristics are derived by testing on a statistical basis.

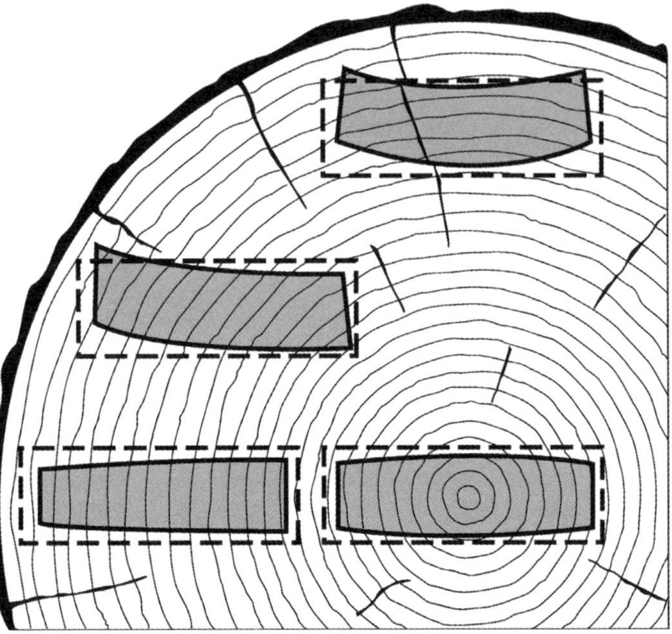

Figure 11.1 *Drying shrinkage of boards from different cross-sectional log areas.*

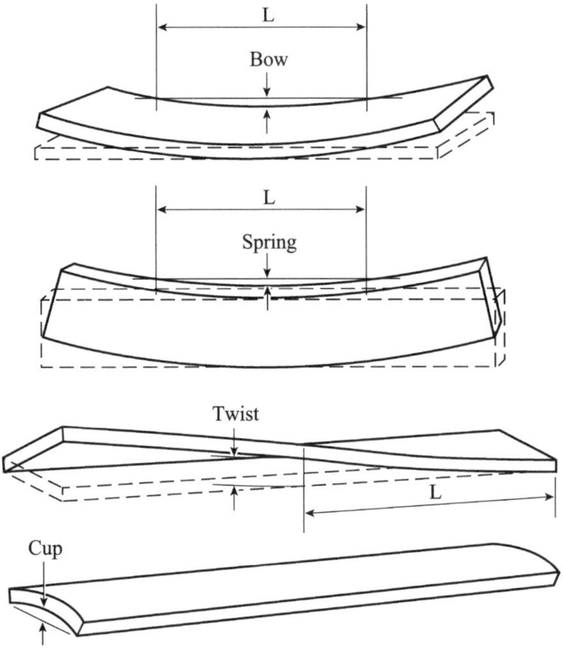

Figure 11.2 *Measurement of board distortions.*

Table 11.5 European exposure hazard classes (solid timber)

Hazard classes	Typical moisture content in timber	Biological agency			
		Fungi	Insects (local)	Termites borers	Marine
1 Above ground under cover, permanently dry exposure	< 20%		x	x	
2 Above ground under cover, high humidity, occasionally wet	> 20%	x	x	x	
3 Above ground not covered, frequent wet exposure	> 20%	x	x	x	
4 Contact with ground or fresh water, permanently wet	> 20%	x	x	x	
5 In salt water, permanently wet	> 20%	x	x	x	x

Timber strength exhibits significant load duration effects, being much stronger for short duration loading. For instance, the resistance against wind gust loading is up to 1.75 times the resistance against permanently sustained loading (BS 5268-2).

BS 5268-2 gives design stresses for certain individual species. It also gives groupings of species of similar characteristics into strength classes (see also BS 338) with common strength values for each class and for timbers graded by reference to specified levels of permissible defects.

Strength classes allow the designer flexibility in selecting different species or a mixture of species according to availability and cost. However, it should be emphasised that different species within the same strength class may differ in natural durability and treatability (see Section 2.1.1). It should also be realised that each strength class adopts the lowest of the strength characteristics within the group of species concerned and will incur a strength penalty for a particular species with actual strength values higher than the strength class minima.

Timber is graded structurally by visual or mechanical means into different strength grades or classes for particular species. Defects affecting strength properties are knot size, wane, fissures and splits, slope of grain, annual growth rings, resin and bark pockets, distortion, and rejection of pieces with compression wood, insect damage, and fungal decay (other than blue stain). Glulam is rated stronger than the grades of individual laminates because strength-weakening defects are randomly distributed across the laminates.

Mechanical stress grading consists of passing pieces of sawn or planed timber of known cross-section continuously through rollers, which apply a mid-span bending load inducing a fixed deflection between two support rollers over a fixed span. The dynamic modulus of elasticity over the span of the piece of specific size is derived from the load recorded. Based on the known correlation between modulus of elasticity and strength characteristics, the strength grade of the piece of timber is determined and marked. Mechanical stress grading can be carried out much faster and more efficiently than visual grading. However, a visual inspection is still required to reject pieces with insect damage, decay and excessive distortion.

11.2.4 Roundwood

Round poles undergo minimal processing and retain the concentric cross section of heartwood surrounded by sapwood. Treatable species of round poles can therefore be preservative treated effectively and more uniformly, usually with creosote under pressure penetrating into the sapwood. However, creosote is now regarded as harmful to operatives and the environment, and is being replaced with other types of preservative. Roundwood is also much stronger because of minimal cutting of wood fibres.

11.2.5 Glue laminated timber

The durability of glue laminated timber (glulam) is related to that of the timber laminates and the adhesive.

To meet the requirements of gluing, timber grades for the laminates are slightly different. For instance, wane is not permitted. Dimensional accuracy of the laminates, surface preparation and moisture content are also critical to develop adequate glue bonding.

Preservative treatment may be carried out on the laminates before lamination, or on the assembled glulam member.

Durability is generally enhanced in glulam due to the more distributed defects and the suppression of splitting by the restraining effect of adjacent laminates. Similarly, the strength properties of glulam are also enhanced compared with the properties of the laminates.

11.2.6 Plywood

The durability of plywood depends upon the timber species and the adhesive used, although it is generally slightly more durable than the timbers species from which the plywood is made. It should be emphasised that the term ‘exterior quality’ or ‘external grade’ frequently used to describe plywood quality relates only to the resistance of the glue lines to exterior conditions of use (WBP type adhesives) and not necessarily to the timber species.

Practically all structural plywoods are imported into the UK, mainly from Canada, USA, Finland and Sweden. All structural plywoods are bonded with WBP type adhesives. Strength and moduli vary with timber species and veneer grading, and ply lay-up.

Nowadays, structural plywoods are usually made from mixed species, with the outer veneers often of more durable species. It is important when specifying plywood to refer to the type, grade, thickness, number of plies and whether a sanded or unsanded finish is required.

Design stresses for imported structural plywoods are given in BS 5268 Part 2. The Standard requires manufacture to be subject to independent quality control by the appropriate inspection and certification body for the exporting country, namely:

- British Standards Institution (BSI);
- American Plywood Association;

- Timberco (TECO) USA;
- Council of Forest Industries (COFI), Canada;
- Technical Research Centre (VTT), Finland;
- National Testing Institute (Statens Provningstansalt), Sweden.

11.2.7 Reconstituted wood-based panels

The specification and durability ratings for various wood-based panel products are given in many recently published ENs. The BS versions of the ENs (which differ only in the UK National Annexes) are listed in the BSI Catalogue.

Chipboard

Chipboard is made from wood chips bonded with an adhesive under heat and pressure. In the 1960s there were some grades of chipboard with inadequate strength, excessive creep, poor moisture resistance and durability, which proved unsatisfactory for structural uses such as floor and roof boarding. Since then, much improved structural grades of chipboard for specified service conditions have been developed under BS and European Standards which are widely used, principally for floor boarding, roof boarding and sheathing in timber frame walls.

Chipboard consist of about 85% randomly orientated wood chips of different sizes, usually from coniferous timber, which are sprayed with adhesive before being bonded and heat pressed. The board is layered, with outer layers of more durable timber and larger amounts of adhesive than core layers. By varying the materials, lay-up and pressing cycle, boards of different physical properties for different service conditions are produced.

The common adhesives used are synthetic resins, such as urea formaldehyde (UF) for boards with limited moisture resistance, and melamine urea formaldehyde (MUF) for greater moisture resistance. The lay-up is usually of outer layers with higher resin content of around 11% and in the core of about 5%. They are made in thicknesses of about 6 mm to more than 40 mm. Densities range from 450 to 750 kg/m^3.

Oriented strand board (OSB)

OSB was developed in the 1970s to replace some plywood applications. The wood flakes, usually of aspen (USA) or Scots pine (Europe), are about 50 to 75 mm long, 20 to 30 mm wide and 0.5 mm thick. As its name indicates, the particles are oriented in the longitudinal direction of manufacture in surface and core layers, taking up some 50% of the board volume. The adhesive is normally phenol formaldehyde.

They are made in thicknesses of about 5 to 25 mm. The as-pressed board surface is slightly irregular, reflecting the matrix of the flakes, and is sanded smooth where required. Board density is around 550 to 750 kg/m^3.

Fibreboards

Fibreboards of various densities and structural properties are made from wood which is first reduced mechanically and by steaming under pressure to its basic fibres. Forestry thinnings, sawdust, and wood waste are used; but not bark. Steaming under pressure softens the natural thermoplastic bonding agent lignin, which provides all or most of the adhesive in fibreboards.

In the traditional wet production method, the fibres are mixed with hot water and additives introduced, then hot pressed. This squeezes out excess water through a lower wire mesh plate. By varying the pressure, boards of different thicknesses and densities are produced, ranging from softboard of up to 25 mm thick with density below 400 kg/m^3; mediumboard of 6 to 13 mm thick with density of 400 to 900 kg/m^3;and hardboard of 3 to 8 mm thick with density of 900 to 1100 kg/m^3. Tempered hardboard is a high density water repellent board in which the material is passed through a hot oil bath and sometimes incorporating additives such as phenol formaldehyde.

Fibreboards have a lower equilibrium moisture content than solid timber in the same hygrothermal condition, but are not suitable for external exposure.

Medium density fibreboard (MDF)

MDF is produced by a dry process. Very fine wood fibres are coated with up to 10% of resin, such as UF, MUF or MSI (isocyanate) and hot pressed to produce sheets with fine, smooth surfaces at thicknesses up to 40 mm and densities of 600 to 1000 kg/m^3.

MDF boards are used mainly in modern furniture and non-structural applications such as skirting boards and architraves. Boards can accept edge profiling and surface routing, revealing a smooth machined surface without grain appearance. They are also increasingly being used structurally where good surface appearance is also required such as in floorboarding and staircases. They are generally not suitable for outdoor exposure, although more moisture-resistant boards have recently become available.

Cement bonded particleboard

Cement bonded particleboard consists of timber shavings and thin chips with about one-third by weight of ordinary Portland cement as binder, pressed and heat cured. The chips are randomly orientated. The water/cement ratio is about 1:1. Board thicknesses are from 6 to 40 mm and density of around 1200 kg/m^3.

Cement bonded particleboards have excellent fire resistance, are durable in exterior use and highly resistant to termite attack.

General characteristics of reconstituted boards (except plywood)

Although reconstituted boards have a lower variability in characteristics because of their more even consistency, they have certain distinctly different characteristics from solid timber. These are influenced by the small timber particle sizes, the greater exposure of their end-grain when wetted, and by the properties of the adhesives as well as their bonding with the timber particles. Such boards are generally more visco-elastic than solid timber: they exhibit much higher creep deformation under stress and more pronounced duration-of-load effects, which increase with moisture absorption.

Water absorbed by the wood particles in the board leads to more pronounced thickness swelling, some of which is irreversible when re-dried. Moisture is absorbed mainly through the large surface areas of the end- and the side-grain of the wood particles, which are only partially sealed off by the binder. Thickness swelling of reconstituted boards may be several times the cross-grain swelling of solid timber. For these reasons, with the exception of cement bonded particleboard, reconstituted structural grade boards should only be used in hazard classes 1 and 2 service conditions of exposure, and only exceptionally in class 3.

In the board plane, longitudinal moisture movement is also greater than that of timber parallel to the grain. Thus when used in long lengths such as for floorboarding, moisture movement gaps should be provided.

Table 11.6 summarises the dimensional changes with relative humidity in timber and reconstituted boards.

11.2.8 Adhesives

Adhesives for wood are required to be at least as strong and as durable as the timber they connect.

In the liquid stage, fresh adhesive must have an 'open time' before the onset of hardening to allow enough time for the adhesive to be mixed, spread and pressed between the timber parts to be connected. It is usually possible to adjust the open time by varying the composition. The open time also varies with ambient temperature.

An adhesive's liquid consistency should be suitable for smooth and even spreading, and it must have adequate gap-filling properties to fill voids in surface irregularities of the connecting timber surfaces. A gap-filling adhesive is one which provides adequate bond strength for loadbearing use in glue lines up to 1.3 mm thick. It should also be sufficiently fluid for excess adhesive to be squeezed out when the connection is compressed before curing.

Structural glues are generally of the formaldehyde thermosetting types supplied as two-part products. The highest glue quality classified in BS 1204 Part 1 is Type WBP

Table 11.6 Dimensional change of certain timber and wood-base panels from 65% to 85% relative humidity

Material	Dimensional change (%)		
	Parallel to grain or board length	**Perpendicular to grain or board length**	**Through thickness**
Solid timber			
Douglas fir	< 0.10	0.80 (R)	1.00 (T)
Beech	< 0.10	1.20 (R)	2.00 (T)
Plywood			
Douglas fir	0.15	0.15	2.00
Chipboard			
Loadbearing – dry	0.25	0.25	7.00
– humid	0.20	0.20	4.00
Cement bonded particleboard	0.18	0.18	0.50
OSB	0.20	0.20	5.00
Fibreboards			
Hardboard	0.15	0.15	3.50
Tempered hardboard	0.15	0.15	3.50
MDF	0.20	0.20	3.50

R = radial direction
T = tangential direction
Source: Timber Engineering STEP1 A11/6 Table 3.

(Weather and Boil Proof) can make joints highly resistant to outdoor exposure, micro-organisms, cold and boiling water, steam and heat. Certain grades of phenol and resorcinol based adhesives satisfy this requirement. They set hard as rigid joints.

Some glues are susceptible to degradation by ultraviolet radiation in exposed exterior conditions.

It should be emphasised that structural gluing should be carried out in humidity and temperature controlled conditions. Gluing should be limited to side-grain joints and finger end-joints. Timber cannot be glued effectively on end-grain faces.

Thermoplastic one-part PVAC (polyvinyl acetate) adhesives have pronounced thermo-plastic behaviour when cured. Joints made with these adhesives exhibit creep under sustained load. They are therefore not generally suitable for structural applications, nor are they resistant to moisture and high temperature.

Timber to be joined must satisfy the following requirements:

a) *Gluability:* The timber species, including any preservative treatment, must be compatible with the adhesive and satisfy the required glued strength specified in the appropriate Standards. Most timbers are gluable but it should be noted that variability can occur within a species; for example, difficulties may be encountered due to particular samples being particularly absorbent or having a high resin content. BRE Digest 314 *Gluing wood successfully* gives guidance on making effective glued joints for a range of species. BRE Digest 340 *Choosing wood adhesives* advises on measures to be taken for gluing timber which has been treated with preservatives or flame retardants. Compatibility of the glue with timber species and additional treatment should be checked with the glue manufacturer.

b) *Dimensional stability:* Timber distortion by moisture variation should be controlled before gluing, to minimise interface tensile stresses being induced in the glue line. In glulam, it is generally necessary to kiln-dry the laminates to around 12% moisture content before planing to specified smoothness. In the case of other glued structural assemblies, timber up to 20% moisture content may be satisfactorily glued provided certain limits of maximum timber or board thickness, and moisture content differences of the pieces, are not exceeded.

c) *Surface condition:* The surface to be glued must be clean, smooth, freshly and lightly sanded. The best bond is achieved on freshly planed surfaces.

11.2.9 Moisture content

The strength, stiffness and creep properties of timber and timber-based products are affected by their in-service moisture content. Timber attains different equilibrium moisture contents when exposed to different environments of temperature and humidity. The equilibrium moisture contents for wood-based panel products are generally slightly lower than for solid timber in the same environment.

For the purpose of assigning strength values under specified environmental conditions, BS 5268-2 and Eurocode 5 define the following service classes:

- Service class 1: Characterised by a moisture content in the materials corresponding to a temperature of 20°C and the relative humidity of the surrounding air only exceeding 65% for a few weeks per year. For most softwoods, the average moisture content will not exceed 12%.

- Service class 2: Characterised by a moisture content in the materials corresponding to a temperature of 20°C and the relative humidity of the surrounding air only exceeding 85% for a few weeks per year. For most softwoods the average moisture content will not exceed 20%.
- Service class 3: Climatic conditions leading to higher moisture contents than in service class 2.

Strength values are generally assumed to be the same for solid timber and plywood in service classes 1 and 2, while for class 3 the strength values are reduced by factors of from 0.6 to 0.9 depending on the type of stress or moduli.

For reconstituted wood-based panels, strength values for service class 2 are about 70% to 80% of those for service class 1. Due to enhanced creep and irreversible dimensional changes at high moisture contents, the use of wood-based panel in service class 3 is not recommended.

To reduce movement and creep under load, and maintain the tightness of mechanical fasteners in structural connections, the moisture content of timber and wood-based panels when installed should be close to that likely to be attained in service conditions. Table 11.7 gives examples of timber moisture contents in buildings.

Timber when sawn from the green log has a high moisture content (about 30%) and, apart from external uses, should be dried before grading and installation. Timber treated with a water-borne preservative should be dried for use in service class 1. To minimise fissures and distortion, timber should be installed with a moisture content as close as possible to the relevant service class condition.

Timber may be dried naturally after sawing by prolonged airing under cover with individual planks separated for ventilation (sticked), until it dries down to the required moisture content. Too rapid a drying rate may exacerbate the development of fissures and distortion. However, the outdoor ambient temperature and humidity may not be sufficient or may take too long to induce natural drying, particularly to moisture levels for timber to be used in service class 1; thus kiln drying is normally employed.

Kiln drying is carried out on parcels of wet timber in closed kilns where air of controlled temperature and humidity is passed through the sticked parcels for specified

Table 11.7 Moisture content of timber related to service class

Service class	Examples of use in building	Average moisture content likely to be attained (%)	Moisture content which should not be exceeded in individual pieces when installed
3	External, fully exposed	20 or more	–
2	Covered and generally unheated	18	24
2	Covered and generally unheated	15	20
1	Internal, continuously heated	12	20

Source: BS 5268 Part 2: 2002.

durations, the schedules or programmes being dependent on the timber species, shape, and initial and final moisture contents. The drying rate is controlled to minimise the occurrence of fissures and distortion; ideally, strength grading should be carried out after drying. Moisture content can be determined by means of electric resistance probe meters, which are convenient for site measurement, or by drying samples in a heated oven and measuring weight loss.

Dried timber should be parcelled, stored, transported and installed with protection against re-wetting and against developing fungal growth.

In the case of wood-based panels, particularly heat-pressed reconstituted products, they generally have a relatively low moisture content at the time of manufacture and usually require conditioning by ventilating them naturally to increase their moisture content before installation. This reduces the expansion with any subsequent uptake of moisture.

11.2.10 Characteristic strength values

Characteristic strength and moduli values for individual species are defined in Eurocode 5 as the population 5-percentile value of results of load testing on small specimens, in accordance with the standard methods of test EN408 and EN1193 and derived in accordance with EN384.

Strength grading for a particular species and permissible defects is carried out according to EN318 and EN519 for visual and machine grading, respectively.

It should be noted that different climatic conditions may affect the rate of growth and strength properties of the same species of tree.

Characteristic values for wood-based panels are determined in accordance with BS EN 12369. Characteristic values are used directly in the limit state design format of Eurocode 5, modified by such other factors as service class, load duration, size effects and partial safety factors for timber. In BS 5268 Part 2, which retains the traditional permissible stress format of design, similar characteristic values are expressed as grade stresses, which incorporate global safety factors.

11.2.11 Time dependent effects

The strength and moduli of timber are markedly influenced by duration of load. Eurocode 5 divides durations of load into 5 classes, as given in Table 11.8.

Table 11.8 Load duration classes

Load duration class	Order of accumulated duration of characteristic load	Example
Permanent	> 20 years	self weight
Long term	6 months – 10 years	storage
Medium term	1 week – 6 months	imposed load
Short term	< 1 week	snow*, wind
Instantaneous		accidental load

*Where heavy snow persists for prolonged periods, part of the load should be regarded as medium-term.

The following are examples of modification factors given in BS ENV 1995-1-1 for time dependent effects on strength and deformation:

- Modification factors (multipliers) k_{mod} for the effect of duration of load on strength in service class 1 for solid and laminated timber, and plywood, ranging in value from 0.60 to 1.10 for permanent to instantaneous load durations, and from 0.40 to 1.1 for certain structural grades of particleboard.
- Modification factors k_{def} to allow for the effect of creep on the final deformation u_{fin} in relation to the instantaneous deformation, expressed by the equation $u_{fin} = u_{inst}(1+k_{def})$ in service class 1 for solid and laminated timber, ranging in value from 0.60 to 0.00 for permanent to instantaneous load durations, and from 1.50 to 0.00 for certain structural grades of higher grades of particleboard.

Values of k_{mod} and k_{def} for other service classes, load durations and other wood-based panels are given in Eurocode 5.

11.2.12 Size effects in solid and laminated timber

The bending, tensile and compressive strengths and the modulus of elasticity of solid and laminated timber increase with reducing section size. For example, BS 5268 Part 2 advises that the design stresses for beams should be increased by a factor of 1.17 when the depth is reduced from 300 mm to 72 mm.

11.3 Factors affecting durability

11.3.1 Exposure hazards

The hazardous conditions of exposure to decay and insect attack in European conditions are defined as hazard classes in BS EN 335 Part 1, summarised for solid timber in Table 5.

Specific guidance on the likely effects of moisture and biological agencies on solid timber and wood-based panels are given in BS EN 335 Part 2.

Wood-based panels differ in their composition, manufacture and attain different equilibrium moisture contents in comparison with solid timber. Consequently their hazard classes are modified, as given in BS EN 335. With the exception of cement bonded particleboards, no known suitable wood-based panel products are manufactured for use in hazard classes 4 and 5.

BS EN 335 enables the end use of timber products to be assigned to one of the five European hazard classes, subject to regional differences which may modify the hazard class for local conditions. It facilitates the choice of timber species or panel product and any preservative treatment, for the appropriate hazard class of exposure in service.

11.3.2 Fungal attack

Timber with a moisture content above about 22% is susceptible to attack by various types of fungi in the presence of oxygen. Fungal attack can cause decay, weakening and staining. Some fungi species may remain dormant during dry periods and attack the timber again when re-wetted.

Wood-rotting fungi develop from minute airborne spores which germinate if they land on damp timber and produce thread-like hyphae which collectively form a mycelium. The hyphae penetrate the wood, breaking down and feeding on the cell walls. The mycelium eventually produces a fruit body, which then release spores to the atmosphere.

In buildings, fungal decay of timber is described as either wet rot or dry rot, the latter having also the ability to grow through masonry.

Wet rot fungi are subdivided into white and brown rots (see BRE Digest 345). White rot destroys both the cellulose and lignin components of the timber causing the timber to be lighter in colour and have a lint-like texture. Brown rot attacks the cellulose and leaves the lignin as a brown residue, causing the timber to crack along and across the grain; very decayed wood will crumble to dust when dry.

Dry rots belong to the group of brown rots. Refer to BRE Digest 299.

Numerous species of fungi exist worldwide, varying in vigour, timber preferences and susceptibility to control by preservatives. Features used in identification include type of fruit body (the most distinctive feature, colour and appearance of the hyphae and mycelium), and form of decay. Identification is essential for determining the correct remedial treatment.

Blue-stain or blue-fungi and mould-fungi cause, respectively, deep seated and superficial discoloration of damp timber of more than 25% moisture content, feeding on the cell contents and stored food and thus confined to sapwood. They do not cause loss of strength but spoil the appearance of the timber. Even if arrested by drying, they can revive when re-wetted and disrupt paint finishes. To avoid this risk, boron-diffusion preservative treated timber, in which the boron penetrates the whole tree log after felling, may be used indoors for visible timber. However, boron treatment leaches in water and the timber should therefore be protected before, during and after installation.

11.3.3 Insect attack

Most wood-boring insects are beetles. On externally exposed timbers rotted by fungal action, certain bees or wasps will attack the softened timber. In marine waters, floated logs or timber used in structures may be attacked by marine borers which are beetles or molluscs. Insect attack is characterised by bored holes of various sizes, depths and concentration. Some of the more common wood-boring insects are described below.

House Longhorn beetle (*Hylotrupes bajulus*) infestation, in softwoods only, has been found in certain areas of SE England and UK Building Regulations have required pre-treatment of roof timber in new buildings in these areas. The Forest Longhorn beetle (*Cerambycids*) attacks the bark by scoring and tunnelling into the sapwood of the freshly felled logs and causes damage to timber throughout the world.

Common furniture beetle (*Anobium punctatum*), commonly called ‘woodworm’, can attack hardwoods and softwoods in buildings in damp conditions such as damp ground floors or roof voids. Infestation tends to die out when the building is dried and heated. Tropical hardwoods, and wood-based structural panel products are generally immune from attack.

Death watch beetle (*Xestobium rufovillosum*) is common in Southern England, rare in Northern England and unknown in Scotland. Infestation is usually found in large size hardwoods such as damp oak and elm where some wood rot has already occurred. Attack continues in dry timber where rot has been arrested and may carry over into softwoods which are adjacent. For these reasons the insect must be considered a primary pest.

Marine borers include wood-boring molluscs, related to oysters and clams, such as the genera *Teredo, Bankia* and *Martesia,* the first two of which are of worm-like appearance and commonly called shipworm. Crustacean borers, related to shrimps and lobsters, include *Limnoria, Chelura* and *Sphaeroma*, which are much smaller than molluscs. Infestation is more severe in warmer saline waters. Marine borers can only exist in a salinity exceeding 5–9 parts per 1000 and will survive out of saline water for up to 40 days. No species of timber is naturally immune from marine borers, but commercially available species with resistant heartwood include:

Basralocus (Angelique) *(Dicorynia guianensis)*
Belian *(Eusideroxylon zwageri)*
Brush box *(Tristania conferta)*
Ekki *(Lophira alata)*
Greenheart *(Ocotea radiaei)*
Manbarklak *(Eschweilera longipes)*
Okan *(Cylicodiscus gabunensis)*
Opepe *(Nauclea diderrichii)*
Pyinkado *(Xylia xylocarpa)*
Red louro *(Ocotea rubra)*
Southern blue gum *(Eucalyptus globulus)*
Turpentine *(Syncarpia laurifolia)*

For marine work, sapwood should be excluded by taking it off from larger logs.

The type of insect-borer may be identified visually by the type and condition of the timber, size and shape of holes, colour, shape of bore dust, and size and shape of tunnels within the timber, and if necessary by further identification of the insect or borer. For further information see BRE Digest 307 and PRL Technical Note 59.

11.3.4 Corrosion of metal fasteners by preservative-treated timber

Zinc-coated steel, stainless steel and aluminium are the usual materials for metal fasteners such as nails, screws, bolts, shear plates, cover plates, and tooth-plate connectors.

It has been found that water-borne preservatives in timber based on copper-chrome-arsenic (CCA) compositions can increase the corrosion of certain grades of aluminium alloy and zinc-coated mild steel where the relative humidity is maintained close to 100% or where free water is present. Furthermore, 'fixation' of the preservative chemicals does not prevent corrosion.

In interior conditions of use, the risk of corrosion of fasteners embedded in timber (such as trussed rafters with galvanised tooth-plate connectors) can be alleviated by:

- ensuring that the moisture content of CCA treated timber is adequately controlled from fabrication to construction and end use;
- preventing conditions from occurring which could lead to high moisture levels (more than 18%) such as by adequate ventilation of the structure.
- where high moisture levels may persist in internal or external conditions, using fasteners of an austenitic grade of stainless steel;
- using steel fasteners with a heavy coating of epoxy resin, which have been found to be suitable for chlorinated environments such as swimming pool enclosures.

Advice on dealing with hazards from certain chemical and high temperature environments is given in references such as *Structural use of wood in adverse environments* (Meyer and Kellogg 1982) and BRE Digest 302.

11.4 Maximising durability

11.4.1 Protection from exposure

Protecting timber from exposure to severe thermal and moisture cycling will prolong its durability. Evidence of extended durability is widely seen in roofs of old churches and other public buildings, as well as in bridges. Small hardwood roundwood piles with reinforced concrete pile-caps supporting low- and medium-rise buildings are widely used in SE Asia; their durability derives from the favourable conditions of constant immersion in ground water and the absence of exposure to erosion and moisture cycling.

Timber frame houses with load bearing studs and trussed rafters are now commonly accepted in the UK to mortgagors and insurers (such as the National House-Building Council) as an alternative to traditional masonry construction. They are detailed to remain dry in service, with sole plates in proximity to the ground being preservative treated. Protection from outdoor exposure is effected by cladding them with roofing slates or tiles and non-load-bearing brickwork.

Exposed unprotected timber end-grain is particularly prone to splitting and decay when exposed to excessive drying and wetting by liquid water. Upper ends of vertical timber exposed externally should be capped with or sealed with resin or epoxy coatings. The lower ends of external timber such as column bases should be raised off the ground by not less than 150 mm to minimise splash back, by means of bolt-on steel shoes (Fig. 11.3). The end-grain should be protected against capillarity by coating with a suitable resin, rubber or bitumen compound. Lapped timber faces and bolt holes should be similarly protected.

The protection of exposed side-grain timber against erosion and moisture changes may be effected with good detailing to minimise water penetration (Figs 11.4 and 11.5), or by applying a suitable coating. Where protection is provided by a protective screen, the timber face should be ventilated.

Other examples of high moisture content situations include:

- moisture penetration from warm, damp air, particularly in unventilated spaces;

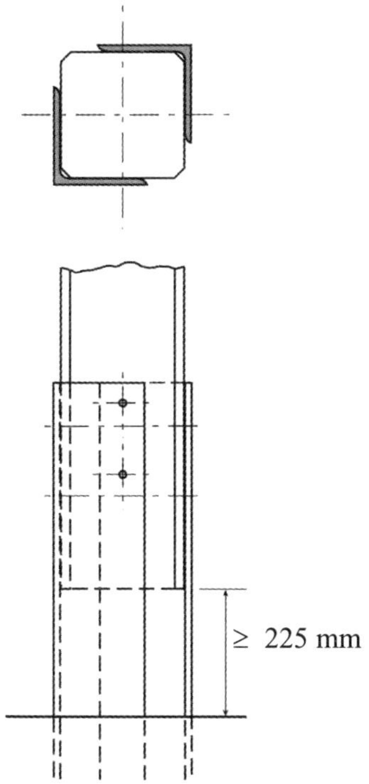

Figure 11.3 *Example of bolted exterior column base.*

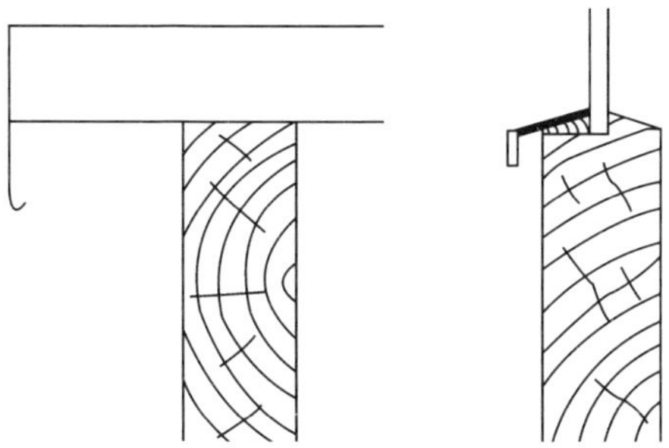

Figure 11.4 *Edge beams detailing for limited penetration of water (Reproduced by permission from* Timber Engineering STEP 1).

- undrained ledges and pockets between timber and metal plates where water can collect for prolonged periods;
- using wet timber in construction which cannot dry out quickly.

11.4.2 Preservative treatment

Preservative treatment should be used sparingly, as it brings adverse environmental consequences by the use of toxic substances. Ideally, durable species of wood should be used (see Table 11.4) without the need for treatment, but this is not always possible.

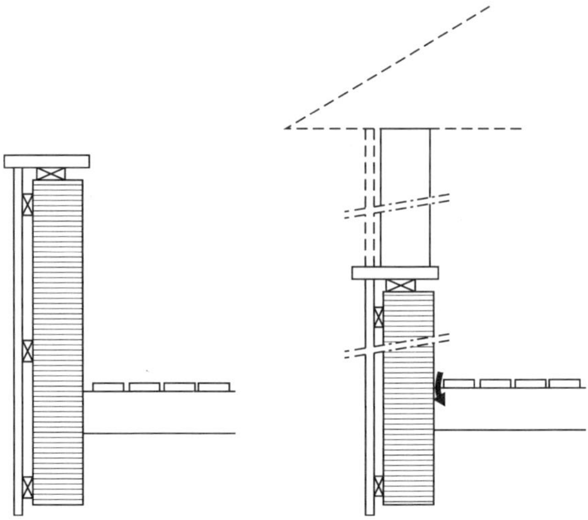

Figure 11.5 *Protection of bridge timber (Reproduced by permission from* Timber Engineering STEP 1).

Fungi and insects are the two main biological agents responsible for timber degradation in service (see Sections 3.1 and 3.2), and suitable chemical preservative treatments can be used to avoid or reduce degradation (see BS 1282 and BS EN 351 Part 1).

It should also be borne in mind that pressure-impregnated preservatives only penetrate the timber by a few millimetres. Fixing holes and cuts along or at the ends of timber, made after preservative treatment, will expose the untreated timber. Liberal brush application of preservative at these exposed points is helpful but will not match the original pressure impregnated preservative, nor can this be always achieved during construction when exposed faces or ends become inaccessible or obstructed. In such cases, periodical renewal of preservative and remedial work will be necessary for continued protection.

The only preservative treated wood-based panel product available from some producers is plywood. However, it is generally pressure-treated after manufacture so that only the outer veneers receive the preservative, the glue lines forming a barrier to the penetration of preservative into inner veneers.

Advice on insecticidal treatments for eradicating infestations by specific wood-boring insects in existing buildings is given in BRE Digests 307 and 327. Treatment types include the application of solvent based liquid, emulsion, paste, smoke, vapour, gas fumigation and heat sterilisation. Safety and environmental aspects of the use of certain treatment chemicals during application, and the type of building occupancy, should be considered when choosing the type of treatment (see BRE Digest 371 and BWPDA code of practice).

11.4.3 Maintenance and repair

As for structures of any material, including timber, periodic inspection and timely repair is essential.

Decayed parts may be cut out and replaced by timber implants with compatible properties, which may be glued or mechanically fastened. Only certain types of adhesives are suitable for external conditions and must be applied under strictly controlled dry conditions and accurate matching surfaces.

Where decayed timber is replaced and re-connected with a combination of gluing and mechanical fasteners, only the glue or mechanical fastener connection strength should be assumed to be acting effectively but not both, since they have different load-displacement characteristics. Where the loss of strength of the decayed timber is critical, the structural member should be shored before cutting and repair.

Timber decayed to a limited extent may be cut out and replaced by appropriate resin based cement mortar fillers.

Bolted connections should be tightened periodically to avoid loosening due to timber shrinkage and the initial creep of the timber, particularly after a new timber structure has had time to settle down to an equilibrium moisture content.

Bibliography

BS EN 335-1:1992, *Hazard classes of wood and wood-based products against biological attack: classification of hazard classes,* London: BSI.

BS EN 335-2:1992, *Hazard classes of wood and wood-based products against biological attack: guide to the application of hazard classes to solid wood*, London: BSI.

BS EN 318:1993, *Fibreboards. Determination of dimensional changes associated with changes in relative humidity*, London: BSI.

BS 1204:1993, *Specification for synthetic resin adhesives (phenolic and aminoplastic) for wood*, London: BSI.

BS EN 350-1:1994, *Guide to the principles of testing and classification of natural durability of wood*, London: BSI.

BS EN 350-2:1994, *Guide to natural durability and treatability of selected wood species of importance in Europe,* London: BSI.

BS EN 336:1995, *Structural timber. Conifer and poplar. Sizes. Permissible deviations*, London: BSI.

BS EN 338:1995, *Structural timber. Strength classes*, London: BSI.

BS EN 384:1995, *Structural timber. Determination of characteristic values of mechanical properties and density*, London: BSI.

BS EN 408:1995, *Timber structures. Structural timber and glued laminated timber. Determination of some physical and mechanical properties,* London: BSI.

BS EN 519:1995, *Structural timber. Grading. Requirements for mechanical strength-graded timber and grading machines*, London: BSI.

BS EN 335-3:1996, *Hazard classes of wood and wood-based products against biological attack: Application to wood-based materials*, London: BSI.

BS EN 351-1:1996, *Durability of wood and wood-based products. Preservative treated solid wood. Part 1 – Classification of preservative penetration and retention*, London: BSI.

BS 4978:1996, *Specification for visual strength grading of softwood*, London: BSI.

BS 6100:1998, *Glossary of building and civil engineering terms. Part 4.1: 1992 (1998): Characteristics and properties of timber and wood-based panel products*, London: BSI.

BS EN 1193:1998, *Timber structures. Structural timber and glued laminated timber. Determination of shear strength and mechanical properties perpendicular to the grain*, London: BSI.

BS EN 1194:1999, *Timber structures. Glued laminated timber – strength classes and determination of characteristic values*, London: BSI.

BS 1282:1999, *Wood preservatives – Guidance on choice, use and application*, London: BSI.

BS EN 13556: 2003, *Round and sawn timber – Nomenclature of timbers used in Europe*, London: BSI.

BS EN 12369-1:2001, *Wood-based panels – Characteristic values for structural design*, London: BSI.

BS 5268:2002, *Part 2: Structural use of timber – Code of practice for permissible stress design, materials and* workmanship, London: BSI.

BS EN 1995-1-1:2004, *Eurocode 5: Design of timber structures – Part 1-1. General – Common rules and rules for buildings*, London: BSI.

British Wood Preserving and Damp-proofing Association (1998) *Safety precautions for the design and operation of timber treatment installations,* BWPDA Code of practice, London: BWPDA.

Building Research Establishment (1985) 'Building overseas in warm climates', *BRE Digest 302*, Watford, UK: BRE.

Building Research Establishment (1986) 'Identifying damage by wood-boring insects', *BRE Digest 307*, Watford, UK: BRE.

Building Research Establishment (1986) 'Gluing wood successfully' *BRE Digest 314*, Watford, UK: BRE.

Building Research Establishment (1989) 'Choosing wood adhesives' *BRE Digest 340*, Watford, UK: BRE.

Building Research Establishment (1989) 'Wet rots: recognition and control', *BRE Digest 345*, Watford, UK: BRE.

Building Research Establishment (1992) 'Remedial wood preservatives – using them safely', *BRE Digest 371*, Watford, UK: BRE.

Building Research Establishment (1993) 'Dry rot: its recognition and control', *BRE Digest 299*, Watford, UK: BRE.

Building Research Establishment (1993) 'Insecticidal treatments against wood-boring insects', *BRE Digest 327*, Watford, UK: BRE.

Building Research Establishment (1998) 'Timbers: their natural durability and resistance to preservative treatment', *BRE Digest 429*, Watford, UK: BRE.

Meyer, R.W. and Kellogg, R.M. (Eds) (1982) *Structural Uses of Wood in Adverse Environments,* New York: Van Nostrand Reinhold.

Prince Risborough Laboratory (1972) *Marine borers and methods of preserving timber against their attack,* Technical Note 59, Building Research Establishment, Watford, UK.

STEP 1, *Timber Engineering*, Edited by H. J. Blaas, P. Aune, B.S. Choo, R. Gorlacher, D.R. Griffiths, B.O. Hilson, P. Racher, G. Street, Pal Centum Hout, The Netherlands, 1995.

Appendix: Glossary of selected timber terms

Blue stain: Blue discoloration of timber, confined to sapwood, principally in green timber, resulting from the growth of certain fungi that invade and digest the contents of the wood cells, but which do not rot the timber.

Equilibrium moisture content: Moisture content at which timber and wood based panel products neither gain nor lose moisture when exposed to given constant conditions of temperature and humidity.

Fibre saturation moisture content (fibre saturation point or fsp): Hypothetical moisture content of timber at which all free moisture has been removed but at which the cell walls are still saturated with bound moisture.

Fissure: A term covering all splits, checks and shakes that result from a longitudinal separation of the wood fibres due to shrinkage caused by the timber drying and may appear on any face, edge or end of the timber.

Heartwood: Inner zone of the wood that, in the living tree, has ceased to contain living cells and reserve materials (e.g. starch). Not to be confused with heart. *Note:* In commercial practice it is usual to restrict the term to the darker coloured wood that is visually distinct from the sapwood.

Sapwood: Outer zone of wood that, in the growing tree, contains living cells and reserve material (e.g. starch) generally lighter in colour than the heartwood though not always clearly differentiated.

Slope of grain: Angle between the direction of the grain and the axis of a piece.

Wane: Original rounded surface of a tree remaining, with or without bark, on any face or edge of square sawn timber.

Wood borer: Wood-boring invertebrate, chiefly insects and their larvae but also marine borers.

Worm holes: Hole, tunnel or channel in timber caused by a wood borer.

12 Durability of FRP Composites for Civil Infrastructure Applications

Professor B Benmokrane, P Wang, T Pavate and M Robert

12.1 Introduction

This chapter deals with the durability of fibre reinforced polymer composite materials used in civil infrastructure. A critical and comprehensive review on the scope of application of fibre reinforced polymer (FRP) in concrete structures, and concerns about the durability of FRP composites, has been presented.

Electrochemical corrosion of steel in various concrete structures has become a serious concern all over the world. Fig. 12.1 shows a typical infrastructure corrosion problem. In Canada, it is estimated that the required repair cost for parking garages alone is in the range of $6 billion (Benmokrane et al. 2001) and in US, for all concrete structures, the rehabilitation cost is estimated to exceed $250 billion per year (Chin 1996). Therefore, it would be logical and highly appropriate to be concerned and to prevent such corrosion-related deterioration, which could in future save the billions of dollars required to deal with steel corrosion problems. A very recent step in that direction, which has received global attention, is the use of FRP composites as a cost effective alternative to traditional construction materials and techniques. FRP materials are composed of synthetic or organic fibres impregnated with resin matrix. They have a successful history of application in the aerospace, marine, space, and chemical industries. The most commonly available FRPs, which can be used for civil infrastructure, are glass (GFRP), carbon (CFRP) and aramid (AFRP). These FRP

Figure 12.1 *Electrochemical corrosion of steel in concrete structures.*

materials offer excellent mechanical properties: weight saving, magnetic neutrality, damage tolerance, freedom from electrochemical corrosion, and real-time monitoring. Recent experimental and analytical research, worldwide, has led to substantial progress in their use in civil infrastructure ranging from internal reinforcement (e.g. reinforcing bars and tendons for concrete reinforcement), external reinforcements (e.g. wraps for seismic retrofit of columns and externally bonded reinforcement for strengthening of walls, beams, and slabs), composites bridge decks, and composites structural systems (e.g. FRP structural shapes, vehicular and pedestrian bridge systems, piling products for marine waterfront structures) (Saadatmanesh and Ehsani 1998; Benmokrane and Rahman 1998; Benmokrane and El-Salakawy 2002; Dolan, Rizkalla and Nanni 1999).

The principle problems confronted by designers and engineers in using FRP composite materials are their durability qualities, and long-term behaviour in general, and particularly in severe environmental conditions. In contrast to the plethora of work originating from the aerospace, chemistry and military industries on the durability of FRP composites, there is a small, but growing, body of research, which is specifically concerned with the durability of FRP composites in civil infrastructure applications. This chapter summarises recent research on various aspects of the durability of FRP composites for use in civil infrastructure. The objectives of this chapter are twofold: firstly, to review recent durability studies on FRP composites, and secondly, to highlight technical issues related to the long-term performance of various structural forms, which include FRP internally reinforced concrete structures, FRP externally reinforced concrete structures, FRP wrapped concrete columns and FRP structural shapes. The main focus is on the effects of environmental conditions such as moisture/solution, alkaline, fire, UV, thermal, and mechanical conditions i.e. fatigue, creep.

12.2 Comparison of properties: FRP versus traditional building materials

Advanced FRP composites are made from different constituent materials, i.e. fibre, resins, interface, fillers, and additives. Higher modulus fibres contribute to the mechanical strength of the FRP, whereas the matrix helps to transfer or distribute the stress from one fibre to another, through interface shear resistance, and to improve the durability of the fibre against environmental and mechanical damage. The interface between the fibre and the matrix is known to significantly affect the performance of FRP composites. In addition to these three basic components (fibres, resins, and interface), the fillers serve to reduce cost and shrinkage. The additives help to improve the mechanical and physical properties of the composites as well as the workability.

12.2.1 Fibres

Fibres control the strength and stiffness characteristics of the FRP composite. There can be one or, in the case of hybrids, more than one type of fibre, in the composite. A great deal of research and development has been conducted with fibres on the effects of fibre types, volume fraction, architecture, and orientations. The fibre generally occupies 30–70% of the matrix volume of the composite (Fig. 12.2). The fibres can be chopped, woven, stitched, and/or braided. The most common types of fibres, used in advanced composites for civil infrastructure applications, are glass, aramid and carbon.

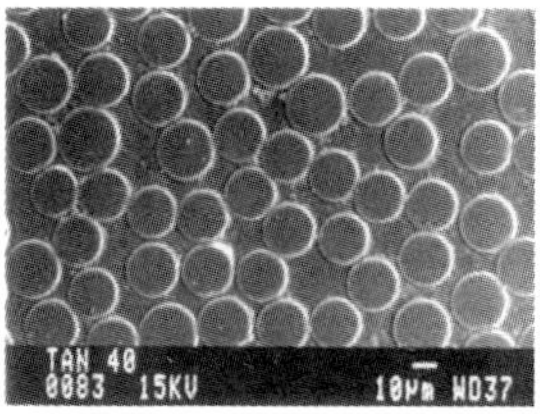

Figure 12.2 *Microstructure of FRP bar.*

Glass fibre reinforced composite is the oldest and most widely used due to its low cost. However, glass fibre has a low E-modulus, and GFRP is known to have a much lower stiffness than that of steel. The performance of glass fibre is prone to deterioration due to alkaline attack, and has become a particularly serious concern where it is used in highly alkaline concrete. However, high-alkali-resistant glass fibres have been developed. Carbon fibre has a high E-modulus and CFRP shows a higher stiffness, which can be comparable to, or even exceed, that of steel. Aramid or Kevlar (trade name) is the least used fibre for concrete structures. It generally offers higher strength and stiffness compared to glass fibre, but cannot match that of carbon fibre.

12.2.2 Resin matrix

Several types of polymer resins are used to impregnate the fibres in FRP composite. Polymer resins can be thermosetting or thermoplastic. Thermosetting matrix resins provide better thermal stability, chemical resistance, reduced creep and stress relaxation than thermoplastic resins. Thermoplastic resins impart better impact strength, fracture resistance and storage life to the FRP composite. The majority of FRP composites to date have been made with thermosetting polymer. Although many types of polymer resins are available, the most common resins used in the construction industry are unsaturated polyesters, epoxies, and vinyl esters. Resin with high molecular weight, and with an additive for minimising moisture absorption, does offer resistance to moisture ingress. Therefore, resins with higher molecular weight and a minimum number of end groups, that modify viscosity by using monobasic acids, improve the durability of a composite.

12.2.3 Interface

The interface is a region in the composite that lies between the resin matrix and its fibres. This region has gradients in physical properties that significantly influence the durability of FRP composites, such as the hygrothermal properties, stress corrosion, and fatigue behaviour. By applying suitable sizing, the performance of FRP composites can be improved. To accurately understand this improvement in a quantitative way, the multilayer nature of the interface/interphase has been taken into account in recent studies (e.g. Verghese 1999).

12.2.4 Filler/additives

As resins are very expensive, it is not cost effective to fill up the voids in a composite matrix purely with resins. Fillers are added to the resin matrix for controlling cost and improving mechanical and chemical properties. The three major types of filler are calcium carbonate, kaolin, and alumina trihydrate. Other common fillers include mica, feldspar, wollastonite, silica, talc, and glass. When one or more fillers are added to a properly formulated composite system, the performances, which include fire, chemical resistance, high mechanical strength, and low shrinkage, can be improved. Other improvements include toughness as well as high fatigue and creep resistance.

A variety of additives are used in the composites to improve the material properties, aesthetics, manufacturing process, and performance. The additives can be divided into three groups; catalysts, promoters and inhibitors; colouring dyes and releasing agents. Their roles are as simple as their names imply, and they need no further discussion.

12.2.5 Manufacturing process

Manufacturing process variables, such as heating and cooling rates, cure temperature and time, have a significant effect on cure performance and hence the chemistry of the composite. Void content, in composites, is a function of the compaction and the consolidation that takes place during cure. A higher degree of compaction serves to eliminate voids and non-wetted fibres in the composites, which could serve as potential stress concentrations and prevent future damage.

In general, unlike traditional construction material, due to the wide range of constituent material properties that exist in FRP composites, which include fibre/resin combinations, fibre to resin volume ratios, fillers, sizing chemistry, interface, additives, and manufacturing processes, an almost endless range of combinations can be achieved. Different combinations and interactions among these properties significantly affect the performance of FRP composites in general, and durability in particular.

12.3 Application scopes

The scope of applications of FRPs in concrete construction is very wide. In fact, the true potential of FRP is yet to be realised. From the basic application point of view, FRPs can be used in concrete in three basic forms:

- internal reinforcement for reinforced concrete structures;
- external reinforcement for strengthening or repairing existing deficient structures;
- FRP structural elements (e.g. beams, girder, and column) in concrete–FRP composite structures.

Fig. 12.3 shows typical varieties of FRP reinforcements for civil infrastructure applications. Fig. 12.4 illustrates GPRP reinforcing bars for use in concrete.

12.3.1 FRP internally reinforced concrete members

The prospect of FRP bars being used as internal reinforcement to concrete is now reasonably well known. Fig. 12.5 shows the construction of a bridge using FRP bars as internal reinforcement. Research on the behaviour of FRP internally reinforced concrete has drawn considerable global interest. Although the basic principles of the analysis of

Figure 12.3 *FRP reinforcements for civil infrastructure application.*

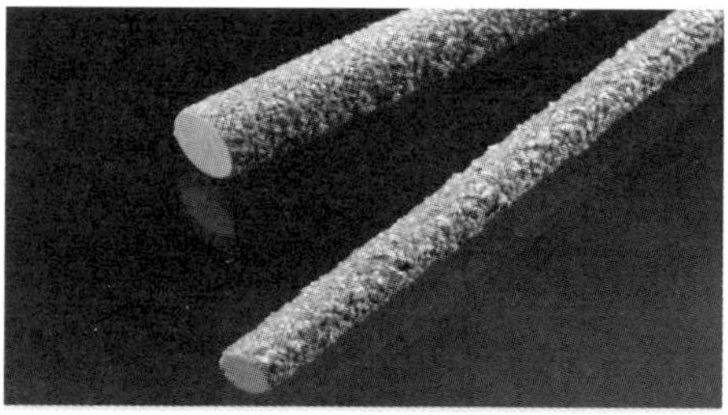

Figure 12.4 *GFRP reinforcing bars.*

Figure 12.5 *Concrete bridge deck slab reinforced with GFRP bars.*

conventional steel reinforced concrete structures remain valid for FRP reinforced concrete structures, some modifications are suggested by researchers, to take into account the fundamental FRP material properties (e.g. stiffness, bond, etc.), while predicting the various parameters of structural behaviour (e.g. deflection, crack width in concrete, etc). Apart from using internal FRP bars in concrete, pre- or post-tensioned FRP tendons can also be used as an effective concrete reinforcement. However, loss of prestress due to creep/relaxation or slip in the anchorage grip is a serious problem that must be considered. It is expected that more research results may become available on the structural behaviour of FRP bar reinforced concrete structures, which will lead to a better understanding and more comprehensive structural design guidelines.

FRP composites are also being investigated for use as internal reinforcement in concrete bridge decks. The deck is the part of a bridge that requires most maintenance, due to the deterioration of the wearing surface, corrosion of the steel, and degradation of the reinforcement. In addition, there is often a need to increase load-bearing capacity, due to increased load and traffic flow. With the successful experience of the

Figure 12.6 *Concrete columns reinforced with FRP sheets.*

application of FRP composites in other concrete structures such as concrete beams, columns and slabs (Benmokrane and Rahman 1998), logical attention is being given to develop FRP reinforced concrete deck systems. A number of field installations have been reported in the USA, Canada and elsewhere (Benmokrane and Rahman 1998).

12.3.2 FRP externally reinforced concrete members

FRP reinforcement, external to a concrete structure, can be effective in the form of external plate bonding or fibre wrapping. External plate bonding by adhesive is a well-established technique to strengthen or repair deficient reinforced concrete beams or slabs. Initially the method was developed with steel plates but, at present, CFRP plates are being widely used all over the world (Benmokrane and Rahman 1998). Recent research has been conducted on the flexural strengthening of concrete beams with prestressed FRP sheets and laminates. This study has involved the development of novel systems for anchoring and applying prestress to post-strengthened, reinforced and prestressed concrete beams. In comparison to non-prestressed bonded FRP, prestressing leads to a more efficient use of FRP reinforcement, since a given FRP strain becomes associated with a lower overall structural deformation in a prestressed beam (Garden, Hollaway and Thorne 1998).

The wrapping of concrete columns by resin impregnated FRP fabrics or straps to improve the strength and ductility is a technique used in seismic retrofitting of concrete structures. This method is an alternative to the steel jacket technique. The application of FRP external reinforcement has a long-term durability advantage, and a great potential to be used extensively to strengthen or upgrade ageing reinforced concrete infrastructure facilities. Fig. 12.6 shows the wrapping technique used for column rehabilitation.

12.3.3 FRP structural members

The use of FRP composites in load-bearing structural components, has been used in a number of projects within the last few years. Buildings with landing pads for helicopters,

Figure 12.7 *Parking garage concrete structural slabs reinforced with GFRP bars.*

pedestrian bridges, bridge decks and cooling towers are some of the examples documented (Benmokrane and El-Salakawy 2002; Dolan, Rizkalla and Nanni 1999). Other potential structural applications include cables, parking garages (Fig. 12.7), chemical plants and wastewater treatment facilities. Moreover, the concept of using FRP as a structurally integrated permanent formwork maximises the advantages of FRPs and concrete, simplifies the construction procedures, and reduces erection time. Additionally, concrete-filled FRP round tubes have been used as non-corrosive piles in marine environments to replace reinforced, prestressed, and concrete-filled steel tube piles (Dolan, Rizkalla and Nanni 1999). The same concept has also been proposed for bridge columns and piers in seismic zones (Benmokrane and El-Salakawy 2002).

Recent research has been conducted to fully develop FRP composite deck systems. Theoretical and laboratory studies have been carried out (Arockiasamy et al. 1999; Benmokrane and El-Salakawy 2002). However, current application of FRP composites in this area is limited to either laboratory research or demonstration projects.

12.4 Durability concerns

Although FRPs are resistant to electrochemical corrosion, which affects steel severely, the performance of FRPs may deteriorate due to environmental, physical or chemical conditions, leading to loss of strength and stiffness. The literature (Taerwe 1995; Saadatmanesh and Ehsani 1996; Saadatmanesh and Ehsani 1998; JCI 1997; Benmokrane and Rahman 1998; Benmokrane and El-Salakawy 2002; Dolan, Rizkalla and Nanni 1999) indicates that the performance of FRPs deteriorates due to certain physical (e.g. cyclic or sustained loading, moisture diffusion, extreme temperature variation) or chemical (e.g. alkalinity) exposure. The degree of damage/deterioration depends on a variety of factors such as the type and volume of fibres, and resin matrix, the exposed environment, and the manufacturing process. Furthermore, due to the addition of FRP composites to concrete structures, the durability performance of FRP reinforced concrete structures becomes more complex due to the combined effect of FRP composites, interface, concrete, and various environmental and mechanical conditions.

Hence, the whole process of assessing the durability of FRPs, in association with concrete structures, is a very complex and multidimensional task. The necessity of a broad assessment of the durability of FRP composites in association with concrete, and as an individual material, cannot be over emphasised. In the following sections, an attempt has been made to look at various durability aspects of FRP composites, FRP reinforced concrete members, and the bond between FRP and concrete. To focus on the complexity of the analysis, attention is restricted to the following durability influence factors:

1. fluids (moisture; chemical solutions);
2. alkalinity;
3. freeze–thaw;
4. ultraviolet radiation;
5. fire;
6. creep/relaxation;
7. fatigue.

12.5 Durability studies on FRP composites

12.5.1 Fluids

The effect of fluids on the performance of FRP composites has been one of the most widely studied subjects during the past decades. The majority of theoretical investigations have been focused on the sorption processes and reporting weight-gain data by using linear or non-linear Fickian's law. In general, the sorption behaviour of fluid into FRP composites depends upon the types of fluid (e.g. water, acid, or base), fluid concentration, temperature, externally applied stress, constituent materials (fibre, resin matrix, and interface), manufacture process, and the state of the material (e.g. extent of damage). A comprehensive review of this can be found in Weitsman and Elahi (2000). Weitsman (1998) identified five different types of fluid sorption behaviour based on weight gain versus fluid sorption time plot:

1. linear Fickian behaviour;
2. 'pseudo-Fickian', where the material does not attain an equilibrium moisture content;
3. 'two-stage diffusion' behaviour, related to changes in environmental conditions, e.g. temperature, applied load, or relative humidity;
4. rapid moisture weight gain, resulting from the large deformation or damage to the material, e.g. fibre–matrix debonding and matrix cracking, which is often irreversible;
5. an irreversible process as a result of leaching out of the material following chemical or physical breakdown.

It has been shown that rate of sorption is controlled by the chemical structure of the resin, (e.g. degree of cross-linking, the type of cross-linking, and the presence of voids), interface, and the manufacturing process. Consequently, researchers have attempted to control the diffusion process by using a resin matrix with a lower permeability (Benmokrane 2000), modifying the interphase region by using suitable sizing chemistry, and selecting a suitable manufacturing process to reduce the void content.

More significantly, fluid ingress can degrade the resin matrix by chemical attack (hydrolysis) or by a drop in the glass transition temperature. As a consequence, fluids affect the matrix-dominant properties, such as the transverse strengths and the shear strengths of the FRP composites. Reduction in these properties becomes more pronounced with increased exposure, temperature, and time (Gao and Weitsman 1998; Liao et al. 1998).

Glass fibres are particularly sensitive to fluids ingress, as glass fibres are subjected to chemical and physical attack. The extent of the degradation of glass fibres depends on their composition, fluid type, fluid concentration, and exposed temperature. Extensive studies have been conducted in this research area (Schutte 1994;Liao 1998; and Weitsman and Elahi 2000). Typically, the tensile strength of E-glass/vinyl ester composites decreases by 40% in a 100% relative humidity environment at 93°C. The flexural and tensile modulus decreases by approximately 10% at 93°C, and at 100% relative humidity for 200 days exposure (Springer, Sanders and Tung 1980). Moreover, fibre orientations are reported to significantly affect the ultimate tensile strength (UTS) and the Young's modulus of composites after moisture exposure (Gopalan et al. 1989). Gopalan et al. (1989) subjected specimens of E-glass/epoxy composites with different fibre architectures to water at 70°C for 20 days, and reported a decrease in the ultimate tensile strength (UTS) of 33.3, 8.0, 13.7 and 20.4% for resin, unidirectional, bi-directional and chopped strand mat respectively. Likewise, for these specimens, the Young's modulus decreased 45.2%, 5.0%, 14.6% and 24%, respectively.

Carbon fibres are not affected by fluids ingress. However, the resin matrix is usually affected, and consequently so are the properties of the composites. For unidirectional carbon composites, this usually leads to a reduction in the compressive strength and shear strength, and to a small effect on the tensile strength (Sen et al. 1996). Hancox and Mayer (1994) reported minimal weight gain and strength loss for carbon/epoxy specimens exposed to 65% relative humidity for over four months, and to boiling water for over three weeks. Rege and Lakkad (1983) tested the compressive, flexural and interlaminar shear strengths of CFRP laminates immersed in salt water and distilled water for 120 hours at various temperatures. The data showed a decrease in strength with increasing temperature for both solutions. It was also observed that, for coupons immersed in fresh water and in salt water at the same temperature, greater decreases in strength and higher equilibrium uptake occurred with the salt water. Water was also found to accumulate at the fibre–matrix interface.

Aramid fibres are affected by fluids, mostly at higher temperatures (Dolan 1993). Saturated AFRP composites have been reported to lose 35% of their flexural strength at room temperature (Allred 1984), and up to 55% if stressed and under wet–dry thermal cycles (Sen et al. 1996). Technora fibres and Kevlar fibres are in the same category of aramid fibres, however, their behaviour is different under the combined effects of fluids and temperatures. For Technora fibre, no degradation was observed in distilled water at any temperature, however, a strength reduction was noticed for the samples immersed in acidic and alkaline solutions. This reduction increased with time and with the temperature. In the case of Kevlar fibres, strength reduction was observed at high temperatures not only in acidic and alkaline solutions, but in distilled water (Uomoto and Nishimura 1999).

Furthermore, studies have been conducted on the combined effect of loading, temperature, and fluids. Buck et al. (1998) reported that the combined effects reduced the ultimate tensile strength of E-glass/vinyl ester composites even in a short time. The effects were more pronounced at higher temperatures. The main failure mode was reported to be delamination and debonding, which were mainly due to pitting and surface damage of the E-glass fibres and matrix micro-cracking. The presence of sustained load played a significant role in the moisture diffusion and subsequent material deterioration at the molecular level by causing such damage as pitting of the E-glass fibres.

In terms of fluids and their effects on fracture and toughness of FRP composites, a detailed review is given by Weitsman (1998). In general, the reported literature is contradictory. On the one hand, due to an increase in compliance through the plasticisation process, increases in fracture energy and fracture toughness are reported. On the other hand, due to degradation in the fibre–matrix interface region, reductions in fracture energy and toughness are reported.

12.5.2. Alkaline reaction

Alkaline resistance of fibres

Alkaline reaction of FRP composite in concrete is one of the major durability concerns for design engineers, since concrete is used in most civil infrastructure structures, which almost inevitably ensures that the FRP composites will be either embedded in concrete (e.g. rebar or tendons) or will be placed around concrete (e.g. column retrofit or beam strengthening). Typically, a concrete environment has high alkalinity, which depends upon the design mixture of the concrete, and on the type of cement. This alkaline environment damages glass fibres through loss in toughness, strength and embrittlement. Glass fibres are damaged due to the combination of two processes (1) chemical attack on the glass fibres by the alkaline cement environment, and (2) concentration and growth of hydration products between individual filaments (Murphy et al. 1999). The embrittlement of fibres is due to the nucleation of calcium hydroxide on the fibre surface. The hydroxylation can cause fibre surface pitting and roughness, which act as flaws severely reducing fibre properties in the presence of moisture. In addition, calcium, sodium and potassium ions found in the concrete pore solution are highly aggressive towards glass fibres. Therefore, the degradation of glass fibres not only depends upon the high pH level, but also the combination of alkali salts, pH, and moisture.

Aramid fibres show strength degradation in an alkaline environment. Kevlar 29 exposed to 10% sodium hydroxide solution for 1000 hours loses 74% of its strength. High modulus aramids such as Kevlar 49 demonstrate better alkaline resistance (Malvar 1998). Kevlar 49 has been reported as having 3% strength loss after 100 hours immersion in a 4% NaOH solution (DuPont 1992). Carbon fibres are commonly believed not to be affected by an alkaline environment. Judd (1971) reported that carbon fibres were resistant to alkaline solutions at all concentrations, and at all temperatures up to boiling. Carbon tows immersed for 257 days in a very basic 50% w/v sodium hydroxide solution showed variations in strength and elastic modulus of around 15%. Carbon strands soaked for 9 months in a pH 13 solution (at 60°C) showed no variation in strength or stiffness (Santoch et al. 1993).

Alkaline resistance of resin matrix

For the use of resin matrix in FRP composites, researchers (Malvar 1998; Benmokrane 2000) suggested that resin must provide high toughness, which must be enough to prevent the development of micro-cracks, and diffusion through the resin matrix must be minimal. Among the commonly used resin matrices for construction (i.e. vinyl ester, epoxy, and polyester) vinyl ester resin was reported to be the most suitable. The reason is that, in vinyl ester resin, the weakest ester linkage is partly replaced by the stronger ether linkage that is highly resistant to alkali. Furthermore, vinyl ester resin is tough because of a longer distance between cross-links. Thus, if flaws/micro-cracks pre-exist in a specimen, their growth under stress would be at a lower rate in vinyl ester (Benmokrane 2000). Additionally, vinyl ester resins are resistant to a wide range of acids (sulphuric, hydrochloric, hydrofluoric, phosphoric, nitric) as well as to chlorine (Malvar 1998).

Although resin matrices provide a certain level of protection to fibres from alkaline degradation, migration of high pH solutions and alkali salts through resin (or through void, crack, interface between fibre and matrix), to the fibre surface needs to be considered. Katsuki and Uomoto (1995) used electron probe microscopy to track the ingress of alkali ions (sodium) into aramid, carbon and glass reinforced vinyl ester rods. Sodium ions penetrated into GFRP in a radial direction with time. No degradation was noticed in AFRP or CFRP rods immersed for 60 days. Hojo et al. (1991) studied the corrosion behaviour of resins in aqueous solutions, and compared it with that of metal. Three forms of corrosion mechanisms were found namely surface reaction, corrosion layer forming, and penetration types. In surface reaction type corrosion (observed in epoxy resin cured with phthalic anhydride in alkaline solution), the ester bonds in the resin were attacked by hydrolytic reaction with liquid and then dissolved from its surface, uniformly. In the corrosion layer forming type (observed in unsaturated polyester in alkaline solution), the ester bonds in the main polymer chain were corroded by the same mechanism, but the cross-linked polymer chain remained. A corrosion residual layer was then formed. In penetration type corrosion, a large quantity of environmental liquid was absorbed until an equilibrium state was reached, and then the strength of the specimen decreased quite remarkably. This type of corrosion was observed in orthophthalic polyester in boiling water and in epoxy resin cured with amine in sulphuric acid. By using corrosion depth, with immersion time, it was found that the concept of corrosion rate could be applied even in polymeric materials. The corrosion rates were dependent upon chemical structure and the reactivity between the resin and the environments.

Effect on strength and stiffness

Aqueous solutions with high pH are known to degrade the tensile strength and stiffness of GFRP bars (Porter and Barnes 1998), although particular results vary tremendously according to differences in test methods. Higher temperature and long exposure time exasperate the problem. Tensile strength reductions in GFRP bars ranging from 0–75% of the initial values have been reported in the cited literature. Tensile stiffness reductions in GFRP bars range between 0–20% in many cases. Tensile strength and stiffness of AFRP rods in elevated temperature alkaline solutions with and without applied tensile stress have been reported to decrease by between

10–50%, and 0–20% of the initial values, respectively (Rostasy 1997; Sen et al. 1998). In the case of CFRP, strength and stiffness have been reported to decrease between 0–20% (Takewaka and Khin 1996).

In order to evaluate the long-term durability performance of FRP in an alkaline environment, extensive studies have been conducted to develop accelerated ageing procedures and predictive models for long-term strength estimates, especially for GFRP rebars. Several models are being developed which include Pilkington Brothers, based on a glass reaction model developed at Iowa State University (Porter et al. 1995), a durability test protocol model, being developed at the University of Sherbrooke (Benmokrane 2000), and diffusion models developed at West Virginia University and at the University of Southern Florida (Faza et al. 1994). Research on the effects of temperature on the durability of FRP bars in a concrete alkaline environment indicates that an acceleration factor for each temperature difference can be defined by using Arrhenius theory. These factors differ for each product, depending upon the type of fibre, type of resin, and the bar size. In addition, these factors are affected by the environmental conditions, such as the surrounding solution media, temperature, pH, moisture, and freeze–thaw conditions (Benmokrane and El-Salakawy 2002). Unfortunately, currently, no accelerated ageing models have been developed for CFRP bars (Porter 1999).

Stress corrosion in an alkaline environment

Besides the effect of pH and alkali salts, there are other important concerns relating to the stress corrosion of FRP bars, especially with regard to long-term loading conditions. Stress corrosion is the process where a material is held under constant load in a corrosive environment until failure occurs. This process is known as stress corrosion (or in the case of constant strain, as strain corrosion). Premature failure (Sen et al. 1993) or rapid loss in tensile strength of GFRP composites in an alkali environment has been reported (Porter and Barnes 1998).

Benmokrane (2001) studied the stress corrosion mechanism of GFRP bars in various alkaline environments (NaOH, simulated pore water, moist concrete) under different stress levels (22–68% of ultimate strength). Accelerated ageing tests have been conducted on more than 400 GFRP rebars. Fundamental material properties of aged and unaged GFRP rebars have been studied, using micro-structural analysis techniques to evaluate the degradation mechanism and the penetration of the chemical solution into the GFRP bars. The results obtained, indicate that the stress corrosion mechanism of GFRP rebars depends upon the environment and the stress level. Under a stress level of about 20–30% of the ultimate strength, in a NaOH solution, the stress corrosion mechanism involves mainly the damage of the fibres. When in a simulated pore water solution and a concrete environment, stress corrosion is caused by interface damage. At high stress levels above 55%, matrix and fibre cracking are the most dominant mechanisms.

Nkurunziza et al. (2004) studied the residual tensile properties (strength and stiffness) of GFRP reinforcing bars under combined effects of temperature with alkaline or water medium, in addition to constant sustained tension. Three bar sizes (16, 12.7 and 9.5mm) were tested under elevated temperature or ambient temperature for different durations. The elevated temperature was used as an accelerated aging factor. By using the Arrhenius model, the elevated temperature used in this test, for 2 and 4 months, simulated field conditions of 40 and 80 years. The alkaline or water medium represent the most common field conditions. The following conclusions were obtained from this research;

- A maximum of 11% reduction in tensile strength (compared to guaranteed strength) was observed in the 9.5mm GFRP bars after exposure to the combined effect of 60°C and alkaline solution under sustained stress level of 29%.
- The residual tensile strength for conditioned specimens was higher (a minimum of 130%) than the specified design strength as recommended by ACI440.1R-03 design guidelines.
- The reduction in the ultimate strain after conditioning is within the acceptable limits. The lowest residual strain was 43% higher than the specified design strain recommended by ACI 440.1R-03.
- No significant loss in the elastic modulus was observed under the severe conditions. This was a general conclusion in the three bar sizes tested. In addition, the smaller bar size showed a slight increase in the modulus after conditioning. This maybe due to more curing of the matrix under the influence of the elevated temperature.

To improve the durability of FRP composites in an alkaline environment, significant attention has been given to the development of newer types of glass fibres such as Adventex and AR-glass fibres (alkaline resistant). The application of special surface coatings, the better selection of a resin matrix, the selection of suitable chemistry sizing and improvements in the manufacturing process, have also helped.

12.5.3 Freeze–thaw

Degradation mechanism

The majority of studies in this subject are being directed towards aerospace materials. The reported literature on the effect of freeze–thaw on pultruded FRP composites is very limited. In terms of material degradation due to low temperature and freeze–thaw cycles, Lord and Dutta (1988), have given an extensive review. In general, at low temperatures, complex residual stresses arise within the FRP composites as a result of matrix stiffening, and the mismatching of thermal expansion coefficients. Residual stresses can cause micro-cracks in the matrix and at the fibre–matrix interface, leading to the degradation of the FRP composites. Micro-cracks can grow under low temperature thermal cycling and coalesce to form transverse matrix cracks and cause fibre–matrix debonding. The severity of cracking is directly related to the amount of resin shrinkage during curing. The authors also found that even if the composites were produced in a stress-free state, the subsequent thermal cycling usually produced micro-cracks. Transverse matrix cracking can affect laminate stiffness, strength, dimensional stability, and fatigue resistance.

Another important research area is the long-term synergistic effects of freeze– thaw, fluids and loading. These synergistic effects may lead to an increased matrix crack density and as a consequence, high permeability, resulting in the FRP composites suffering reduced corrosion resistance. Hence, a comprehensive understanding of these synergistic effects is very important for design purposes. However, the reported literature is very limited in this area.

The effect of freeze–thaw cycles on the mechanical properties of FRP composites

In general, the literature show that unidirectional tensile strengths decrease in the –10 to –40°C temperature range, whereas the off-axis and transverse strengths may increase due to matrix hardening. Increasing freeze–thaw cycles accentuate residual stresses resulting in an increased severity and density of cracks. An apparent increase in matrix brittleness and a decrease in tensile strength was also reported (Lord and Dutta 1988; Benmokrane and Rahman 1998; Dolan Rizkalla and Nanni 1999).

Kuz'min et al. (1989) reported an increase in the ultimate tensile strength in wound GFRP tubular specimens. Weiss (1982) studied the effects of different temperatures (from 293 K to 77 K) on the mechanical properties of CFRP composites. The Young's modulus showed a 7% increase. Poisson's ratio showed a 6% reduction. Fracture stress in the fibre direction increased from 1700 MPa to 2010 MPa, and fracture strain increased from 1.22% to 1.34%. Allred (1984) subjected Kevlar fabric laminates to moisture and 2 hour temperature cycles from –20°F to 125°F. The results indicated that the moisture cycles, without freeze–thaw cycles, had little effect on the strength of the specimens. However, after 360 freeze–thaw cycles (4 months), the ultimate tensile strength of the laminates decreased by 23%. After 1170 freeze–thaw cycles, the strength was degraded by 63%.

Gomez and Casto (1996) conducted freeze–thaw tests on isoptholic polyester and vinyl ester, pultruded glass reinforced composites. Samples were subjected to freeze–thaw cycles (–17.8°C to 4.4°C) after initially being submerged in 2% sodium chloride and water. The results showed significant loss in flexural strength and toughness after 300 freeze–thaw cycles.

Specimens made of a carbon FRP grid and a glass–carbon FRP grid with junctions at a 100 mm spacing were exposed to various environmental conditions for one year. These conditions included exposure to salt and alkali, UV radiation with wet–dry cycling, and freeze–thaw cycles (Rahman et al. 1996). The results indicated that the tensile strengths of the specimens were not affected by these conditions, whereas the junction strength was found to be sensitive to salt, alkali, and freeze–thaw cycles.

12.5.4 Fatigue

Whilst FRP composites are not usually isotropic, they are certainly non-homogeneous and non-uniform. They may not be continuous (due to imperfect manufacturing), and may display a non-linear response, especially under dynamic loading. Therefore, the fatigue behaviour of FRP composites is complex in the way that different degradation modes can be developed during fatigue loading. The development of the degradation mode is strongly influenced by material properties (e.g. fibre orientation, matrix, fibre–matrix interface, ply orientation and stacking sequence, fibre geometries), environment (e.g. moisture, temperature), and loading history. Therefore, it is important to understand the fatigue behaviour of FRP composites in the civil engineering infrastructure environment for design purposes. Over the previous decades, extensive fatigue data about FRP composites has been generated for other industrial applications (Liao et al. 1998; Weitsman and Elahi 2000). However, limited studies have been conducted on the fatigue performance of pultruded FRP composites (e.g. bars, tendons, sheets, plates, and FRP structural shapes) for civil infrastructure applications.

From previous research (Benmokrane and Rahman 1998), some general tendencies have been observed in unidirectional FRP coupons with 60% fibre volume fraction, using tension–tension fatigue testing in an ambient laboratory environment. Of all the types of FRP composites for civil infrastructure applications, CFRP composite is generally reported to be least prone to fatigue loading. At one million cycles, the fatigue strength is generally between 50% and 70% of the initial static strength and is relatively unaffected by realistic moisture and temperature exposure of concrete structures unless constituent materials are substantially degraded by the environment. GFRP composites were reported to have approximately a 10% loss in the initial static capacity per decade of logarithmic lifetime (Mandell and Meier 1983).

Although aramid fibres show poor durability in compression, the tension–tension fatigue behaviour of an impregnated aramid fibre bar is excellent. Two million-cycles endurance limits of commercial AFRP bars for concrete applications have been reported in the range of 54% to 73% of the initial tensile strength (Odagiri et al. 1997). Based on these observations, Odagiri et al. (1997) suggested that the maximum stress to be set is between 0.54 to 0.73 times the tensile strength.

Studies on the fatigue behaviour of FRP tendons indicated that the fatigue life of FRP tendons in pretensioned beams is shorter than that of tendons tested in air. For loaded post-tensioned beams, the curvature of the tendon profile causes the tendon to rub against the edge of the crack instigating failure point. Special attention needs to be given to the fatigue resistance of the anchorage. Such devices can usually develop the full tendon strength under monotonic load conditions, but less than this value when a cyclic load is applied. Rostasy and Budelmann (1991) evaluated the S-N curves for GFRP tendons. They reported that the fatigue strength of the GFRP tendons is influenced by the anchorage properties. However, the fatigue strength of GFRP is markedly below than that of wedge-anchored pre-stressing wire.

Saadatmanesh and Tannous (1999) studied relaxation, creep, and fatigue behaviour of commercially available CFRP tendons (Leadline 8 mm diameter PC-D8 and 7.5 mm diameter CFCC). For the fatigue tests, the investigators used ambient laboratory conditions with sinusoidal load control at 3–5 Hz, at seven different minimum loads corresponding to 30%, 40%, 50%, 60%, 70%, 80%, and 90% of ultimate strength and three different stress ranges corresponding to approximately 5%, 10%, and 20% of ultimate strength. The results indicated that the residual strength of Leadline and CFCC decreased less than 10% of virgin ultimate strength. Little change was observed in the modulus of both tendons at the lower stress ranges, but some increase was noted at the higher stress ranges. No change in Poisson's ratio was noticed in Leadline.

Certain exposure environments (e.g. elevated temperatures, humidity, and corrosive fluids) reduce the fatigue life of FRP composites. These effects are dependent upon the fibre and matrix types, laminate lay-up, pre-conditioning methods, solution content, and the environmental conditions during fatigue, especially the interface between the fibre and the matrix. In general, the effects of environmental conditions on the tensile fatigue behaviour is well documented for aerospace composite materials, but there is a lack of data on the effect of the exposure environment on the flexural and the compression fatigue behaviour for pultruded FRP composites. There is very limited long-term fatigue data, especially beyond 10^7 cycles (Liao et al. 1998).

Pre-conditioning under relatively low humidity (65% R.H.) may not have a significant effect on the fatigue behaviour of FRP composites (Jones et al. 1983). However, pre-conditioning at elevated temperatures in water (>75°C) always has a deleterious effect on the fatigue performance of GFRP and AFRP composites. The authors noticed that the fatigue degradation rate for GFRP pre-conditioned in boiling water was slower when compared to that of dry specimens. The authors attributed this phenomenon to several factors, firstly, plasticisation of the resin may result in an increased fatigue strain of the resin, and secondly, boiling permitted relaxation of the thermal strains introduced during processing. Also, the stress transfer capacity of the fibre–matrix interface will be reduced by boiling water, which in turn will reduce the stress concentration in the vicinity of the broken fibres or the resin micro-cracks.

Hayes et al. (1998) studied the effect of moisture on the fatigue of pultruded glass/ vinyl ester composites. The results indicated that the quasi-static tensile strength was reduced by 26% at a moisture concentration of 0.95% (by weight). This reduction in strength was not recovered when the material was desorbed, suggesting that exposure to moisture caused permanent damage to the material system. The fatigue life of the wet material was reduced by a factor consistent with its reduction in fracture strength, i.e. the dry and wet S-N curves have the same slope but different intercepts.

McBagonluri et al. (2000) studied the effect of fresh and salt water on the tension fatigue in pultruded glass/vinyl ester composites under both isothermal and cyclic ageing conditions. The results indicated that the temperature effect on the fatigue response did not depend upon the presence or absence of the ambient fluid. The test frequency effect was insignificant on the fatigue response.

Other parameters such as alkalinity, salinity of the concrete, temperature, surface deformation, ratio of maximum to minimum cyclic stress and loading frequency were also found to affect the fatigue life of FRP composites (Rahman et al. 1997). The fatigue strength of CFRP bars encased in concrete has been observed to decrease when the environmental temperature increased from 68 to 104°F (20 to 40°C) (Adimi et al. 1998). In the same investigation, the endurance limit was found to be inversely proportional to the loading frequency. It has also been found with CFRP bars, that the endurance limit depends also upon the ratio of the maximum to the minimum cyclic stress. A higher mean stress or lower stress ratio (minimum divided by maximum) will cause a reduction in the endurance limit (Saadatmanesh and Tannous 1999). The addition of ribs, wraps, and other types of deformations are commonly used to improve the bond behaviour of FRP bars. Such deformations, however, have been shown to induce local stress concentrations that significantly affect the performance of the CFRP bar under fatigue loading (Katz 1998). Local stress concentrations degrade fatigue performance by imposing multiaxial stresses that serve to increase matrix-dominated damage mechanisms.

12.5.5 Creep/relaxation and creep rupture

Creep/relaxation of FRP composites

When a material is subjected to a constant applied stress, the material will exhibit a time-dependent strain, known as creep. The associated time-dependent stress change exhibited by the material is called stress-relaxation. Although the short-term properties of FRP composites are excellent, the behaviour under long-term sustained loading may exhibit time-dependent response due to viscoelastic behaviour of the polymeric matrix. Thus, the apparent stiffness and strength of FRP composites will decrease slowly over time. The time temperature superposition (TTSP) principle, which is a widely accepted accelerated test method, has been utilised successfully in extrapolating short-term creep data over many decades. Details about this technique can be found from many text books (e.g. Ferry 1980).

Creep behaviour of FRP composites is strongly dependent upon the structure (e.g. fibre orientation), and load condition of the material. For unidirectional FRP composites, the creep compliance is less affected by the creep behaviour of the polymer matrix when the material is loaded along the fibre direction than in any other direction. For off-axis loading, the creep behaviour is strongly dependent upon the creep of the matrix. In FRP composites reinforced with discontinuous randomly orientated, or with continuous bi-directional woven fibres, the properties of the material are matrix dominated, therefore the creep of the polymer matrix is the main reason for the creep behaviour. In addition, physical ageing, temperature, moisture, stress level, and ultraviolet radiation affect the creep behaviour of FRP composites. Some data related to these influencing parameters can be found in recent reviews (e.g. Scott et al. 1995; Liao et al. 1998; Weitsman and Elahi 2000). Data on the effect of moisture absorption on creep behaviour are scarce (Mohan and Adams 1985). The moisture absorption level is history-dependent, therefore sorption behaviour under temperature cycles is not the same as under a constant temperature and humidity (Dillard 1991). The combined effects of temperature, aggressive agents (acid, base, moisture, etc) and stress level, are extremely important. The literature on this research topic is limited.

Most of the literature relates to the results for uniaxial tension testing. Relatively little work has been conducted on compression and bending tests. This is mainly due to the difficulty of setting up such tests and accurately controlling the test data (Liao et al. 1998). There is less literature available concerning creep relaxation and cure kinetics in the natural service environments of ambient cured vinyl esters, polyester and epoxies (Morgan 1999). In general, the creep behaviour of carbon fibre reinforced composites is well documented. However, only limited data exists for pultruded GFRP composites and structural members. A few long-term creep test data (with creep time greater than 10 000 hours) is available (Bhatnagar et al. 1981).

The effect of fluids and temperature

Fluids and time can often affect the creep behaviour of FRP composites in a manner that is similar to the effect of temperature and time. Fluid absorption in FRP composites will lead to residual stress and plasticising of the resin, which can accelerate the time-dependent behaviour of FRP composites. Liao et al. (1998) schematically described the effect of time, temperature, and fluids on the creep behaviour of FRP composites. The author concluded that creep compliance was increased with the increase in fluid content and temperature over time.

Bhatnager et al. (1981) studied the creep strain of unidirectional and bi-directional E-glass/epoxy laminates at 75, 100 and 150°C under stress levels of 40%, 60% and 80% of the short-term ultimate tensile strength. The results indicated that the elastic behaviour of the specimens was linear, except at the highest temperature and stress level. The observed viscoelastic behaviour in each test was definitely non-linear with respect to stress. Increasing temperatures resulted in greater departure from linearity for the viscoelastic behaviour of both the bi-directional and unidirectional specimens. The viscoelastic deformations were hardly recovered after unloading. Wang and Wang (1980) studied the combined effect of moisture, temperature and stress on the tensile creep behaviour of glass/epoxy composites. Unidirectional laminates were loaded at angles of 0°, 45°, and 90° with respect to fibre orientation. At room temperature with a low stress level (6.2 MPa) and moisture content (0.5–0.94% by mass), 0° laminate exhibited minimal creep when compared to the 90° and 45° laminates. When the temperature was increased to 100°C, a significant increase in creep of the 0° laminate was noticed. It was also found that the creep of the 45° and 90° laminates was strongly influenced by moisture and temperature even in low stress levels.

Although recent studies have been conducted on the effect of environmental conditions on creep/relaxation behaviour of FRP reinforcement, the literature is limited in this research area. The relaxation, fatigue, and creep behaviour of two types of commercial CFRP tendons, namely Leadline PC – with 8.8 mm diameter and 1×7 – 7.5 mm diameter CFRP cable, were examined by Saadatmanesh and Tannous (1999). Twelve Leadline, and 12 CFCC, tendon specimens were tested in air at temperatures of 130°, 25° and 60°C to determine their relaxation behaviour. In addition, the relaxation behaviour of 24 Leadline, and 24 CFCC, specimens were examined in a chemical solution. The loss in tensile force for the 3000 hour test duration at stress ratios of 0.4 and 0.6, was generally less than 10%, and depended primarily on the initial stress ratios and the type and temperature of the environment. Leadline and CFCC exhibited good fatigue strength for a stress range of 100 and 107 MPa, respectively. Both Leadline and CFCC exhibited good creep behaviour with limited creep strains at a sustained load of approximately 40% of their ultimate tensile strength for the 3000 hour test duration.

Creep rupture

Creep rupture is another important concern, when FRP composites are subjected to long-term loading. Creep rupture is the tensile fracture of a material subjected to sustained high stress levels over a period of time and occurs when the material's strain capacity is reached. As the ratio of the sustained tensile stress to the short-term strength of the FRP bar increases, endurance time decreases. The creep rupture endurance time can also decrease irreversibly under sufficiently adverse environmental conditions, such as high temperature, ultraviolet radiation exposure, high alkalinity, wet and dry cycles, or freeze–thaw cycles.

In general, carbon fibres are the least susceptible to creep rupture, aramid fibres are moderately susceptible, and glass fibres are the most susceptible. Creep-rupture tests have been conducted on GFRP, CFRP, and AFRP bars with a diameter of 0.25 in. The bars were tested at different load levels at room temperature. The results indicated that a linear relationship exists between creep-rupture strength and the logarithm of time for all load levels. The ratios of stress level at creep rupture after 500 000 hours to the initial ultimate strength of the GFRP, CFRP, and AFRP bars were extrapolated and found to be 0.3, 0.47, and 0.91, respectively (Yamaguchi et al. 1997).

In another study, the stress at creep rupture at the ultimate strength after 50 years calculated using a linear relationship, was found to be 0.66 for AFRP and 0.79 for CFRP (Ando et al. 1997). Tests were also conducted to study the environmental effects on the creep-rupture strength (Dolan et al. 1997) and to evaluate the creep-rupture strength of GFRP (Seki et al. 1997). Short-term (48 hours) and long-term (1 year) sustained loads corresponding to 50% of the initial ultimate strength were applied to GFRP and CFRP rebars at room temperature. The specimens showed little creep or reduction of tensile modulus. The ultimate strength did not change significantly after conditioning (Lyer and Anigol 1991).

12.5.6 Ultraviolet rays

The shorter wavelength of ultraviolet rays, present in sunlight, is known to initiate chemical reactions within polymeric materials. FRP composites are polymeric and therefore prone to the same photochemical damage as unreinforced polymers. Photochemical reactions in polymers potentially lead to chain scission or cross-linking reactions. Chain scission reactions decrease the molecular weight of the surface polymers, allowing erosion of the low molecular weight fragments (Monney 1998). Continued exposure and subsequent erosion, results in a substantial loss of resin from the polymer surface and, in the case of FRP composite, leads to exposed fibre and increased moisture permeation, which may lead to a change in mechanical properties (Startsev et al. 1999).

The UV effect is primarily at the surface of the composites to a depth of about 10 micrometres (Deler and Miller 1988). Therefore, for FRP composites used as internal reinforcement in concrete structure, this effect would not have any consequence, but for FRP composites used as external reinforcement, ultraviolet rays would be likely to have a negative influence on the long-term structural performance. By using UV absorbing additives, the rate of photo-oxidation reactions can be minimised. However, the protective coating will be degraded eventually by the UV radiation and will need to be maintained. Periodic inspections of the protective coating should be performed to ensure that the composite surface does not become visible.

Published literature on the effect of UV radiation on FRP composites indicates that only minor changes in the mechanical properties occur following UV exposure either in a laboratory setting or in the field. This is particularly true for thick sections, in which the percentage of the thickness affected by UV damage is relatively small. Thinner specimens exhibit greater changes in mechanical properties. Larsson (1998) reported that the UV effect was a surface phenomenon in Kevlar 49 composites, thus any influence on the mechanical properties are more pronounced for thinner samples.

For thick samples, no mechanical property degradation was observed. Chin et al. (1997) observed surface chemical reactions in vinyl ester and isopolyester neat resins. No changes were observed in the glass transition temperature of the materials, which indicated that the damage was confined to the surface region. Isopolyester was shown to have a better UV resistance than vinyl ester. Polyesters were reported to be prone to surface photo-oxidation reactions.

Degradation in tensile strength and stiffness of FRP composites are found to be fibre dependent as a result of UV exposure. The literature indicates that after 1000 hour cycles of ultraviolet radiation, Aramid fibres show a 50% reduction in tensile strength, glass fibres show a 20% reduction, and carbon fibres show no reduction. Little or no decrease in the tensile strength of neat resin samples was observed. In terms of FRP rods, after 1250 cycles of ultraviolet radiation, AFRP rods show a 13% decrease in tensile strength. GFRP rods show an 8% reduction in tensile strength. CFRP rods show no reduction in tensile strength. These results indicate that the decrease in tensile strength as a result of UV exposure is detrimental to seismic retrofitting of columns, since decreases in tensile strength will reduce the confining capacity of the wraps (Kato et al. 1998).

The effect of UV on the creep behaviour of a variety of polymeric materials was investigated by Regel et al. (1967). In this study, the loaded specimen was irradiated with UV radiation from a lamp, at an intensity of about 0.84 J/cm^2 on the target. In the wave length interval from the short time period that the UV radiation was turned on, the creep strain increased sharply. When the UV radiation was switched off, the creep rate changed to the starting value suggesting that the creep-rate change was reversible. The sharp increase occurred independent of whether the material was in the initial or the steady state of creep. This was observed in all the nine different polymer materials tested (natural rubber, polyacrylonitrile, polymethylmethacrylate, cotton, celluose triacetate, and polystyrene). Similar procedures using infrared radiation (IR) did not yield similar results, suggesting that the effect of IR on the creep rate was insignificant. The authors attributed bond breakage as the intrinsic mechanism. Experiments of Regel et al., involving stress relaxation, indicated that UV also increased the rate in stress-relaxation.

12.5.7 High temperature

It is well known that organic matrix based fibre reinforced materials exhibit viscoelastic transitions followed by reversible and irreversible thermal degradation when exposed to elevated temperature, especially at a temperature close to or above, the glass transition temperature (Tg). Beyond Tg, the elastic modulus of a polymer will be significantly reduced due to changes in its molecular structure. The value of Tg depends upon the type of resin, and is normally in the range of 200 to 300°F in a composite material. The fibres, which exhibit better thermal properties than the resin, can continue to support some load in the longitudinal direction until the temperature threshold of the fibres is reached. This can occur at temperatures near 1800°F for glass fibres and 350°F for aramid fibres. Carbon fibres are capable of resisting temperatures in excess of 3000°F. However, due to a reduction in force transfer between fibres through bond to resin, the tensile properties of the composites are reduced. Test results have indicated that 480°F (much higher than the glass transition temperature) reduced the tensile strength of GFRP and CFRP composites by at least 20% (Kumahara et al. 1993). Other properties more directly affected by shear transfer through the resin, such as the bending strength, were reduced significantly at higher temperatures.

Based on the available literature, Kodur and Baingo (1999) compared the behaviour of FRP composites with that of traditional building materials at elevated temperature. The rate of strength loss was much greater for FRP than for concrete or steel. FRP composites lost little strength when the temperature was up to 100°C. After that temperature, the strength degradation became much faster, resulting in a 50% strength loss at 200°C. In the case of concrete, a 50% strength loss did not occur until about 700°C, whereas for steel, the corresponding temperature was 500°C. The critical temperature for FRP composites was much lower than that of steel. The effect of high temperature on the properties of FRP rebars were studied by Kumahara et al. (1993), who found a reduction of 20% in the tensile strength of glass and carbon fibre FRP rebars at a temperature of 250°C. The reduction of tensile strength of aramid was 60% at 250°C.

Research is still continuing to improve the performance of FRP composites at high temperatures.

12.5.8 Fire

Another significant concern in organic polymer matrix composites is the possibility of fire, resulting in the spread of flame, the release of heat and toxic byproducts, and potential structural collapse. Fire-related issues associated with composite materials are more severe in confined spaces (such as tunnels and buildings) as opposed to open spaces (such as roads and bridges). Interior applications of composites are likely to come under existing building, or construction code, requirements. Compared to non-filled plastics, composites have an advantage in their high non-combustible fibre content (up to 70% by wt.), which makes less fuel available to the fire. When the outermost layers of resin are lost, the fibres serve as an insulating layer, slowing heat penetration and the evolution of gases from the interior of the composite.

Published literature indicates that the most important issues in this area for FRP composite are flame spread, fire endurance, smoke, toxicity, and heat release. These subjects have been studied extensively. Recent developments in fire research have highlighted the importance of the heat release rate (HRR) as the primary fire hazard indicator. The rate of heat release, especially the peak amount, is the primary characteristic in determining the size, growth, and suppression requirements of a fire environment. It is also important to understand the structural resistance of building structures to different fire threats.

Glass reinforced phenolic composites were reported to have inherent low flame rate characteristics, low smoke generation, and a low heat release rate (Sorathia et al. 1995). Therefore, phenolic resin based composites offer benefits in fire performance over many existing polymer composites such as polyester and vinylester based composites. Egglestone and Tourley (1994) investigated the burning characteristics of glass reinforced composites panels with an isophtholic polyester resin, vinyl ester resin and phenolic resin as the matrices. The results showed that phenolic composites have a longer ignition time, reduced heat output, less contribution to a low level sustained fire and a lower smoke yield. Besides the use of phenolic resin, fire resistance can also be increased through special fire-resistant additives, intumescent coatings, and addition of inorganic fillers (microfine clays, alumina trihydrate, etc.) into the resin during processing.

12.6 Durability studies on FRP – Concrete structures and FRP structural members

There is a growing number of studies on the durability performance of FRP (internally and externally) reinforced concrete structures and FRP structural members. Although several durability parameters have been investigated for some structural forms, the literature is very limited.

12.6.1 Freeze–thaw

Recent research has been conducted on the behaviour of concrete structures (internally and externally) strengthened with FRP composites subjected to cold climate conditions. Freeze–thaw cycling is an issue of particular concern. Tests on FRP wrapped concrete cylinders exposed to freeze–thaw action have revealed that freeze–thaw cycles lead to more brittle failure of FRP wrapped concrete cylinders than similar specimens kept at room temperature. Moreover, FRP wrapped cylinders exposed to freeze–thaw cycles showed a significant increase in strength over unwrapped cylinder exposed to the same freeze–thaw cycles. Increasing the number of FRP layers leads to a further increase in strength (Soudki and Green 1997).

Research on the effect of freeze–thaw exposure has also been conducted on FRP externally strengthened beams. The literature indicated that freeze–thaw exposure led to a decrease in the moment capacity and ultimate deflection of the FRP strengthened beams. The decrease rate was larger for precracked specimens than for uncracked specimen (Del Mar Lopez 1999). Arntsen and Pedersen (1999) examined the effect of freeze–thaw cycles on beams strengthened with CFRP sheets. Some specimens were moist-conditioned at 80% RH while the others were water-saturated before strengthening and subsequently sealed against evaporation. The beams were subjected to 56 freeze–thaw cycles (–22°C to +17°C), each of which were of 24 hours duration (9 and 15 hours above and below 0°C, respectively). The results showed that the deviation between the load-deflection curves was negligible for different exposures. Nollet et al. (1999) studied the short-term flexural performance of beams strengthened by CFRP strips at room temperature (25°C) and at a freezing temperature (–25°C). It was observed that due to freezing, both the load and the deflection increased. A thirty-five percent increase in flexural capacity was noted under the freezing temperature.

Some studies have been conducted on the effect of freeze–thaw exposure on FRP bar reinforced concrete structures. Challal et al. (1991) conducted freeze–thaw tests on three 75 × 100 × 400 mm concrete specimens reinforced with an ordinary steel bar (diameter 11.3 mm), an epoxy coated steel bar (diameter 11.3), and a glass fibre rod (diameter 9.4 mm). Flexural tests were performed on the specimens at 0, 200, 400, and 600 freeze–thaw cycles. The specimens with the GFRP rod showed a similar behaviour to other specimens, as the strength decay curves were almost parallel.

Benmokrane (2000) investigate the effect of freeze–thaw cycles on concrete beams reinforced with GFRP bars. Freezing and thawing tests were carried out, according to ASTM C666 procedure B, on three 75 × 100 × 400 mm mm concrete beams reinforced with an ordinary steel rebar (diameter = 11.3 mm) and with a GFRP rebar (diameter = 9.4 mm). Flexure tests were performed on these specimens at 0, 200, 400 and 600 cycles. The results indicated that the freeze–thaw cycles did not affect the flexural behaviour of the GFRP reinforced RC beams. Vijay and GangRao (1999) immersed GFRP bar reinforced concrete beams in salt (pH 7) and alkaline (pH 13) solutions respectively, and subjected them to freeze–thaw cycles for 12 months. No stiffness losses were noticed after the freeze–thaw cycles. The crack patterns of the conditioned beams increased less (from 2.54 to 5.08).

12.6.2 Fatigue

Studies on prestressed concrete members indicated that the overall fatigue characteristics of GFRP bar pretensioned concrete beams matched those of similarly loaded steel pretensioned beams. The fatigue behaviour of the GFRP pretensioned beams in the post-crack range resulted in much higher deflection and crack widths than similar beams reinforced with pretensioned steel (Mikami et al. 1990). Nanni et al. (1995) investigated the effect of cyclic loading (20–80% of the maximum beam capacity) on concrete beams reinforced with two-dimensional FRP grids. The beams demonstrated cyclic bending load behaviour similar to that of beams with steel reinforcement.

Test results have been reported recently for T-shaped and rectangular reinforced concrete beams post-strengthened for flexural fatigue by bonding CFRP sheets to the tension faces (Dolan, Rizkalla and Nanni 1999; Benmokrane and Rahman 1998). The results show that the fatigue life of RC beams could be enhanced significantly through the use of externally bonded FRP reinforcement. Analytical methods for fatigue performance have been proposed and compared with experimental results. Design recommendations have been proposed for the use of FRP sheets to reduce fatigue failure. The analytical model provides accurate results for the fatigue life prediction of reinforced concrete beams reinforced with CFRP sheet. Maruyama et al. (1997) studied the flexural fatigue resistance of concrete beams reinforced with CFRP strands and subjected to 1,200,000 cycles at a peak load higher than the cracking load (60% of the static ultimate load). No significant reduction in bending strength was observed. Grace and AbdEl-syed (1996) tested double T-bridge (externally prestressed with carbon fibre reinforced cables and internally reinforced with GFRP rebars) loaded up to 60% of the static ultimate load. There were no significant changes in the static and dynamic characteristics of the specimens after 7 million cycles.

12.6.3 Creep

FRP (internally or externally) reinforced concrete members can exhibit a complex viscoelastic behaviour due to the combination of the viscoelastic behaviours of the concrete, the adhesive, and the FRP composite. Although extensive studies have been conducted on the individual viscoelastic behaviour of FRP composites, concrete, and adhesive, very limited studies have been conducted on the viscoelastic behaviour of composites systems.

For externally reinforced RC members, studies have been conducted on small-scale and large-scale FRP laminate bonded RC beams both experimentally and theoretically. Plevris and Triantafillou (1994) reported an analytical procedure to predict the time-dependent behaviour of RC beams strengthened with FRP laminates. Findley's model (Findley 1960) was used to predict the behaviour of FRP composites. The addition of CFRP plates reduced the rate of compressive creep in the concrete of the strengthened systems. The response of the steel was shown to be time-dependent. Barnes and Garden (1997) reported the tension creep test on 1 m CFRP reinforced concrete beams. The samples were subjected to various environmental conditions, which included 22°C, 40°C and 60°C temperatures, with 50% relative humidity. At the end of the creep test, the creep deflection of the unbonded beam was 3.7% of the total deflection, which was found to be higher than 2.4% in the CFRP plated beams. This indicates a reduction in the time-dependent deformation due to the addition of the bonded FRP plate.

In terms of FRP bar reinforced concrete beams, the available literature (Kage et al. 1995) indicates that the creep curve of FRP bar reinforced concrete beams is similar to that of steel reinforced members. The creep deflection of FRP bar reinforced concrete beams can be predicted using the existing theory for steel reinforced concrete beams with modification of creep coefficients for FRP reinforced sections. Brown and Bartholomew (1996) indicated that long-term deflection varied on the basis of stress levels. This is not addressed by the equations in ACI 318, which only multiply the initial deflection by the time dependent factor, ξ. Brown concluded that the creep coefficient should be adjusted twice; first, to account for the different stress level in the concrete, and second, to account for the larger initial deflection. From the available data (Brown and Bartholomew 1996), the ratio of ξ_{Frp}/ξ_{steel} varied from 0.46 for AFRP and GFRP to 0.53 for CFRP. In another study, the modification factor, ξ, based on the compression-controlled failure varied from 0.75 after one year to 0.58 after five years (Vijay and GangaRao 1998).

Vijay and GangaRao (1999) subjected GFRP rebar reinforced concrete beams to sustained loads between 20% to 50% of the ultimate load, for up to 847 days. The concrete creep strain curves for GFRP reinforced concrete beams were similar to those of the steel reinforced concrete beams. The creep coefficient in GFRP reinforced concrete beams was less than that of the steel reinforced concrete beams. This was due to the initial strains and deflections of the GFRP reinforced concrete beams being higher than that of the steel reinforced beams. The creep coefficient was found to be 32.8 mm under low sustained loading (20% of ultimate sustained stress) as compared to 39.1 mm for beams under high-sustained loading (50% of the ultimate sustained stress).

12.6.4 Fire/high temperature

For FRP internal reinforced concrete structures, FRP rebars are embedded in a non-flammable concrete substrate. There is currently no adhesive connecting rebar to the concrete, so the flammability of the interface is not an issue. The issue here is that of structural integrity during a fire. The properties of the polymer at the surface of the bar are essential in maintaining a bond between the FRP and the concrete. At a temperature close to the glass-transition temperature, however, the mechanical properties of the polymer are significantly reduced, and the polymer is not able to transfer stresses from the concrete to the fibres. One study carried out with bars with glass-transition temperatures of 140–255°F (60–124°C) reports a reduction in the pullout (bond) strength of 20% to 40% at a temperature of approximately 212°F (100°C), and a reduction of 80% to 90% at a temperature of 390°F (200°C) (Katz et al. 1999). In a study on the flexural behaviour of beams partially pretensioned with AFRP tendons and reinforced with either AFRP or CFRP bars, beams were subjected to elevated temperatures under a sustained load. Failure of the beams occurred when the temperature of the reinforcement reached approximately 392°F (200°C) and 572°F (300°C) in the carbon and aramid bars, respectively (Okamoto et al. 1993). Another study involving FRP reinforced beams reported reinforcement tensile failures when the reinforcement reached temperatures of 482–662°F (250–350°C) (Sakashita et al. 1997). Locally, such behaviour can result in increased crack widths and deflections. Structural collapse can be avoided if high temperatures are not experienced at the end regions of FRP bars, allowing anchorage to be maintained. Structural collapse can occur if all the anchorage is lost due to the softening of the polymer, or if the temperature rises above the temperature threshold of the fibres (1800°F (980°C) for glass fibres).

The effects of fire on external FRP reinforced concrete members, depend upon whether the structures are in a confined space (interior of a building, or parking garage, tunnel, etc.) or in open spaces (such as road and bridges). Fire-related issues are more severe in confined spaces. In a tunnel, for example, the developing fire can produce copious quantities of potentially toxic smoke. In open spaces, such as bridges, a developing fire increases the chance of structural collapse. If the structure is one in which the composite form has only a reinforcing or repair role, the consequences of local heat-induced composite failure is not likely to be serious. In this scenario, time is available to repair the damage. However, if the affected composite component is part of the primary structure, the structure may collapse. For earthquake reinforcement, the problem is somewhat more complex. If an earthquake-induced fire did destroy the composite reinforcement on the structure, the structure might survive the initial earthquake only to fall victim to an aftershock occurring after the fire.

Currently, the application of FRP composites is associated with concrete structures. There is insufficient information on the relationship between standard, bench-scale fire test performance, and full-scale fire behaviour for all types of composites and FRP reinforced concrete members and so precludes an extensive fire hazard analysis.

12.7 Fatigue behaviour of FRP bridge decks

Cyclic loading promotes the creation and development of flaws and micro-cracks in FRP materials. Also, as loading frequency increases, the stress difference between fibre and resin increases, resulting in interfacial damage. The cyclic loading category in bridges is called high cycle, or fatigue loading, where the load history contains cycles in the range of two million at a low bond stress range. Fig. 12.8 shows the setting used for fatigue testing of an all composite bridge deck. Such high cycle

Figure 12.8 *Laboratory fatigue testing of FRP composite bridge deck.*

loading is a problem at service load levels. The most significant effect of high level repeated loads is the reduction of bond strength at failure. Moreover, since bridge decks are subjected to the combined effects of extensive cyclical loads, and long-term aggressive environments, their endurance limits under varying temperatures and aggressive environments have to be established for design purposes. Literature in this research area is very limited.

Sanjeev and GangaRao (1998) studied the fatigue behaviour of concrete decks reinforced with GFRP rebars. The decks were subjected to 2 million fatigue loading cycles. Stiffness reductions of 30% to 40% were found in the concrete decks after cracking, compared to the uncracked sections. Decks transversely post-tensioned were found to have a degradation rate one-fifth of those not transversely post-tensioned. Furthermore, it was found that the degradation rate of GFRP reinforced decks compared well with that of steel reinforced decks in the fatigue crack propagation zone. The residual deflection zone was found to be nearly zero and a possible reason is that the stress induced in the rebar is less than the endurance limit of the FRP rebars when under tensile fatigue.

In terms of the application of FRP in FRP modular decks, the reported design method is based on the American Association of State Highway and Transportation Officials (AASHTO) LRFD specifications (1994) and the AASHTO standard specifications (1996). The LRFD specification defines the reliability based limit states to be considered in bridge deck design and provides load combinations and load factors. However, neither reliability calibration, nor the material resistance factors for FRP decks are available in the current specification. Mertz and Kulicki (1997) proposed that for the extension of LRFD to FRP decks, three limit states should be considered: service, strength and fatigue. The fatigue limit state was discussed by Lopez-Anido, R. et al. (Benmokrane and Rahman 1998), and the fatigue performance criteria were proposed based on assessing the stiffness degradation and the strength reduction. The test results indicated that the FRP deck performed well under fatigue loading without major degradation in stiffness and strength.

12.8 Durability studies on the bond between FRP and concrete

Bonding between FRP and concrete is important for the structural integrity of both FRP internally reinforced and FRP externally reinforced concrete structures. Although attention has been given to this research area, the reported literature is very limited. A review has been given by Nanni et al. (1995) with the emphasis on the test methods of FRP–concrete bonding. Other reviews have been given by Chin et al. (1996) and Liao et al. (1998).

12.8.1 The bond between FRP internal reinforcement and concrete

Effect of thermal cycles

In conventional steel reinforced concrete structures, the coefficients of thermal expansion for steel rebar and concrete are almost similar. However, the coefficients of thermal expansion of FRP rebars vary in the longitudinal and transverse directions, depending upon the types of fibre, resin, volume fraction of the fibre, and the orientation of the fibre. FRP rebars generally consist of unidirectional fibres, except when the surface configuration is modified through the use of veils or fibre types. An absence of transverse fibres results in a transverse coefficient of thermal expansion higher than the longitudinal coefficient of thermal expansion and is about 5 times larger than that of concrete.

Longitudinal thermal expansion of FRP rebar is low, since it is fibre dominated. However, the transverse expansion coefficient of FRP rebar is matrix dominated, and is 3 to 5 times larger than that of concrete. For FRP internal reinforcement applications, it is expected that the longitudinal strength of FRP rods will not be affected significantly in a temperature range of up to 300°C. However, the bond strength of FRP rebars to the concrete is mainly the function of the resin polymer at the surface of the rods (Bank et al. 1998). Therefore, when dealing with the effect of temperature on FRP internal reinforced concrete, the bond strength of FRP rebars will be affected first as the temperature changes.

High temperature has a detrimental effect on bond due to the reducing shear stiffness in FRP. Tsuji et al. (1991) conducted pullout tests on CFRP, AFRP, GFRP rod internal reinforced samples, which were exposed to 80°, 140°, 200°, and 260°C. The reported residual bond strength was in the range of 56% to 86% of the initial values. Katz et al. (1999) studied the bond properties of FRP rebar with different surface treatment at temperatures ranging from 20°C to 250°C. The results showed a 36% to 90% reduction in bond strength as the temperature increased from 20°C to 250°C.

Transverse thermal expansion of FRP rebars may lead to a splitting-type cracking. Gentry and Bank (1994) reported splitting-type cracking that paralleled the reinforcement in precast concrete reinforced with a glass/vinyl ester rebar. Matthys et al. (1996) conducted numerical simulations to predict thermal-induced transverse cracking in concrete reinforced with aramid composite bars and strips. Strain gauge techniques were used by Gentry and Bank (1994) to measure the coefficients of thermal expansion of unidirectional FRP bars. The transverse coefficient of thermal expansion was found to be 3 to 4 times higher than that of plain concrete. The authors also noted that the helical wrapping of the glass roving appeared to constrain the transverse thermal expansion of the bars.

Gentry and Husain (1998) presented a thermoelastic solution of plain and spirally wrapped FRP bar, embedded in concrete, subjected to a uniform temperature increase. The results showed that spiral wrapping reduced the thermal expansion of the rebar. With relatively small temperature increases, tensile stresses in the concrete surrounding the rebar exceed the tensile strength of concrete. Non-linear thermoelastic simulation and finite element simulation demonstrated that cracks decreased with a decrease in rebar diameter and increase in the confining pressure of the concrete. Another durability concern, which is receiving much attention, is the effect of low temperature and freeze–thaw cycles on the bond between the FRP and the concrete. The literature indicated that 200 freeze–thaw cycles do not change the bond behaviour of GFRP and CFRP bar reinforced concrete structures; AFRP bars (both braided and coiled types) show up to a 20% reduction in the bond strength (Mashima and Iwamoto 1993).

Accelerated ageing tests

To study the long-term bond behaviour of FRP rods embedded in concrete, accelerating ageing methods were commonly employed before or after embedding the rods in the concrete. The common accelerated ageing conditions include mechanical loading, moisture, chemical exposure, and temperature.

It was observed that bond strength either increased or did not change when GFRP/concrete blocks were cast, conditioned, and then subjected to direct pullout tests. This may be due to the swelling of the GFRP and a consequent increase in the frictional component of the bond or slower degradation due to the limited access of moisture in the bond region (Porter and Barnes 1998). Al-Dulaijan et al. (1996) observed significant lower bond strengths when commercial GFRP rods were subjected to alkaline conditioning before casting in concrete. The literature suggested that conditioning GFRP rods before embedment in pullout specimens was more deleterious to bond than conditioning after embedment. It is still not possible to exactly relate the acceleration method to the real-time long-term bond behaviour of FRP reinforced concrete structures.

Bakis et al. (1998) compared the bond and tensile strength of various GFRP rods after an accelerated ageing test. The rods were made of three different resin blends: 100% vinyl ester; 50% vinyl ester and 50% iso-polyester; and 20% vinyl ester and 80% iso-polyester. Weight and cross-sectional area measurements, direct pullout bond tests in concrete, and stand-alone tensile tests were used to evaluate the environmental effects. The conditioning consisted of the immersion of GFRP rods in an alkaline solution (pH 12–13), followed by a dry period. The rods were then cast in concrete for pullout tests. Weight and cross-sectional area changes were not dependent upon resin type. Tensile strength degradation was the greatest in the rods containing increased polyester. The ultimate bond strength did not change significantly for the 100% vinyl ester rods. With the blended resins, however, the ultimate bond strength actually increased after conditioning. This increase was more evident as the proportion of the polyester in the resin increased.

Nanni et al. (1998) investigated the acceleration of FRP bond degradation under two conditions: sustained loading and alkaline solution. Two types of FRP rods were used: glass vinyl ester (MGV) and carbon vinyl ester (MCV). After direct exposure to either of the conditioning regimes, residual bond strengths were measured through direct pullout testing, to assess the acceleration of the bond degradation. The results indicated that sustained load conditions were not an accelerator of FRP–concrete bond degradation, whereas, environmental conditions were. The pullout tests were performed on two types of rods with an identical surface configuration. The pullout specimen consisted of an FRP rod embedded concentrically in a concrete cube, with a side dimension of 15 cm. The embedded length of the FRP rod within the concrete cube was adjusted by sliding plastic tubing along both ends of the rod; this tube served as a bond-breaker. All rods used in this study were nominally 12.7 mm in diameter, and had an embedded length of 5 diameters. The environmental conditions included unembedded and embedded FRP rods in a saturated calcium hydroxide solution, which had a pH of 12–13, at 26°, 60°, and 80°C. In the first condition scheme, unembedded and embedded FRP (MGV) rods were immersed directly in the solution. After conditioning, the rods were allowed to dry in an ambient environment for 10 to 14 days, before they were cast into the pullout specimens. They were called condition-then-cast specimens. The second scheme involved the immersion of FRP–concrete pullout specimens in the calcium hydroxide solution. It was called the cast-then-condition conditioning scheme, and included both MGV and MCV specimens. After conditioning, cast-then-condition specimens were allowed to dry in an ambient environment for 10 to 14 days, before being subjected to pullout testing.

Kojlma et al. (1998) studied static and dynamic bond characteristics of FRP tendons embedded in pretensioned, prestressed concrete members. Braided AFRP and twisted CFRP rods were used. The bond behaviour was affected significantly by prestress, embedded length, repeated loading as well as the rod type. By the introduction of prestress, the slipping load and the maximum load of the specimens with AFRP tendons increased. Furthermore, cycling loading was found to affect the bond behaviour.

Sen et al. (1998) conducted a three-year exposure study, to evaluate the bond behaviour between the CFRP and the concrete for the CFRP pretensioned elements after exposure to wet–dry cycles. The bond was evaluated by testing pretensioned beam specimens in flexure. The beams were designed to fail by the rupture of the CFRP material, so that both material and bond degradation were due to the changes in the ultimate capacity. Thus, the results provide a quantifiable measurement of degradation. Thirteen beams were used in this study: Three unexposed specimen, and ten exposed specimens. The beam dimensions were 114 mm × 150 mm × 2.44 mm. These dimensions ensured flexural failure by rupture of the CFRP rod. Exposed specimens were periodically removed and tested to failure to determine their service and their ultimate response. The failure modes were sensitive to the extent of the precracking damage sustained prior to exposure. The bond between CFRP and concrete showed evidence of degradation in severely damaged specimens, possibly due to the moisture absorbed by the epoxy matrix. In less damaged specimens, there was no significant bond degradation. Thus, if CFRP is used to replace steel in pretensioned piles, driving stresses should be carefully monitored to prevent any damage.

12.8.2 Bond between FRP external reinforcement and concrete members

The durability of FRP external reinforcement (e.g. sheets, plates) can basically be determined in the same way as for FRP external reinforcements (e.g. FRP rods). However, durability related to bond performance in concrete is different. The adequacy of the properties of the polymer between the FRP and the concrete interface is essential in maintaining the bond between FRP and concrete. It is well known that the performance of the polymer will be affected by a variety of environmental and mechanical conditions. Therefore, it is essential to understand the durability performance of the bond between the FRP and the concrete for the successful application of this technique. In general, although the literature is limited in this research area, and conclusive results have not been achieved, some studies are being conducted to identify the effects of the environmental conditions (e.g. thermal cycles, fluids), mechanical conditions (e.g. cycling loading) and their combined effects on bond behaviour.

The effects of acidic and alkaline conditions under varying and constant temperatures, on bond in concrete beams wrapped with carbon fibre sheets were studied (Javed 1996). The results showed acidic exposure led to a 17% decrease in bond shear strength while alkaline exposure led to a 24% decrease in bond shear strength. Hygrothermal conditions led to a 29% decrease in bond shear strength. Toutanji and Gomez (1997) studied the bond durability of concrete beams externally bonded with GFRP and CFRP sheets. The specimens were exposed to two different environments: room temperature (20°C), and 300 wet–dry cycles (salt water was used for the wet cycles and the dry cycles were at 35°C and 90% humidity). Wet–dry cycles led to the degradation of epoxy, which reduced the bond between the FRP sheet and the concrete.

Ferrier et al. (1998) noted that the shear modulus of resin at the interface between the FRP and concrete is critical to the durability of FRP externally reinforced concrete. The authors studied the effect of temperature on the shear modulus of the resin at the interface between the FRP and the concrete. The results indicated that with exposure to a 20°C dry environment an 80% decrease in shear modulus was recorded, while a 70% decrease in the shear modulus was reported for the samples exposed at 60°C in the dry environment.

Freeze–thaw and wet–dry cycles exposure was reported to have a significant effect on the bond transfer length, shear stress distribution, and differential strain between the concrete and the GFRP plate (Mukhopadhyaya et al. 1998). Green and Bisby (1998) studied the effects of freeze–thaw cycles on the bond between the CFRP sheet and the concrete beams. After 50, 150, and 300 freeze–thaw cycles with temperatures ranging from –18°C to 15°C, no degradation of the bond of the CFRP strengthened beams, was noted.

Warren et al. (1998) conducted a durability study on CFRP external strengthened concrete beams under marine conditions. Unreinforced concrete beams, 90 mm × 90 mm × 600 mm, were prepared. In one group of beams, CFRP rods were embedded in epoxy resin in a longitudinal slot cut in the concrete. In the second group, CFRP sheets were glued to the surface of the beams with epoxy. One group was exposed to one cycle at –23°C. Other groups were exposed to cyclical temperature in an environmental chamber from –23°C to 60°C every 4 hours. In addition, two groups of samples were exposed to a continuous salt fog of 46°C for 1 year. Exposure of the test specimens in a differential temperature of 83°C for 762 thermal cycles did not lead to visible cracking. After 1188 cycles, all of the samples strengthened with CFRP rods contained visible cracks in the concrete adjacent to the groove filled with epoxy. Exposure to the 46°C salt fog conditions did not result in any visible deterioration of the bond in any of the samples.

Green et al. (2000) studied the effect of freeze–thaw cycles on the bond durability between FRP plate and concrete. The experimental programme had two phases. One portion of the work, used single lap CFRP–concrete joints loaded in shear. The other portion used modified flexural specimens to examine similar bond parameters. The two different specimen types were crucial in order to first examine the FRP–concrete joints loaded in pure shear and then the joints loaded in both shear and flexure, so that the results would be applicable to both shear and flexural strengthening applications. Freeze–thaw cycles of all the specimens were conducted using a cold climate testing facility. Freeze–thaw cycles were applied to the blocks at a rate of one cycle per day, in accordance with ASTM C310 (1971) with 16 hours of freezing in cold air at –18°C followed by 8 hours of thawing in a warm bath at 15 °C. Controlled specimens were stored at room temperature and relative humidity. The specimens were divided into groups, which were subjected to 0, 50, 150, or 300 freeze–thaw cycles.

In the pullout tests, the average shear stress between the two strain gauges was determined using following equation:

$$f_b = \frac{\varepsilon_{n+1} - \varepsilon_n}{\Delta x} E_p t_p \tag{1}$$

where; f_b is the bond stress (MPa), ε_n is the strain at a particular strain gauge, t_p is the thickness of the sheet and is the distance between the gauges.

In the flexural tests on the beams, the average bond stress at failure was calculated using the following equation:

$$\tau_{bond,ave} = \frac{\varepsilon_{p\max} E_p A_p}{L b_p} \tag{2}$$

where; $\tau_{bond.ave}$ is the average bond stress at failure, $\varepsilon_{p\max}$ is the maximum strain in the plate at failure, E_p is the elastic modulus of the plate material, A_p is the cross-sectional area of the plate material, L is the total bond length and b_p is the plate width.

The results indicated that freeze–thaw cycling does not reduce the load-carrying capacity of the joint between the concrete and the FRP plates when the joints were loaded either in pure shear or in combination of shear and flexure. Nevertheless, the adhesive may be affected slightly by the freeze–thaw exposure resulting in changes in the failure modes. The authors recommended that the experimental work represented the first step in the study of FRP–concrete bond durability. Studies considering a variety of other factors are required to establish adequate durability and design data. For instance, studies using deteriorated (or extensively precracked) members would be helpful. An investigation into the effects of freeze–thaw cycling under sustained load was recommended. Tests on different epoxy binding agents under the effects of freeze–thaw cycling were also recommended.

12.9 Development of design guidelines and design codes

The most important development in the use of FRP composites for civil infrastructure is the development of design guidelines. During the past decades, substantial progress has been made in the development of various design guides, draft codes and experimental guidelines for the application of FRP composites in civil engineering infrastructure. Attention has been given to the durability parameters in these draft codes (e.g. environmental reduction factors, resistance factors, creep stress levels, fatigue stress limits, and fire endurance). A summary of the draft guidelines is given in the appendix. The results of these international efforts on developing design guides are expected to materialise in the near future.

12.10 Summary and concluding remarks

A number of technical issues, concerning the durability of FRP composites and FRP reinforced concrete structures exposed to civil infrastructure environments have been reviewed. These issues included fluids, alkalinity, freeze–thaw, UV rays, fire, fatigue, and creep/relaxation. Categories studied included the constituent materials of FRP composites, FRP external reinforcement (sheet, laminates, tubes), FRP internal reinforcement (rods, tendons), FRP composite decks, FRP structural shapes, FRP internal reinforced concrete members, FRP external reinforced concrete members and FRP bridge decks. From these reviews, the following conclusions were obtained.

Development on design guidelines

Substantial progress has been made in the development of various draft codes for the application of FRP composites in civil engineering infrastructure. Attention has been given to durability and damage tolerance in these draft design guides. Some durability parameters have been introduced in the draft codes (e.g. environmental reduction factors, resistance factors, creep stress levels, fatigue stress limits, and fire endurance). More durability parameters must be included in the design codes, and, further studies must be conducted to calibrate these parameters.

Durability studies on FRP composites

In terms of the effect of fluids, extensive data have been documented from the aerospace industry, however, there is a lack of long-term data on the effects of fluids on pultruded FRPs. The effects of fluids on FRP reinforcement were studied in terms of mechanical properties. Fick's law was used to simulate fluid diffusion into FRP reinforcement, and to predict the strength properties. Further study needs to be conducted on the combined effect of temperature, fluids and loading patterns. With respect to the effects of alkaline environments, extensive studies have been conducted concerning the effect of alkaline environments on FRP composites and their constituents, involving alkali ion penetration, simulation of the phenomenon of penetration of alkali, the effect of alkalinity on structural engineering properties, the effect of alkali on resin matrix, and fibres. Studies on FRP reinforcements exposed to an alkaline environment have been reported for the engineering properties such as tensile, compressive, fatigue, creep and bond. Only epoxy and vinyl ester resins are suggested for use in FRP reinforcement for construction. Polyester resin is not recommended. From the literature, most of the researchers indicated that further study needs to be conducted to establish a standard test method. Furthermore, approaches to the long-term strength prediction needs additional study.

The behaviour of FRP composites under long-term real-life sustained loading may exhibit time-dependent responses due to the viscoelastic behaviour of the polymeric matrix especially under conditions of high temperature and moisture. Hence, creep is considered to be the critical design issue. Limited data exists for long-term creep behaviour. Most of the existing data refer to tensile creep. There is a lack of data concerning the synergistic effects of moisture, alkalinity, and temperature. Furthermore, the creep behaviour of FRP reinforcement in concrete under sustained loads needs additional study to determine the creep rates for various levels of sustained stress. In such tests, the effect of sustained load on the residual tensile strength, the influence of temperature on the time-deformation relationship, creep coefficients, moisture impact on creep performance, and endurance time should be considered. The accelerated ageing test may be used to assess the materials specifications for FRP reinforced rebars, to minimise the ageing of the materials and to establish the remaining life.

Fatigue behaviour is one of the important parameters that needs to be considered for bridge decks, parking areas and other structures subjected to long-term cyclic loading. Various aspects of fatigue behaviour of FRP composites have been studied, which involve fatigue mechanisms, the effect of the constituent materials, environmental condition effects, frequency to fatigue testing, fatigue life prediction methodologies, stress corrosion in the resin, stress corrosion in the fibres, and the stress rupture of the composites. The literature indicates that a reasonable amount of data exist concerning the fatigue of FRP composites produced from low cost fabrication methods. However, these data are available for only a limited set of fatigue conditions. More data are needed about the fatigue of composites in conjunction with combined environment conditions such as temperature, alkalinity, moisture, and UV.

There is a lack of long-term data concerning the freeze–thaw effect on pultruded FRP composites. Freeze–thaw in the presence of salt can result in accelerated degradation due to the formation and expansion of salt crystals in addition to the moisture induced swelling and contraction. Combined thermal and moisture effects can be significant in the case of freeze–thaw and needs to be considered at the design stage.

The effect of UV radiation on thick FRP composites has been reported to cause only minor changes in the mechanical properties. Thinner FRP specimens exhibited greater changes in their mechanical properties. It is difficult to assess the effects of UV only as most studies are carried out in an outdoor environment or in an artificial weathering device where moisture was also present. Limited studies have been conducted on the effect of ultraviolet light on FRP reinforcements (e.g. rods, tendons, rebars). More studies need to be conducted to determine the effect of UV on FRP products.

There is very limited literature on the properties of FRP at high temperatures, and on the fire performance of FRP–concrete systems. The literature shows that a temperature of 250°C will reduce the tensile strength of GFRP and CFRP rebars by 20%, and by 60% in AFRP rebars. Phenolic resin based composites offer better fire performance than many existing polymer composites such as polyester and vinyl ester based composites. However, the mechanical properties of phenolic resin based composites are not well documented.

Durability studies on FRP reinforced concrete structures

Studies on the durability performance of FRP (internally and externally) reinforced concrete members have received attention recently. Several durability parameters have been investigated such as fatigue, creep, freeze–thaw cycles, and the effect of high temperature and fire. Literature is very limited in this emerging research area.

The literature indicated that freeze–thaw exposure led to a decrease in the moment capacity and in the ultimate deflection of FRP externally strengthened beams. The decrease rate was larger for precracked specimens than for uncracked specimens. Freeze–thaw exposure did not affect the flexural behaviour of FRP bar reinforced RC beams, however, the combined effect of freeze–thaw cycles and an aggressive environment increases the crack-patterns. Tests on FRP wrapped concrete cylinders exposed to freeze–thaw action have revealed that freeze–thaw cycles lead to more brittle failure of FRP wrapped concrete cylinders than similar specimens kept at room temperature. Moreover, FRP wrapped cylinders exposed to freeze–thaw cycles show a significant increase in strength over unwrapped cylinders exposed to the same freeze–thaw cycles. Increasing the number of FRP layers led to a bigger increase in strength

Studies on prestressed concrete members indicated that the overall fatigue characteristics of GFRP bars pretensioned concrete beams matched those of similarly loaded steel pretensioned beams. Fatigue behaviour of GFRP pretensioned beams in the post-crack range resulted in much higher deflection and crack widths than similar beams reinforced with pretensioned steel. The literature showed that the fatigue life of RC beams could be enhanced significantly through the use of externally bonded FRP reinforcement.

Limited studies have been conducted on the viscoelastic behaviour of FRP (internally and externally) reinforced concrete structures. For externally reinforced RC members, experimental studies have been conducted on both small scale and large scale FRP laminate bonded RC beams. Externally bonded FRP plates lead to a reduction in the time-dependent deformation, and reduce the rate of compressive creep in the concrete of the strengthened systems. In terms of the FRP bar reinforced concrete beams, the literature indicates that the creep curve of FRP bar reinforced concrete beams is similar to that of steel reinforced members. The creep deflection of FRP bar reinforced concrete beams could be predicted using the existing theory for steel reinforced concrete beams with a modification made for the creep coefficients of the FRP reinforced sections. Although theoretical analyses have been conducted in this area, the literature is inconclusive.

Currently, the application of FRP composites is associated with concrete structures. There is insufficient information on the relationships between standard, bench-scale fire test performance and full-scale fire behaviour of all types of composites and FRP reinforced concrete members. This limits the validity of the fire hazard analysis.

The fatigue behaviour of FRP reinforced concrete decks or FRP modular decks is an important durability issue. The literature is very limited in this area. Future studies need to be conducted to establish the endurance limit of FRP reinforcement in tension-fatigue loading. In addition, the endurance limits under varying temperatures and aggressive environments need to be established to arrive at the threshold limits for design purposes.

Durability studies on the bond between FRP and concrete

Limited theoretical and experimental studies have been conducted on the durability of the bonding between FRPs and the concrete in reinforced (externally and internally) concrete structures. The parameters studied were thermal effects, accelerated ageing tests, and an alkaline environment. For FRP internal reinforced concrete members, high temperature has a detrimental effect on bond due to the reducing shear stiffness in the FRP. Freeze–thaw and wet–dry cycles exposures were reported to have a significant effect on the bond transfer length, shear stress distribution, and different strains between GFRP and the concrete, with less effect on the bond between CFRP and the concrete. Further studies need to be conducted to establish the accurate reduction factors in the bond strength equation for design purposes.

Bibliography

State-of-art reports, draft codes and design guidelines

ACI 318-05 (2005) *Building Code Requirements for Structural Concrete and Commentary*. American Concrete Institute Committee 318, 430pp.

ACI 440 (2001) (a) *Guide for the Design and Construction of Concrete Reinforced with FRP Bars*, draft, American Concrete Institute Committee 440H, 97 p.

ACI 440 (2001) (b) *Guide for the Design and Construction of Externally Bonded FRP Systems for Strengthening Concrete Structures*, draft, American Concrete Institute Committee 440F, 78 p.

ACI 440R (1996) *State-of-the-art Report on Fibre Reinforced Plastic Reinforcement for Concrete Structures*, American Concrete Institute Committee 440, 67 p.

AIJ (1991) *Evaluation items and evaluation methods for new reinforcement materials and new reinforced concrete* (in Japanese), Report, Task Committee on Continuous Fibre Composite Materials, Architectural Institute of Japan, Tokyo, Japan.

ASTM C666-97 (1997) *Standard Test Method for Resistance of Concrete to Rapid Freezing and Thawing* (now superseded) ASTM International.

BRI (1993) *Design guidelines of FRP reinforced concrete building structures* (in Japanese), Building Research Institute, Japanese Ministry of Construction, Tokyo, Japan. A translation is published as: Sonobe et al. (1997) 'Design guidelines of FRP reinforced concrete building structures', *Journal of Composites for Construction*, Vol. 1, No. 3, August, ASCE, USA, pp. 90–115.

BRI (1995) *Design guidelines for FRP prestressed concrete members* (in Japanese), Building Research Institute, Japanese Ministry of Construction, Tokyo, Japan.

CHDBC (1998) *Canadian Highway Bridge Design Code*, First Edition.

CSA (2000) *S806-00 Design and Construction of Building Components with Fibre Reinforced Polymers*, Canadian Standards Association, Ontario, Canada, Draft 7, 192 p.

CUR (1996) *Kunststof wapeningselementen in beton - Preadvies*, CUR report 96-9 (in Dutch), CUR preliminary advisory Committee PC 97, Centre of Civil Engineering Research and Codes (CUR), The Netherlands, 66 p.

JSCE (1992) *Application of continuous fiber reinforcement materials to concrete structures* (in Japanese), Concrete library No. 72, Japanese Society for Civil Engineering, Tokyo, Japan.

JSCE (1993) *State-of-the-art report on continuous fiber reinforcing materials*, Concrete Engineering Series 3, Japanese Society for Civil Engineering, Tokyo, Japan, 164 p.

JSCE (1997) *Recommendation for design and construction of concrete structures using continuous fibre reinforcing materials*, Concrete Engineering Series 23, Ed. A. Machida, Research Committee on Continuous Fibre Reinforcing Materials, Japan Society of Civil Engineers, Japan, 325 p.

JCI (1998) *Technical Report on Continuous Fibre Reinforced Concrete*, Technical Committee on Continuous Fibre Reinforced Concrete (JCI TC952), Japan Concrete Institute, Japan, 151p.

SINTEF (1998) *Modifications to NS3473 when using fibre reinforced plastic (FRP) reinforcement*, SINTEF Report STF22A98741, Sintef Structures and Concrete, Norway, 40 p.

IStructE (1999) *Interim guidance on the design of reinforced concrete structures using fibre composite reinforcement*, Institution of Structural Engineers, UK, 116 p.

ISIS Canada (2000) (a) *Reinforcing Concrete Structures with Fibre Reinforced Polymers*, The Canadian Network of Centers of Excellence on Intelligent Sensing for Innovative Structures, ISIS-M04-00, University of Winnipeg, Manitoba, Canada, 40 p.

ISIS Canada (2000) (b) *Strengthening Reinforced Concrete Structures with Externally-Bonded Fibre Reinforced Polymers*, The Canadian Network of Centres of Excellence on Intelligent Sensing for Innovative Structures, ISIS-M05-00, University of Manitoba, Manitoba, Canada, 156 p.

Japan Society of Civil Engineers (1997), *Recommendation for Design and Construction of Concrete Structures Using Continuous Fiber Reinforced Materials*, Concrete Engineering Series 23, Ed. by A. Machida, Research Committee on Continuous Fibre Reinforcing Materials, Tokyo, Japan, 325 p.

Japan Concrete Institute (1998) *Continuous Fiber Reinforced Concrete*, Technical Report, JCI TC952 on Continuous Fiber Reinforced Concrete, Published by Japan Concrete Institute, Tokyo, Japan, 163 p.

Machida, A. (Ed.) 1993, *State-of-the-Art Report on Continuous Fiber Reinforcing Materials*, Concrete Engineering Series 23, Published by Japan Society of Civil Engineers (JSCE), Tokyo, Japan, 164 p.

References

Adimi, R., Rahman, H., Benmokrane, B. and Kobayashi, K. (1998) 'Effect of temperature and loading frequency on the fatigue life of a CFRP rebar in concrete', *Proceedings of the Second International Conference on Composites in Infrastructure (ICCI–98), Tucson, Arizona*, Vol. 2, pp. 203–210.

Arntsen, B. and Pedersen, V. (1999) 'Freeze–thaw durability of concrete beams strengthened with CFRP', *Proceedings of the Eighth International Conference on Structural Faults and Repair, London*, pp. 125–131.

Al-Dulaijan, S.U. (1996) 'Effect of environmental and mechanical conditioning on bond between FRP reinforcement and concrete', PhD dissertation, The Pennsylvania State University, University Park, PA, pp. 319.

Allred, R.E. (1984) 'The effects of temperature and moisture content on the flexural response of Kevlar/epoxy laminates: part II (45/0/90) filament orientation', *Environmental Effects on Composites Materials*, Vol. 2, Ed. G. Springer, Lancaster, PA: Technomic Publishing Company, Inc.

American Association of State Highway and Transportation Officials (1994) *LRFD Bridge Design Specifications*, Washington, DC.

American Association of State Highway and Transportation Officials (1996) *Standard Specifications for High Bridges*, Washington, DC.

Arockiasamy, M., GangaRao, H., Benmokrane, B. and Shahawy, M. (1999) 'Fibre reinforced polymer composites for concrete bridge deck reinforcement', Interim report, prepared for National Cooperative Highway Research Program, Transportation Research Board, National Research Council, pp. 190.

Bank, L.C. and Mosallam, A.S. (1990) 'Creep and failure of a full size fiber reinforced plastic pultruded frame', *Composites Material Technology* 1990, Vol. 1, pp. 49–56.

Bank, L.C., Puterman, M. and Katz, A. (1998) 'The effect of material degradation on bond properties of FRP reinforcing bars in concrete', *ACI Material Journal*, Vol. 95 No. 3, pp. 232–243.

Bakis et al. (1998) 'Effect of cyclic loading on bond behaviour of GFRP rods embedded in concrete beams', *Journal of Composites Technology and Research*, Vol. 20 No. 1, pp. 29–37.

Barnes, R., and Garden, H.N. (1997) 'Time-dependent behaviour and fatigue', *Strengthening of Reinforced Concrete Structures: Using Externally-Bonded FRP Composites in Structural and Civil Engineering*, Eds L.C. Hollaway, and M.B. Leeming, Woodhead Publishing Limited.

Benmokrane, B. and Rahman, H. (Eds) (1998) 'Durability of fibre reinforced polymer composites for construction', *Proceedings of the CDCC98, Sherbrooke, Québec, Canada*, 692 p.

Benmokrane, B. and El-Salakawy, E. (Eds) (2002) 'Durability of Fiber Reinforced Polymer (FRP) Composites for Construction', *Proceedings of the CDCC02, Montreal (Québec), Canada*.

Benmokrane, B. (2000) 'Improvement of the durability performance of GFRP reinforcements for concrete structures', Research report, Department of Civil Engineering, University of Sherbrooke, Sherbrooke, Québec, Canada, 50 p.

Bhatnagar, A., Lakkad, S.C. and Ramesh, C.K. (1981) 'Creep in high glass content unidirectional and bi-directional GRP laminates', *Fibre Science and Technology*, Vol. 15, pp. 13–25.

Brown, V. and Bartholomew, C. (1996) 'Long-term deflections of GFRP reinforced concrete beams', *Proceedings of the 1st International Conference on Composites in Infrastructure (ICCI– 96), Tucson, Arizona*, pp. 389–400.

Buck, S.E., Lischer, D.W. and Nemat-Nasser, S. (1998) 'The durability of E-glass/vinyl ester composite materials subjected to environmental conditioning and sustained loading', *Journal of Composites Materials*, Vol. 32, pp. 874–893.

Challal, O., Houde, J., Benmokrane, B. and Aitcin, P.C. (1991) 'Use of a new glass-fibre rod as reinforcement for concrete structures', *Evaluation and Rehabilitation of Concrete Structures and Innovations in Design, Proceedings, ACI International Conference, Hong Kong*, 1991, Sp.–128, Vol. I, pp. 515–528.

Chin, J. (1996) 'Materials aspects of fibre reinforced polymer composites in infrastructure', NISTIR 5888, National Institute of Standard and Technology, August.

Chin, J.W., Nguyen, T. and Aouadi, K. (1997) 'Effects of environmental exposure on fiber-reinforced plastic (FRP) materials used in construction', *Journal of Composites Technology and Research*, Vol. 19 No. 4, pp. 205–210.

Coomarasamy, A. and Goodman (1999) 'Investigation of the durability characteristics of fiber reinforced plastic (FRP) materials in concrete environment', *Journal of Thermoplastic Composite Materials*, Vol. 12, No. 3, pp. 214–226.

Conrad, J.O., Bakis, C.E., Boothby, T.E. and Nanni, A. (1998) 'Durability of bond of various FRP rods in concrete', *Proceedings of the First International Conference on Durability of Fibre Reinforced Polymer (FRP) Composites for Construction*, Eds B. Benmokrane, and H. Rahman, pp. 299–310.

Daminan, K., Thomas, S.O. and Derek, J.H. (1998) 'Newly developed manufacturing process enhances durability of hollow FRP rebar', *Proceedings of the First International Conference on Durability of Fibre Reinforced Polymer (FRP) Composites for Construction, Sherbrooke, Québec, Canada*, pp. 311–316.

Deler, L.C. and Miller, R.W. (1988) 'Characterization and weather aging and radiation susceptibility', *Engineering Material Handbook: Vol. 2 – Engineering Plastics*, Metals Park, OH ASM International. pp. 576–580.

Del Mar Lopez, M., Naaman, A.E. and Till, R.D. (1999) 'Bending behaviour of reinforced concrete beams strengthened with carbon fibre reinforced polymer laminates and subjected to freeze–thaw cycles', *Proceedings of the Fourth International Symposium: Fibre Reinforced Polymer Reinforcement for Reinforced Concrete Structures*, (Eds) Dolan, C.W., Rizkalla, S.H. and Nanni, A., ACI SP–188, pp. 559–575.

Dillard, D.A. 1991 'Viscoelastic behaviour of laminated composite materials', *Fatigue of Composite Materials*, Ed. K.L. Reifsnider, Elsevier Science Publishers B.V., New York, pp. 339–384.

Dolan, C.W., Rizkalla, S. and Nanni, A. (Eds) (1999) 'Fiber reinforced polymer for reinforced concrete structures', *Proceedings of the Fourth International Symposium: Fibre Reinforced Polymer Reinforcement for Reinforced Concrete Structures*, Eds C.W. Dolan, S.H. Rizkalla, and A. Nanni, ACI SP–188, 1182 p.

Dolan, C., Leu. B.L. and Hundley, A. (1997) 'Creep-rupture of fiber reinforced plastic (FRP) reinforcement for concrete structures: Properties and Applications', *Developments in Civil Engineering*, Vol. 42, pp. 129–163.

DuPont de Nemours & Co. (1992) *Kevlar Data Sheet*, Wilmington, DE.

Faza, S., GangaRao, H. and Ajjarapu, S. (1994) *Strength and stiffness degradation of glass reinforced polyester and vinyl ester structural plates*, West Virginia University, Constructed Facilities Centre, Morgantown, WV.

Ferrier, E., Lagarde, D. and Hamelin, P.P. (1998) 'Durability of reinforced concrete beams repaired by composites', *Durability of Fiber Reinforced Polymer (FRP) Composites for Construction*, CDCC 98, Eds B. Benmokrane, and H. Rahman, pp. 217–228.

Ferry, J.D. (1980) *Viscoelastic Properties of Polymers*, Third edition, Wiley-Interscience.

Findley, J.D. (1980) 'Mechanism and mechanics of creep of plastics', *SPE, J.* pp. 57–65.

Garden, H.N., Hollaway, L.C. and Thorne, A.M. (1998) 'The strengthening and deformation behaviour of reinforced concrete beams upgraded using prestressed composites plate', *Materials and Structures* Vol. 31, pp. 247–258.

Gates, T.S. (1993) 'Effect of elevated temperature on the viscoplastic modelling of graphite/polymeric composites', *High Temperature and Environment Effects on Polymer Composites*, ASTM STP 1174, pp. 201–221.

Gentry, T.R. and Husain, M. (1998) 'Thermal compatibility of concrete and composite reinforcement: analytical and numerical simulation', *Proceedings of the durability of fibre reinforced polymer (FRP) composites for construction*, Eds B. Benmokrane and H. Rahman, pp. 203–216.

Gentry, T.R. and Bank, L.C. (1994) 'Thermal compatibility of plastic composite reinforcements and concrete', *Proceedings of 3rd Material Engineering Conference ASCE, Reston*, pp. 575–582.

Gerritse, A. and Den Uij, J.A. (1995) 'Long-term behaviour of Arapree', *Non-Metallic (FRP) Reinforcement for Concrete Structures – Proceedings of the Second International RILEM Symposium (FRPRCS–2), Gent, Belgium*, Ed. L. Tarewer, pp. 57–66.

Givler, R.C., Gillespie, J.W. and Pipes, R.B. (1982) 'Environmental exposure of carbon/epoxy composite materials systems', *Composites for Extreme Environments*, ASTM STP 768, American Society for Testing and Materials, pp. 137–147.

Gomez, J.P. and Casto, B. (1996) *Freeze–thaw durability of composite materials*, Virginia Transportation Research Council.

Gopalan, R., Somashekar, B.R. and Dattaguru, B. (1989) *Polym. Degrad. Stability*, Vol. 24, pp. 361–371.

Green, M.F. and Bisby, L.A. (1998) 'Effects of freeze/thaw action on the bond of FRP sheets to concrete', *Proceedings of the First International Conference on Durability of Fibre Reinforced Polymer (FRP) Composites for Construction*, Eds B. Benmokrane and H. Rahman, pp. 179–190.

Guetta, B., Bunsell, A.R., Belinski, C. and Lalau Keraly, F. (1989) *Composites*, Vol. 20, No. 5, pp. 461–465.

Hancox, N.L. and Mayer, R.M. (1994) *Design data for reinforced plastics*, Chapman & Hall, New York, NY, pp. 202–204.

Hawkins, G.F., Steckel, G.L., Bauer, J.L. and Sultan, M. (1998) 'Qualification of composites for seismic retrofit of bridge columns', *Proceedings of the First International Conference on Durability of Fibre Reinforced Polymer (FRP) Composites for Construction*, Eds B. Benmokrane and H. Rahman, pp. 25–36.

Hayes, M.D., Garcia, K., Verghes, N. and Lesko, J. (1998) 'The effect of moisture on the fatigue behaviour of glass\vinyl ester composites', *Proceedings, Second International Conference on Composites in Infrastructure, Tucson, Arizona (ICCI–98)*, Eds H. Saadatmanesh and M.R. Ehsanni, V.1, pp. 1–12.

Hojo, H., Tsuda, K., Ogasawara, K. (1991) 'Form and rate of corrosion-resistance FRP resins', *Advanced Composites Materials*, Vol. 1, No. 3, pp. 55–67.

Holmes, M. and Rahman, T.A. (1980) 'Creep behaviour of glass-reinforced plastic box beams', *Composites*, pp. 79. April.

Iyer, S.L. and Anigol, M. (1991) 'Testing and evaluating fiberglass, graphite and steel prestressing cables for pretensioned beams', *Proceedings of the Specialty Conference on Advanced Composite Materials in Civil Engineering Structures, ASCE, New York*, pp. 44–56.

Javed, S., Kumar, S.V. and GangaRao, V.S. (1996) 'Experimental behaviour of concrete beams with externally bonded carbon fibre tow sheets', The 51st Annual Meeting of SPI/CI Conference and Exposition.

Jones, F.R., Rock, J.W. and Wheatley, A.R. (1983) 'Long term durability of stressed GRP in acid environments', *Journal of Material Science Letters*, Vol. 2, pp. 519–521.

Judd, N.C.W. (1971) 'The chemical resistance of carbon fibers and a carbon fiber\polyester composites', *The First International Conference of Carbon Fibers*, Plastics Institute, pp. 1–8.

Kage, T., Masuda, Y., Tanano, Y. and Sato, K. (1995) 'Long term deflection of continuous fibre reinforced concrete beams, Non-Metallic (FRP) Reinforcement for Concrete Structures', *The Second International RILEM Symposium, Ghent, Belgium*, Ed. L. Tarawer, Aug. pp. 251–258.

Karbbari, V.M., Rivera, J. and Dutta, P.K. (2000) 'Effect of short-term freeze–thaw cycling on composites confined concrete', *Journal of Composites for Construction*, Nov. Vol. 4. No. 4, pp. 191–197.

Kato, Y., Nishimura, T., Uomoto, T. and Yamaguchi, T. (1998) 'The effect of ultraviolet rays to FRP rods', *Proceedings of the First International Conference: Durability of Fiber Reinforced Composites for Construction, Sherbrooke, Quebec*, Eds B. Benmokrane, and H. Rahman, pp. 487–499.

Katsuki, F. and Uomoto, T. (1995) 'Prediction of deterioration of FRP rods due to alkali attack', *Non-metallic (FRP) Reinforcement for Concrete Structures, FRPCS–2, Ghent, Belgium*, pp. 108–115.

Katz, A. (1998) 'Effect of helical wrapping on fatigue resistance of GFRP', *Journal of Composites for Construction*, Vol. 2, No. 3, pp. 121–125.

Katz, A., Berman, and Bank, L.C. (1999), 'Effect of high temperature on bond strength of FRP rebars', *Journal of Composites for Construction*, Vol. 3. No. 2, May, pp. 73–81.

Katz, A., Berman, N. and Bank, L.C. (1998) 'Effect of cyclic loading and elevated temperature on the bond properties of FRP rebars', *Proceedings of the First International Conference: Durability of Fiber Reinforced Polymer Composites for Construction, Sherbrooke, Quebec*, Eds B. Benmokrane and H. Rahman, pp. 403–413.

Kodur, V.K.R. and Baingo, D. (1999) *Fire resistance of FRP reinforced concrete slabs*, IR758 Ottawa. Canada National Research Council.

Kojima, T., Inoue, S. and Matsuo, M. (1998) 'A study of static and dynamic bond characteristics of non-metallic fiber tendons in pretensioned concrete', *Durability of Fiber Reinforced Polymer Composites for Construction*, CDCC–98, Benmokrane, B. and Rahman, H., pp. 415–421.

Kumahara, S., Masuda, Y. and Tanano, Y. (1993) 'Tensile strength of continuous fiber rebar under high temperature', *Fiber-Reinforced Plastic Reinforcement For Concrete Structures*, Eds C.W. Dolan, S.H. Rizkalla, and A. Nanni, American Concrete Institute, Detroit, pp. 731–742.

Kuz'min, S.A., Bulmanis, V.N. and Struchkov, A.S. (1989) 'Experimental investigation of the strength and deformability of wound fibre glasses and organoplastic under low climatic temperatures', *Mekhanika Kompozitnykh Materialov*, 1 (Jan./Feb.), pp. 57–61.

Larsson, F. (1988) The effect of ultraviolet light on mechanical properties of Kevlar 49 composites', *Environmental Effects on Composite Materials*, Vol. 3, Springer, Ed., Technomic Publishing Company, pp. 132.

Liao, K., Schultheisz, C.R., Hunston, D.L. and Brinson, L.C. (1998) 'Long-term durability of fibre-reinforced polymer-matrix composite materials for infrastructure application: a review', *Journal of Advanced Materials* Vol. 30. No. 4, 3, pp. 3–40.

Lord, H.W. and Dutta, P.K. (1988) 'On the design of polymeric composite structures for cold regions application', *Journal of Reinforced Plastics and Composites*, Vol. 7. Sept., pp. 435–459.

Malvar, L.J. (1998) 'Durability of composites in reinforced concrete', *CDCC–98 Durability of Fibre Reinforced Polymer Composites for Construction*, Eds B. Benmokrane, and H. Rahiman, Sherbrooke, Canada, pp. 361–365.

Mandell, J.F. and Meier, U. (1983) 'Effects of stress ratio, frequency, and loading time on the tensile fatigue of glass reinforced epoxy', *Long-term Behaviour of Composites*, ASTM STP 813, Ed. T.K. O'Brien, American Society for Testing and Materials, pp. 55–77.

Mashimu, M. and Iwamoto, K. (1993) 'Bond characteristics of FRP rod and concrete after freeze–thaw deterioration', ACI SP–138, *International Symposium on Fiber Reinforced Plastic Reinforcement for Concrete Structures, Vancouver, BC, Canada*, (March) Eds C.W. Dolan, S.H. Rizkalla, and A. Nanni, pp. 51–69.

Matthys, S. et al. (1996) 'Influence of transverse thermal expansion of FRP reinforcement on the critical concrete cover', *Proceedings of 2nd International Conference on Advanced Composites Materials for Bridge and Structure, Canadian Society for Civil Engineering, Montreal*, pp. 665–672.

McBagonluri, F., Garcia, K., Hayes, M., Verghese, K.N.E. and Lesko, J.J. (2000) 'Characterization of fatigue and combined environment on durability performance of glass/vinyl ester composite for infrastructure applications', *International Journal of Fatigue* Vol. 22, pp. 53–64.

Mertz, D.R. and Kulicki, J.M. (1997) 'The application of the AASHTO LRFD bridge design specifications for highway bridge to composite material bridges', *Advanced Rehabilitation, Durability Materials, Non-Destructive Evaluation and Management*, Eds U. Meier and R. Betti, CH. Zurich, pp. 178–184.

Mikami, H. et al. (1990) 'Fatigue characteristics of concrete beams reinforced with braided FRP rods', *Transactions of the Japan Concrete Institute*, Vol. 2, pp. 223–230.

Mohan, R. and Adhams, D.F. (1985) 'Nonlinear creep–recovery response of a polymer matrix and its composites', *Experimental Mechanics* Vol. 9, pp. 262–271.

Monney, L. (1998) 'Ablation of the organic matrix: fundamental response of a photo-aged epoxy–glass fiber composites', *Polymer Degradation and Stability*, Vol. 56, No. 357.

Morgan, R.J., Dunn, C. and Edwards, C. (1999) 'Research panel report: creep and relaxation', Civil Engineering Research Foundation Manufacturing Development Association Study on Durability of Composites in Civil Engineering.

Mukhopadhyaya, P., Swamy, R.N. and Lynsdale, C.J. (1998) 'Durability of adhesive bonded concrete–GFRP joints', *Proceedings of the First International Conference on Durability of Fibre Reinforced Polymer (FRP) Composites for Construction (CDCC–98), Sherbrooke, Canada,* Eds B. Benmokrane and H. Rahman, pp. 373–380.

Nanni, A., Bakis, C.E. and Boothby, T.E. (1995) 'Test methods for FRP-concrete systems subjected to mechanical loads: state of the art review', *Journal of Reinforced Plastics and Composites*, Vol. 14, pp. 524–558.

Nanni, A., Bakis, C.E. and Mathew, J.A. (1998) 'Acceleration of FRP bond degradation', *Proceedings of the First International Conference on Durability of Fibre Reinforced Polymer (FRP) Composites for Construction (CDCC–98), Sherbrooke, Canada,* (Eds) Benmokrane, B. and Rahman, H., pp. 45–56.

Nkurunziza, G., Benmokrane, B., Debaiky, A. S. and Masmoudi, R. (2004) 'Effect of creep and environment on long-term tensile properties of glass FRP reinforcing bars.' *Proceedings of the 4th International Conference on Advanced Composites in Structures and Bridges*, Calgary, Ed Mamdouh El-Badry, pp 145–153.

Nollet, M.L., Perraton, D. and Chaallal, O. (1999) 'Flexural behaviour of CFRP strengthened RC beams under room and freezing temperatures', *Proceedings of the Eighth International Conference on Structural Faults and Repair, London,* pp. 16–19.

Odagiri, T., Matsumoto, K. and Nakai, H. (1997) 'Fatigue and relaxation characteristics of continuous aramid fiber reinforced plastic rods', *The Third International Symposium on Non-Metallic (FRP) Reinforcement for Concretes Structures (FRPRCS–3), Japan Concrete Institute, Tokyo, Japan*, Vol. 2, pp. 227–234.

Okamoto, T., Matsubara, S., Tanigakai, M. and Jasuo, K. (1993) 'Practical application and performance of PPC beams reinforced with braided FRP bars', *Fiber-Reinforced Plastic Reinforcement For Concrete Structure*, Eds C.W. Dolan, S.H. Rizkalla and A. Nanni, American Concrete Institute, Detroit, pp. 875–894.

Pantusco, A., Spadea, G. and Swamy, R.N. (1998) 'An experimental study on the durability of GFRP bars', *Fibre Reinforced Composites in Infrastructure*, Vol. II. ICCI `98, Tucson, Arizona, USA, 5–7 Jan., pp. 476–487.

Plevris, N. and Triantafillou, T.C. (1994) 'Time-dependent behaviour of RC members strengthened with FRP laminates', *Journal of Structural Engineering*, Vol. 120 No. 3, pp. 1016–1042.

Porter, L.M. (1999) 'Durability issues for FRP reinforcement to concrete structures', *SAMPE* Vol. 44, pp. 2246–2254.

Porter, M., Mehus, J. and Young, K. (1995) *Aging degradations of fiber composites reinforcement for structural concrete*, Iowa State University, Engineering Research Institute, Ames, IA.

Porter, M.L. and Barnes, B.A. (1998) 'Accelerated aging degradation of glass fiber composites', *2nd International Conference on Composites in Infrastructure*, Vol. II, Eds H. Saadatmanesh and M.R. Eshanni, University of Arizona, Tucson, pp. 446–459.

Rahman, A.H., Kingsley, J. and Crimi, J. (1996) 'Durability of a FRP grid reinforcement', *Proceedings of the Second International Conference on Advanced Composite Materials in Bridges and Structures, Sherbrooke, Canada*, pp. 681–690.

Rahman, H., Adimi, M.R. and Benmokrane, B. (1996) 'Fatigue behaviour of FRP reinforcements encased in concrete', *Proceedings of the Second International Conference on Advanced Composite Materials in Bridges and Structures, Sherbrooke, Canada*, pp. 691–698.

Rahman, H., Adimi, R. and Crim, J. (1997) 'Fatigue behaviour of carbon FRP grid encased in concrete', *Non-Metallic (FRP) Reinforcement for Concrete Structures. Proceedings of the Third International Symposium*, Vol. 2, Oct. pp. 219–226.

Rege, S.K. and Lakkad, S.C. (1983) 'Effect of salt water on mechanical properties of fibre reinforced plastics', *Fibre Science and Technology*, 19, pp. 317–328.

Regel, V.R. et al. (1967) 'Effect of ultraviolet radiation on the creep rate of polymers', *Mekhanika Polimerov*, Vol. 3, No. 4, pp. 615–618.

Rostasy, F.S. (1997) 'Durability of FRP in aggressive environments', *Non-Metallic (FRP) Reinforcement for Concrete Structures*, FRPRCS–3, Vol. 2, pp. 113–128.

Rostasy, F.S. and Budelmann, H. (1991) 'FRP-tendons for the post-tensioning of concrete structures', *Advanced Composite Materials in Civil Engineering Structures, Proceedings of the Specialty Conference, ASCE, Las Vegas*, Feb., pp. 156–166.

Saadatmanesh, H. and Tannous, F.E. (1997) 'Durability of FRP rebar and tendons', *Non-metallic (FRP) reinforcement for concrete structure: Proceedings of the Third International Symposium*, Vol. 2, Oct., Sapporo, Japan, Japan Concrete Institute, pp. 147–154.

Saadatmanesh, H. and Tannous, F.E. (1999) 'Long-term behaviour of aramid fiber reinforced plastic tendons', *ACI Material Journal*, May–June, pp. 297–305.

Saadatmanesh, H. and Ehsani, M.R. (Eds) (1996) *International Conference on Composites for Infrastructure, Proceedings of ICCI–96, Tucson, Arizona, USA*, 1231 p.

Saadatmanesh, H. and Ehsani, M.R. (Eds) (1998) *International Conference on Composites for Infrastructure, Proceedings of ICCI–98, Tucson, Arizona, USA*, (Vol. 1) 723 p., (Vol. 2) 783 p.

Sakashita, M., Masuda, Y., Nakamura, K., Tanano, H., Nishida, I. and Hashimoto, T. (1997) 'Deflection of continuous fibre reinforced concrete beams subjected to loaded heating', *Non-Metallic (FRP) Reinforcement for Concrete Structures*, Japan Concrete Institute, Vol. 2, pp. 51–58.

Sanjeev, V.K. and GangaRao, H.V.S. (1998) 'Fatigue response of concrete decks reinforced with FRP rebars', *Journal of Structural Engineering*, January.

Santoch, N. et al. (1993), 'Report on the use of CFCC in prestressed bridges in Japan', *Fiber Reinforced Plastic Reinforcement for Concrete Structures. International Symposium*, ACI SP–138, pp. 895–911.

Scheibe, M. and Rostasy, F.S. (1998) 'Stress–rupture behaviour of AFRP rebars in concrete and under natural environment', *Conference of Fiber Reinforced Composites in Infrastructure*, Vol. II. ICCI98, Tucson, Arizona, USA, 5–7 January, pp. 138–151.

Schutte, C.L. (1994) 'Environmental durability of glass fibre composites', *Material Science Engineering,* R13, N.7, November, pp. 265–323.

Scott, D.W, Lai, J.S. and Zureick, A.H. (1995) 'Creep behaviour of fiber-reinforced polymeric composites: a review of the technical literature', *Journal of Reinforced Plastics and Composites*, Vol.14, pp. 588–617.

Sen, R., Mariscal, D., Issa, M. and Shahawy, M. (1993) 'Investigation of S-2 glass/epoxy strands in concrete', *ACI SP–138*, pp. 15–33.

Sen, R., Ross, J., Sukamar, S. and Snyler, D. (1996) 'Durability and bond of AFRP pretensioned piles', Research Report, Department of Civil Environmental Engineering, University of South Florida.

Sen, R., Shahawy, M., Sukumar, S. and Rosas, J. (1998) 'Durability of carbon pretensioned elements in a marine environment', *ACI Structural Journal*, November–December, pp. 716–724.

Shih, G.C. and Ebert, L.J. (1997) 'The effect of fiber/matrix interface in the flexural fatigue performance of unidirectional fiber-glass composites', *Composites Science and Technology*, Vol. 28, pp. 147–161.

Springer, G.S., Sanders, B.A. and Tung, R.W. (1980) *Journal of Composites Materials* Vol. 14, pp. 213–232.

Taerwe, L. (Ed.) (1995) 'Non-metallic (FRP) reinforcement for concrete structures', *Proceedings of the Second International RILEM Symposium* (FRPRCS–2), E. & F.N. Spon, pp. 714.

Takewaka, K. and Khin, M. (1996) 'Deterioration of stress-rupture of FRP rods in alkaline solution simulating as concrete environment', *Advanced Composite Materials in Bridges and Structures*, El-Badry, M.M. (Ed.) Canadian Society for Civil Engineering, Montreal, Quebéc, pp. 649–664.

Toutanji, H.A. and Gomez, W. (1997) 'Durability characteristics of concrete beams externally bonded with FRP composite sheets', *Cement and Concrete Composites*, Vol. 19, 351–358.

Uomoto, T. and Nishimura, T. (1999) 'Deterioration of aramid, glass and carbon fibers due to alkali, acid and water in different temperatures', *The Fourth International Symposium: Fiber Reinforced Polymer Reinforcement for Reinforced Concrete Structures*, SP–188, Eds C.W. Dolan, S.H. Rizkalla and A. Nanni, pp. 515–522.

Uomoto, T. and Katsuki, F. (1994) 'Properties of fibre reinforced plastic rods for prestressing tendons of concrete: Deterioration of GFRP rods by alkaline solution', *Seisan-Kenkyu* (Journal of Institute of Industrial Science, University of Tokyo) (Japanese), Vol. 46, No. 8, pp. 24–27.

Verghese, K. (1999) 'Durability of polymer matrix composites for infrastructure: the role of the interphase', Ph.D thesis. West Virginia Tech. University, 258 p.

Vijay, P.V., Derel, A.S. and GangRao, H.V.S. (1995) 'Durability of glass composites under alkaline and prestress environment', *Fibre Reinforced Structural Plastics In Civil Engineering, Proceedings, Indian Institute of Technology, Madras, India*, pp. 371–376.

Vijay, P.V. and GangaRao, H.V.S. (1998) 'Creep behaviour of concrete beams reinforced with GFRP bars', *Durability of Fibre Reinforced Composites (FRP) For Construction*, CDCC–98, Eds B. Benmokrane and H. Rahman, pp. 661–667.

Vijay, P.V. and GangaRao, H.V.S. (1999) 'Development of fiber reinforced plastics for highway application: aging behaviour of concrete beams reinforced with GFRP bars', CFC–WVU report No. 99 – 265 (WVDOH RP T – 699 – FRP1).

Vistasp, M.K., Chin, J.W. and Reynaud, D. (2000) 'Critical gaps in durability data for FRP composites in civil infrastructure', *The 45th International SAMPE Symposium*, pp. 549–563.

Walzak, T.I. (1994) 'Adhesion loss mechanicals of epoxy coatings on rebar surface', Final report for the Ministry of Transportation Ontario and Concrete Reinforcing Institute, Ontario, Canada.

Wang, C.S. and Wang, A.S. (1986) 'Creep behaviour of glass–epoxy composites laminates under hygrothermal conditions', *Proceedings of the Third International Conference on Composites Materials* (ICCM3), Ed. A.R. Bunsell, et al. Vol. 1, pp. 569–583.

Warren, G., Burke, S., Harwell, S., Inaba, C. and Hoy, D. (1998) 'A limited marine durability analysis of CFRP adhered to concrete', *The Second International Conference on Concrete under Severe Conditions, Environment and Loading, CONSEC–98, Tromso, Norway*, June, pp. 1351–1362.

Watanabe, M. (1979) 'Effect of water environment on fatigue behaviour of fiberglass reinforced plastics', *Composite Material: Testing and Design (the 5th Conference)*, ASTM STP 674, Ed. S.W. Tsai, American Society for Testing and Materials, pp. 345–367.

Weiss, W. (1982) 'Low temperature properties of carbon fiber reinforced epoxy resins', *Non-metallic Materials and Composites at Low Temperature*, Eds G. Hartwig, and D. Evans, Plenum Press New York, pp. 293–309.

Weitsman, Y.J. and Elahi, M. (2000) 'Effects of fluids on deformation, strength and durability of polymeric composites – an overview', *Mechanics of time-dependent materials* 4: pp. 107–126.

Weitsman, Y.J. (1998) 'Effects of fluids on polymer composites – a review', Report MAES 95 – 1.0 CM. University of Tennessee, July.

Yamaguchi, T., Kato, Y., Nishimura, T. and Uomoto, T. (1997) 'Creep rupture of FRP rods made of Aramid, carbon and glass fibres', *Non-metallic (FRP) reinforcement for concrete structure, Proceedings of The Third International Symposium Vol. 2, Oct., Sapporo Japan*, Japan Concrete Institute, pp. 179–186.

13 Fibre Reinforced Composites in the Building and Construction Industry

Dr LS Norwood

13.1 Introduction

The discovery of room-temperature curing for unsaturated polyester resin during the early 1940s laid the foundation for an industry based on a new, artificial structural composite material[1]. However, it took several decades before the building and construction industry acknowledged its potential. The potential and durability of fibre reinforced composites (FRP) was first recognised by the marine industry during the 1950s and, today, small craft construction in many parts of the world is based on glass fibre reinforced plastic (GRP), where the material is now established as the first choice replacing the dominance of wood[2].

The attractiveness of FRP for small volume, complex shape fabrication of structural parts results from the ease with which cheap moulds/tools can be made – and with which the composite can be manufactured. In addition, the nature of the reinforcement enables the load-bearing fibres to be laid accurately along the directions of the principle stresses allowing considerable weight saving when compared with many of the more traditional construction materials. Of course, ease of fabrication, structural performance and low weight are just three of a long list of advantages attributed to FRP. Of the other advantages, the most significant are:

- good corrosion and weather resistance, especially to UV radiation and water degradation;
- the ability to introduce fire-resistant characteristics;
- low thermal conductivity;
- almost 100% retention of mechanical properties after long-term use;
- low maintenance requirement;
- excellent freeze–thaw resistance; and
- creep resistance under design loading conditions, provided appropriate factors of safety have been employed.

The earliest use of glass fibre reinforced plastic (GRP) in buildings was in the fabrication of translucent sheeting for lightweight, non-structural roofing and at the time of its introduction, the lack of codes of practice, the lack of textbooks about the material, the lack of consultants with experience of the material, and the lack of reliable data in the desired format made it difficult for the material to gain acceptance. However, over time, as the number of case histories grew, reputable fabricators emerged and proved that stringent specifications could be met. Hence, confidence in the material and its acceptance as a construction material became inevitable.

Whilst the construction industry had lived with the variability and limitations of natural fibre reinforced composites such as wood (with their unacceptable performance under fire conditions) more stringent requirements were placed on artificial composites and their abilities to resist ignition and limit surface spread of flame. The resultant standards and codes of practice encouraged major developments of resins and composites and now many composite products easily achieve the requirements of the standards – which many types of wood would undoubtedly fail and yet wood is still more readily accepted than GRP by the public and professionals alike[3,4,5,6].

Nearly a quarter of a century ago a British Plastics Federation (BPF) investigation of plastics and fire concluded that:

> *fires in buildings originate in their contents and, in the vast majority of circumstances, the structure does not contribute to loss of life*[7].

It is not surprising that fire situations have been difficult to simulatc in thc laboratory since fire is defined as:

> *a process of combustion characterised by heat or smoke or flame or any combination of these*

and the nature of a fire is critical to the outcome of a fire[8]. The fire standards applied to GRP and the performance of the material in fire situations will be discussed later in more detail with reference to the fire design requirements for actual structures.

In many applications the structural performance of GRP has been shown to be outstanding – evidenced by 30 year old ocean-going minesweepers[9], highly impact-resistant, high-speed train cabs[10], chemical process plant[11] and underground petroleum storage tanks[12]. This highlights just a few of the numerous, diverse examples of a structural material adaptable to use in a variety of markets for a wide range of products from small to very large components.

However, in the building and construction industry the structural benefits of GRP have been slow to be accepted – most likely because, unlike steel, concrete and timber, there was little guidance available on the allowable working stresses that should be used. The reason for this surely lies within the great diversity of the material, with the material property characteristics of GRP depending on fibre type (glass, carbon, polyaramid, etc.), fibre orientation, fibre content (by weight or volume), the matrix characteristics and the matrix cure and post-cure conditions. In addition, the material is made by the fabricator as the component is manufactured, which can lead to material variability from fabricator to fabricator for nominally the same component. Over the years the skill and competence of fabricators has increased such that good quality FRP components today are the norm rather than the exception.

One industry, which has valued the unique advantages that GRP has to offer, decided in 1973 to produce a code of practice for its use. This British Standard Code of Practice has since been revised and will provide the foundation for the new European Standard on 'Specification for Vessels and Tanks in Reinforced Plastics'[13]. Whilst this standard was developed for special applications within a specific market area, there is no reason why a similar industry-specific code could not be devised for fibre reinforced composites within the building and construction market.

In 1977, Leggatt discussed, in detail, the issues to be addressed when considering GRP as a structural material. All the points are as valid today as they were in 1977[14].

There have been many books and articles written about FRP but for general background on material properties, FRP fabrication processes and applications the BPF Handbook is a useful starting point[15].

Before discussing applications of GRP in building and construction projects, consideration must be given to the material, its manufacture and design issues.

13.2 Materials used for the manufacture of fibre reinforced plastic (FRP)

The most cost effective fibres, with the desired level of performance and durability in use as a reinforcement for thermosetting resins are based on E-glass. The most cost effective matrices are based on unsaturated polyester resins.

13.2.1 Glass fibre reinforcement

High quality glass is needed in order to obtain the required performance from glass fibre. The filaments, which form glass fibre reinforcement strands, are 13 μm in diameter and are drawn at high speed from temperature controlled platinum bushings. Before the individual filaments come together they are coated with an emulsion containing a number of chemicals, which each perform a specific task during the subsequent processing:

- a linking or coupling agent to aid the bonding process to the matrix;
- a film former to protect the glass filaments from damage during subsequent processing and moulding, and to give integrity to the bundles of glass filaments;
- a lubricant to aid the passage of fibre strands through and over guide points; and
- an antistatic agent to reduce static build up and problems during chopping processes.

The importance of the coupling agent cannot be over emphasised since it contains reactive chemical groups, some capable of bonding to the glass surface and others capable of bonding to the thermosetting resins. When using unsaturated polyester resin matrices it is essential to ensure that the glass fibre reinforcement is polyester-compatible to ensure the best performance from the composite.

The properties of glass fibre are compared with other fibres in Table 13.1. Glass fibre fabrics are available in a large variety of forms from lightweight tissue for improving surface finish to random, chopped fibre mats for producing finished laminates with isotropic properties in the plane – to balance and unbalance multi-axial, continuous fibre reinforced cloths which can be woven or stitched. In addition, combination materials are available as well as fabrics designed for specific fabrication processes.

On a weight-for-weight basis, reinforcing fibres exhibit exceptional performance when compared to steel with, for example, the unit tensile strength of glass fibre (tensile strength divided by specific gravity) being around three times as high as steel.

13.2.2 Thermosetting resin

The most common resins used for the manufacture of fibre reinforced composites are based on unsaturated polyester resins because they have good general properties, excellent resistance to many environments and are the least expensive. For special

Table 13.1 Typical properties of reinforcing fibres

Property		Fibre					
		Glass		Polyaramid	Carbon		
		E	S		High Strength	High Modulus	Jute
Specific gravity		2.56	2.49	1.44	1.81	1.86	1.20
Tensile strength	(MPa)	1700	3100	3600	3300	2600	100 (*600)
Tensile modulus	(GPa)	70	86	124	237	345	12 (*57)
Tensile strength to failure	(%)	3.5	4.0	2.5	1.34	0.74	–
Electrical resistivity	(Ωm)	10^{13}	–	–	16×10^{-6}	16×10^{-6}	–
Specific heat capacity	(J/kg°C)	800	740	1420	710	–	–
Refractive index		1.547	1.523	2.0	–	–	–

* from flexural tests

applications vinyl ester resins and epoxy resins can be used but consideration must be given to the material and processing requirement costs of these resins in relation to the advantages they offer.

Unsaturated polyester resins are long chain polymers, which are supplied ready dissolved in a reactive solvent. The curing process of these resins can be activated in a number of ways, but by far the most common is with the use of a peroxide catalyst. As the peroxide breaks down, it forms free radicals which are extremely reactive chemical species and readily activate the unsaturation (carbon–carbon double bonds) in both the polyester resin chain and the monomer. This results in co-polymerisation of the resin and monomer to form a three-dimensional crosslinked structure.

Since the curing process takes time, which can be adjusted to suit the needs of different moulding processes and different component sizes, fibres can easily be wetted and layers laid one upon the other to build up a structural component. Finally, gelation occurs and within hours the material is solid although, for adequate curing, it is essential to store components at room temperature for a few days. Many resins are only fully cured if exposed to elevated temperature (60–80°C is generally sufficient) for about 24 hours after gelation. A post-cure stage is unnecessary for many applications, for example in the boat building industry GRP has been found to be durable in a very harsh environment without the need of post-cure. However, for chemical plant applications, elevated temperature post-cure is essential in order to meet the level of chemical and temperature resistance often demanded.

Typical mechanical properties of cured cast thermosetting resins are shown in Table 13.2 and typical electrical and thermal properties of cast polyester resins are shown in Table 13.3.

The properties of reinforcing fibres and resins are generally of little importance on their own but the synergistic effects obtained by combining them to form laminates or

Table 13.2 Typical properties of cured cast thermosetting resins

Resin	Property				
	Tensile Strength (MPa)	**Tensile Modulus (GPa)**	**Tensile Elongation (%)**	**HDT(°C)**	**Barcol Hardness**
Medium reactivity orthophthalic acid based polyester	64 *68	3.0 *3.6	4.5 *2.4	57 *75	42 *46
Low reactivity orthophthalic acid based polyester	60 *60	3.2 *3.8	3.0 *2.0	55 *65	43 *45
Medium reactivity isophthalic acid based polyester	55 *66	2.7 *3.5	4.8 *2.4	59 *95	35 *46
Low reactivity isophthalic acid based polyester	57 *75	2.6 *3.3	7.5 *4.5	54 *75	35 *43
Medium reactivity isophthalic-NPG polyester	48 *60	2.6 *3.2	4.7 *2.5	62 *120	32 *40
Low reactivity clear-HET acid based fire retardant polyester	52 *50	3.2 *3.5	2.0 *1.5	58 *70	41 *45
Filled fire retardant polyester based on dibromo-NPG	49	6.3	1.0	68	55
General purpose, epoxy based vinyl ester resin	*80	*3.4	*4.0	*100	*48
Novolac based vinyl ester resin	*75	*3.5	*3.1	*135	*48
Diethylene triamine cured epoxy resin	*70	*3.4	*5.3	*95	–
Diaminodiphenyl methane cured epoxy resin	*80	*2.8	*5.2	*155	–
Tetrahydrophthalic anhydride cured epoxy resin	*81	*2.7	*4.5	*120	–

* Elevated temperature cure (for epoxy resin) and post cure for polyester resin

Table 13.3 Typical electrical and thermal properties of fully cured cast unsaturated polyester resin

Specific heat	kJ/kg°K	2.3
Thermal conductivity	W/m°K	0.2
Coefficient of linear expansion	$\times 10^{-6}$/°C	100
1000 Hz dielectric constant		2.75
50 Hz power factor		0.008
1000 Hz power factor		0.004
5 MHz power factor		0.019
50 Hz permittivity		3.7
5 MHz permittivity		3.2
Voltage breakdown	kV/mm	22
Volume resistivity	TΩm	1

composites produce materials which can be tailored to meet a range of structural, environmental, electrical, thermal, and chemical conditions. It is by 'gluing' together the flexible high strength fibres with a matrix and holding them in directions where the principle stresses are known to occur that enables useful components to be fabricated from composite materials. Once cured, the matrix is able to transfer load to the stiffer fibres via the fibre–matrix interfacial bond. The matrix also offers protection to the fibres from environmental damage and physical erosion.

In addition, the composite can be further protected by in-mould surface coatings, referred to as gelcoats.

13.2.3 Coating composite components

The most commonly used coating for glass fibre reinforced polyester resin is an in-mould coating generally described as the 'gelcoat'.

Gelcoats perform a variety of functions and for many applications the most important is the protection of the structural laminate from liquid (especially water) ingress by capillary wicking via the ends of exposed fibres. They also provide a decorative finish, which is resistant to environmental attack resulting in long-term retention of gloss and colour with minimal maintenance.

Over the years gelcoats based on unsaturated polyester resins have been specifically developed to meet individual market requirements. Gelcoats can differ for external and internal use for the same market. For example, in the building and construction industry both durability and fire resistance are important requirements which dictate the choice of material. Of course, fire-retardant gelcoats are available but their durability to external weathering is inferior to that of non-fire-retardant gelcoats[16]. Hence, the use of a non-fire-retardant gelcoat in combination with a highly fire-retarding laminate provides both the best weather resistance and the best fire resistance. For internal applications, fire-retardant gelcoats can be used with less expensive laminating resins to achieve the desired resistance to fire.

In order to obtain the best performance from a gelcoat it should be brushed or sprayed onto the mould at 500 g/m^2 to achieve a thickness of 0.4 to 0.5 mm. After curing for between 1 to 2 hours, with a peroxide catalyst, the gelcoat is ready for the application of the laminate. Just like paints, gelcoats can be pigmented but it must be remembered that some pigments have long-term durability limitations and are not advised for long-term exposure to high levels of humidity and UV light. Generally, though, the long-term durability of gelcoats is excellent as shown by over 35 years of use in marine applications and over 25 years of use in exterior cladding applications for buildings (see Section 13.4.1).

In recent years, process technology development has resulted in equipment permitting simultaneous spraying of different coloured gelcoats onto the mould to achieve a granite effect, as shown in Fig. 13.1 – the refurbishing cladding used to

Figure 13.1 *The gelcoated granite effect of a GRP clad stanchion on the Roman Catholic Cathedral in Liverpool. Reproduced by permission of: Viv Wiliams Varley (Architects); Smart Crosby International (GRP Consultants); Minster Composite Products Ltd (GRP Fabricators).*

reclad the Liverpool Roman Catholic Cathedral. A similar effect can be achieved with the use of either inorganic granite powder filler in unsaturated polyester resin or polyester resin based 'chips' and alumina trihydrate fillers in unsaturated polyester resin. Either a granite effect gelcoat or thick section granite effect solid surface can be achieved[17]. An example of this surface effect is shown in Fig. 13.2 on a blastproof composite panel, which will be discussed in more detail later. At present, this type of coating is undergoing extensive weathering tests on natural weathering sites but the types of base resin and filler used in its manufacture have, in previous exposure tests, proved extremely resistant to UV light and water degradation.

Typical gelcoat types recommended for use in the building and construction industry and their properties are shown in Table 13.4.

The use of intumescent coatings can impart excellent reduced surface spread of flame characteristics to non-fire-retardant laminates, resulting in Class 1 surface spread of flame when tested according to BS 476 Part 7 and achieving the Class 0 requirement for use in the construction of non-combustible buildings as defined in Scction B of thc 1985 UK Building Regulations[4,6].

Figure 13.2 *A blastproof panel showing the granite effect surface finish.*

Table 13.4 Typical gelcoats used in the building and construction industry (room temperature cured)

Type of Gelcoat	Property				
	HDT (°C)	Tensile Strength (MPa)	Tensile Modulus (GPa)	Elongation to Break (in tension) (%)	24 Hours Water Absorption (mg)
General Purpose (interior applications)	38	31	1.5	5.5 (peak)	28
Isophthalic acid based (brush/spray)	75	75	3.5	3.0	18
Iso-NPG based (brush/spray)	96	57	3.9	1.6	16
Isophthalic acid based, filled, fire retardant	62	59	4.1	2.2	24
Filled, matt finish orthophthalic (internal applications) fire retardant	73	35	7.8	0.4	13
Isophthalic - NPG granite effect coating	100	60	3.3	2.5	16.5

13.2.4 Adhesives

Often it is simpler, more cost effective and structurally better to join composite materials using adhesives. For many years the industry has relied upon thixotropic, gap filling, bonding pastes manufactured from standard isophthalic acid based or orthophthalic acid based unsaturated polyester resins. Although such materials give reasonable bonds for GRP to GRP, GRP to wood and GRP to steel they have their limitations, particularly with respect to toughness and resilience. Recently, a range of urethane acrylate polymers modified with fumarate unsaturation have became available, which enable extremely tough, gap-filled bonds, up to 15 mm thick, to be achieved[18]. These materials have been proven by years of use in minesweeper construction for ensuring top hot stiffener to hull adhesion and integrity whilst under mine explosion, shock wave loading conditions[19].

Under test these adhesives have been shown to have higher interlaminar shear properties than the materials they are joining and hence, substrate failure, rather than adhesive bond or adhesive/cohesive failure, is the typical mode of failure[20].

13.3 Fibre reinforced composites

As mentioned previously, when reinforcing fibres and matrices are combined to form composites, then the resultant material exhibits characteristics which neither the fibre nor resin components possess on their own. However, the stiffness of the composite is very much a function of the reinforcing material, both as a result of the fibre properties, their orientation and the proportion present. Fibre content is usually expressed in terms of a weight fraction, a volume fraction or a resin:fibre ratio. The relationship between weight fraction and resin:fibre ratio depends upon both the specific gravity of the fibre and the resin and is well documented in the literature[15].

13.3.1 The properties of composites

Prediction of mechanical properties

The accumulation of measured data for composite materials has aided the development of theoretical models, which have been refined so that correlation between predicted and measured data is more reliable than it has ever been. Many of the methods are described in Chapters 5 and 6 of the BPF Handbook[15, 21, 22].

Minimum properties of reinforced laminate plies are given in Table 13.5, which can be used to predict the minimum tensile properties of laminates as shown in Figs 13.3 and 13.4. The typical level of correlation between predicted and measured properties for a laminate containing woven roving glass fibre, unidirectional glass fibre and chopped strand mat are shown in Table 13.6. More detail of the theoretical property determination of composite materials is given in the literature[15].

Measured mechanical properties

Fibre and matrix manufacturers have, over many years, generated large databases of material properties for a wide range of resin types, fibre contents, fibre types and cure/post-cure conditions. Some typical data are given in Table 13.7.

In the building and construction industry the need for fire-retardant materials often requires the incorporation of fillers in the resin. Generally, the presence of fillers will affect the mechanical performance of the composite especially when the filler level is high. This increases the modulus of the resin, which in turn contributes to the modulus of the laminate. In addition, filled resins are more difficult to process, which results in lower fibre content and reduced mechanical performance, but the compromise in mechanical performance is balanced by the improvement in structural integrity under fire situations.

Table 13.5 Minimum properties of fibre reinforced polyester resin based laminate plies

Reinforcement	Property	
	Unit Tensile Strength (N/mm width per kg/m² of reinforcement)	Extensibility (unit modulus) (N/mm width per kg/m² of reinforcement)
Glass		
Chopped Strand Mat (CSM) – random short fibres	200	14,000
Woven Roving (WR) (balanced 0° / 90°)	250	16,000
Unidirection (UD) roving	430	25,000
UD continuous filament (0°)	500	28,000
UD continuous filament (90°)	0	8,400
Polyaramid		
Woven (balanced – 0°/90°)	600	35,000
UD (0°)	1,200	70,000

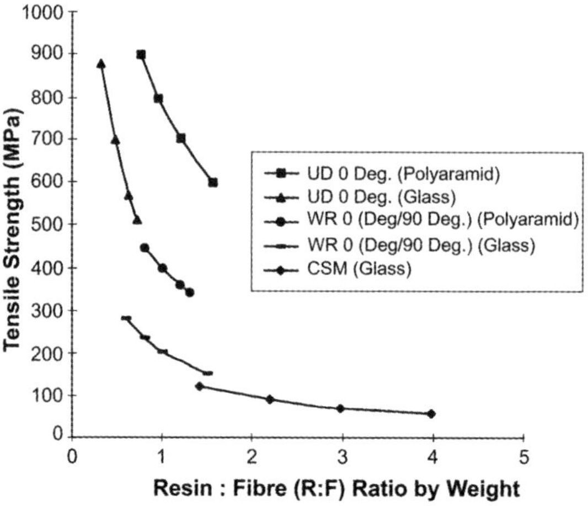

Figure 13.3 *Graph showing the predicted effect of fibre type and content on tensile strength using property data given in Table 13.5.*

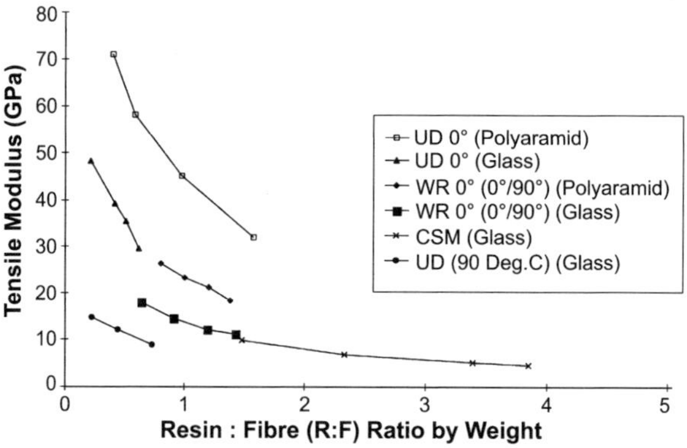

Figure 13.4 *Graph showing predicted effect of fibre type and content on tensile modulus using property data given in Table 13.5.*

Although the presence of filler will limit the fibre content and hence the mechanical properties of fire-retardant laminates, there is no reason why a more structural, non-fire-retardant laminate cannot be protected with either a fire-retardant resin based laminate or an intumescent coating. The specific property benefits (see Table 13.8) of fibre reinforced plastics can then be realised along with their other desirable, decorative and processing advantages[23].

Of course, where structural performance is not a requirement then fibre reinforced fire-retardant unsaturated polyester resins have more than adequate properties to

Table 13.6 Comparison between predicted and measured laminate properties for aglass reinforced polyester resin laminate

Property		**Predicted Data**		**Measured Data**	
		0°	**90°**	**0°**	**90°**
Unit Tensile Strength	N/mm	1095	795	1350	874
Tensile Stiffness	kN/mm	72.1	60.4	72.2	65.5
Tensile Strength	MPa	198	144	231	144
Tensile Modulus	GPa	13.0	10.9	12.3	10.8
Glass Weight	kg/m²		4.05		3.74
Resin Weight	kg/m²		4.86		5.38
Laminate Thickness	mm		5.53		6.00

Table 13.7 Typical measured properties of fibre reinforced composite materials

Property		**Material**			
		Chopped Strand Mat (CSM)	**Woven Roving Glass**	**Satin Weave Glass Cloth**	**Continuous Glass Rovings**
Fibre Content	% weight	30	45	55	70
	% volume	18	29	38	54
Specific Gravity		1.4	1.6	1.7	1.9
Tensile Strength	MPa	100	250	300	800
Tensile Modulus	GPa	8	15	15	40
Compressive Strength	MPa	150	150	250	350
Flexural Strength	MPa	150	250	400	1000
Flexural Modulus	GPa	7	15	15	40
Izod, unnotched, impact strength	kJ/m²	75	125	150	250
Coefficient of linear expansion	$\times 10^{-6}/°C$	30	15	12	10
Thermal Conductivity	(W/mK)	0.20	0.24	0.28	0.29

support the typical loading requirements for cladding and curtain walling, as discussed in the following section.

13.3.2 Fire-retardant composite systems

It is probably impossible to make any material, and especially those with an organic component, totally resistant to the effects of fire whilst retaining reasonable and acceptable costs. However, it is possible with reinforced plastics to reduce the risk, in the event of fire, to an acceptable level without compromising safety and yet still provide a cost effective option. In fact, it has been known for many years that 'fires in buildings originate in the contents and in a vast majority of circumstances the structure does not contribute to loss of life!'[24] D A Purser in an article entitled, 'Toxicity

Table 13.8 Mechanical properties of FRP laminates with other structural materials

Material	Grade	SG	Tensile Mod (GPa)	Ultimate or Proof Strength		Izod Impact Strength (kJ/m²)	Specific Strength (MPa)	Specific Modulus (GPa)
				Tensile (MPa)	Compression (MPa)			
Mild Steel	BS 15	7.80	207	240	240	50	31	27
Aluminium Alloy	HE15 WP	2.70	69	417	417	25	154	26
Stainless Steel	316	7.92	193	241	241	1356	30	24
Random (CSM) Reinforced Plastic	*33	1.47	8	100	150	75	68	5
Unidirectional (Glass) FRP	*82	2.16	53	900	450	250	417	25

* % glass fibre content by weight

Assessment of Combustion Products in Fires' provided a significant change in the views commonly held about death in fire situations[25]. Although he found that both fatal and non-fatal casualties were 'overcome by smoke and toxic gases, rather than by heat or burns', the important factor in most cases was the presence of carbon monoxide in large quantities. From these studies the following significant results emerged:

- 'Carbon monoxide is the single most important narcotic gas present in a fire situation and is the major ultimate cause of death.'
- 'The liberation of hydrogen cyanide from some plastics, wool and foams can cause death at levels of 200 ppm within two minutes.'
- 'Carbon dioxide is always present and, although not toxic below 5% concentration in air, stimulates breathing and the intake of other toxic gases.'
- 'The reduction of available oxygen (hypoxia) impairs mental and physical ability at 15% concentrations and becomes serious at 10% levels.'
- 'Other products of combustion can be considered as irritants and only when the presence of carbon monoxide, hydrogen cyanide, carbon dioxide and low oxygen content in the air are not at critical levels, as to be lethal, are irritants a major problem.'

Even with this understanding of the causes of death in fires, it is still generally accepted that toxic emissions should be reduced to as low a level as practically possible without impairing fire resistance and structural integrity during fires. It is also very important to reduce fume emissions from the non-burning surface of composite components and to ensure structural integrity for a sufficient length of time to ensure escape is possible. The low thermal conductivity of FRP, shown in Table 13.7 with other typical data on FRP, is a real advantage in fire situations – as highlighted in a fire that occurred on an FRP minesweeper some years ago, where it was reported that 'temperatures were high enough to melt aluminium fittings. The GRP laminate's surface was charred but mostly undamaged. The ship was returned to service after repair. With a metallic structure the consequences would have been far worse'[26]. Similar benefits have been realised in the building industry with many examples, some highlighted later, dating back to the 1970s[14].

Fire legislation and standards

Over the last 25 years composite structures for use in buildings have been subjected to a variety of fire performance standards, of which the most well known in the UK is BS 476 (Parts 6 and 7)[3, 4]. There are a number of ways in which resins can be modified to meet the ignition and flame propagation requirements of this standard and each has its benefits and limitations. For example, fire-retardant fillers limit the processability of resins and hence, limit fibre content and structural performance. In addition, some fillers can produce toxic fumes during a fire. Alternatively, unfilled fire-retardant resins rely on technology to incorporate chlorinated and brominated raw materials in the unsaturated polyester resin polymer chain. As a result many composites based on these types of resin can achieve Class 1 surface spread of flame ratings when tested to BS 476 Part 7 and ignitibility resistance when tested to Part 6, which enables them to meet the UK building Regulations for use in the construction of non-combustible buildings[3,4,6,1]. Of course, requirements in other parts of Europe can vary, although there is general correlation between the various surface spread of flame tests. The French NFF-16–101 test requires an assessment of smoke and toxic fumes with the F (smoke) rating made up from three components:[5]

- smoke density measurement;
- opacity (light obscurity) measurement;
- toxic fume measurement.

Few unsaturated polyester resin based laminates can achieve better than the F2 requirement with most only able to achieve F3. However, in recent years, developments in resin technology have resulted in M1 (the best surface spread of flame requirement in the French test, which is equivalent to BS 476 Part 7 Class 1), and F0 classifications with some filled resin systems[27] . The major limitation is a maximum fibre content, by weight of just 20%, which limits applications to semi- and non-structural components only. Hence, such materials can be used for many decorative internal components and cladding panels supporting their own weight.

It is proposed that new European legislation will be introduced to reclassify all building materials with respect to fire resistance and ability to reduce the spread of flame[28]. Thus, plastics will be classified in terms of their fire-resistance performance alongside the more traditional building materials such as wood, concrete, brick, tile, steel, aluminium, etc. Whilst this will, at first sight, show plastics up as being inferior to many of the more traditional building materials in terms of surface spread of flame resistance, it will in effect have changed nothing that is not already known and is simply a reclassification in one common list of materials. When the advantages of plastics over other materials are taken into consideration there will still be many instances when, on balance, their unique advantages will win through and good design will reduce the impact of their limitations.

Fire-retardant resins and composites

The effects of adding fillers to unsaturated polyester resins have already been mentioned but whilst some mechanical performance characteristics are compromised the fire-resistant characteristics are greatly enhanced and, as for many construction materials, a cost effective compromise must be sought. It is possible to achieve excellent resistance to ignitibility and surface spread of flame without the use of fillers and, generally, without seriously reducing mechanical performance but usually at considerable increase in cost because of significant increases in raw material prices.

Typical properties of unfilled fire-retardant cast resins and laminates are shown in Table 13.9 and can be compared with the properties of filled resin systems in Table 13.10. Most notably, the unfilled cast resins are more elastic and the composites made from them, invariably, have higher strength. However, the major benefits of using unfilled resin systems are in terms of ease of processing, for the manufacture of translucent laminates and for achieving high fibre content, if required, for improved structural performance.

For many cladding applications the properties given in Tables 13.9 and 13.10 are more than adequate and have proven track records over the past 25 years, as will be discussed in later sections with reference to specific buildings. Where improved structural properties are necessary, directional reinforcement can be introduced into the structure as and where required to meet the specific stress or stiffness requirements; this, of course, is a major benefit of composites enabling local strengthening or stiffening with accompanying weight savings when compared with many traditional constructional materials.

13.3.3 Processing glass fibre reinforced plastics

Contact moulding

In principle, moulding with thermosetting resins can be a simple operation. Once the resin is mixed with a curing agent, in the correct proportion, it is just a matter of laying reinforcements in the mould, wetting it with resin, consolidating to remove air voids, allowing cure to take place and then demoulding. This basic process which made fibre reinforced plastics such an attractive option for manufacturing structural composites because inexpensive tooling and unskilled labour could be used has also been a barrier to exploiting the full potential of the material since, all too often, failures as a result of unfamiliarity in processing the material, have led to setbacks and questions over its viability as a structural material.

Fibre reinforced plastic is one of the few materials where the moulder/fabricator both manufactures the component and the material simultaneously. Hence, more care than is generally thought necessary has to go into mould preparation, component design, workshop conditions and raw material control. Thorough mixing of curing agents is critical to successful moulding as is the temperature control of all the systems, including the mould and the reinforcement.

Much has been written about the basics of contact moulding which includes both brush application and spray application of resin[29,30,31]. There are many responsible and reliable moulders who mould both large and small components from minesweepers to specialist road vehicles to railway vehicles to chemical plant to

Table 13.9 Typical mechanical properties of room-temperature cured fire-retardant cast resins and laminates (unfilled resins)

Property		Resin Type				
		HET Acid Based	HET-iso Based	Dibromo-NPG Based (Sheeting)	Dibromo-NPG Based (Structural)	
Cast Resin						
Tensile Strength	MPa	52	55	47*		43
Tensile Modulus	GPa	3.2	2.6	2.3*		2.4
Elongation	%	2.0	4.0	3.5*		2.3
HDT	°C	58	57	66*		69
7 Days Water Absorption	mg	28	64	65*		61
Barcol Hardness		41	40	36*		40
Laminate (using filled resins)						
Glass Weight/area	kg/m²	1.8	1.8	1.8	1.8	1.8
Glass Type		4 × 450 CSM (EB)	4 × 450 CSM (EB)	4 × 450 CSM (PB)	4 × 450 CSM (PB)	4 × 450 CSM (EB)
Resin : fibre ratio	By Wt.	3.15:1	2.39:1	2.26:1	2.62:1	2.04:1
Glass Content	% By Wt	24.1	29.5	30.4	27.6	32.9
Tensile Strength	MPa	68	88	118	107	110
Tensile Modulus	GPa	6.3	5.6	6.4	6.2	7.7
Compressive Strength	MPa	159	99	165	192	147
Compressive Modulus	GPa	8.3	8.1	8.4	6.2	5.7
Flexural Strength	MPa	132	166	212	159	243
Flexural Modulus	GPa	5.9	5.9	6.5	6.2	8.1
In-Plane Shear Strength	MPa	45	42	52	50	55
Interlaminar Shear Strength	MPa	4.0	4.1	5.1	7.5	5.5
Water Absorption (7 days)	mg	61	60	44	60	91

* Fully post-cured

simple permanent shuttering. If there is one criticism of contact moulding it concerns control of thickness, which will vary depending upon the skill of the laminator, even across a material containing the same number of plies of reinforcement at, nominally, the same resin content throughout. For many applications this level of thickness variability is of little consequence and is not considered to be particularly relevant in many building and construction applications. However, where structural performance is a major requirement and design demands either well defined stress or stiffness characteristics then control of thickness can be critical.

Of course, care during contact moulding does result in good control of thickness but there will always be an element of variability especially between horizontal and

Table 13.10 Typical mechanical properties of room-temperature cured fire-retardant cast resins and laminates (filled resins)

Property		Resin Type			
		Ortho Based	**Ortho Based**	**Non Fire Retardant Ortho & Prefil F**	**Dibromo-NPG Based**
Cast Resin					
Tensile Strength	MPa	36	48	–	49
Tensile Modulus	GPa	3.0	5.2	–	26.3
Elongation	%	1.8	1.7	–	1.0
HDT	°C	54	63	–	68
7 Days Water Absorption	mg	54	34	–	26
Barcol Hardness		38	53	–	55
Laminate (using filled resins)					
Glass Weight/area	kg/m^2	1.8	1.8	1.8	1.8
Glass Type		4 × 450 CSM (EB)	4 × 450 CSM (EB)	4 × 450 CSM (PB)	4 × 450 CSM (PB)
Resin : fibre ratio	By Wt.	2.14:1	3.04:1	2.23:1	2.38:1
Glass Content	% By Wt	31.9	28.6	31.0	29.5
Tensile Strength	MPa	94	82	80*	77
Tensile Modulus	GPa	5.2	6.9	7.0*	8.1
Compressive Strength	MPa	136	169	151*	165
Compressive Modulus	GPa	8.7	10.5	7.6*	10.2
Flexural Strength	MPa	158	151	128*	168
Flexural Modulus	GPa	6.0	6.0	6.0*	6.9
In-Plane Shear Strength	MPa	51	47	–	54
Interlaminar Shear Strength	MPa	4.9	3.7	4.7*	4.5
Water Absorption (7 days)	mg	101	156	–	38

* Fully post-cured

vertical surfaces. It is important to beware of the cost implications of moving to alternative processing, which will only generally become financially attractive when many, usually hundreds, of the same component are required because tool manufacturing costs are higher. As with so many manufacturing situations there is a cost–performance exercise which has to be carried out then, and only then, can the rationale for changing the manufacturing process be justified. In recent years, legislation against volatile organic component (VOC) emissions, has also been a driving force to move away from the open/contact moulding process[32,33,34]. Here the rationale for change must be based on the cost of meeting the legislation using an open mould process against the cost of using a closed mould process and/or even being in the FRP moulding business at all. Whilst closed mould processing often appears more expensive at the outset, savings in processing time, improvements in productivity, reductions in raw material waste and reductions in rework can compensate for the costs of the change from an open mould process.

There are also several methods of processing FRP mouldings using machine processing, which will be briefly discussed in the following section.

Filament and spray winding

These processes are generally used for the manufacture of large cylindrical components such as tanks and pipes for chemical plants, potable water, sewage applications and components having an internal smooth mould surface.

Centrifugal casting

Generally, used for the production of large hollow components, with the mould surface on the outside. Pipes and tanks are the main components manufactured using this process.

13.3.4 Sheeting manufacture

Most corrugated and flat sheeting is manufactured by machine[29]. The process involves 'sandwiching' glass fibre and resin between two continuous lengths of release film. The sandwich then passes through a series of rollers to expel air bubbles and consolidate the sheet to the required thickness. It is then passed through an oven, where the laminate can be corrugated by rollers and dies prior to curing. Roof sheeting was one of the earliest uses of FRP in the building industry; over the years resins have been developed and modified to provide the specific curing characteristics necessary for the process and to provide translucency in the final product.

Sheeting manufactured with the correct resin will not yellow significantly after years of service even when exposed to tropical UV conditions, provided the resin has been fully cured, the resin content is at least 75% by weight of the sheeting and the release film is removed prior to installation. Resin rich surfaces achieved by the use of surfacing tissue improve the quality of the sheeting and its durability[29].

The typical thermal characteristics of glass reinforced plastic sheeting are:

Thermal conductivity	k	0.2 W/mK
Thermal transmittance	U value	6 W/m²K
Thermal transmittance (double skin with 25 mm spacing)		2 W/m²K
Solar heat transmission		80%

13.3.5 Closed mould processes

Closed mould and matched die moulding are by far the most environmentally friendly means of processing unsaturated polyester resin and produce components with precise, reproducible cross-section[29]. In addition, two mould finish surfaces can be achieved on components, and moulding can be carried out in a variety of ways.

Hot press moulding

Hot press moulding is only cost effective where high volume production runs are required, often for runs of tens of thousands of identical components, because of the cost of presses and matched metal tooling necessary for high pressure processing. Cycle times are fast, usually a matter of just a few minutes and surface finish is excellent. The lengths of identical component runs for building applications are generally too short for this process to be cost effective.

Cold press moulding

Simple presses can be used to impregnate reinforcement with catalysed resin at room temperature under pressures of just a few bar. Cold press moulding requires matched tooling but because of the low pressures involved the moulds do not have to be made of steel and, often, FRP and resin concrete tools are used[30]. It is possible to mould components with a fibre content of at least 50% by weight to produce structural composites.

Resin transfer moulding (RTM)

Resin transfer moulding or resin injection moulding can be used to produce FRP components at low initial costs[29,30]. Matched tools are subjected to only moderate pressures, usually 2–3 bars, and can be made of composite, metal lined composite or light metals. The process involves laying dry reinforcement, core materials (where required) and any other inserts into one of the tools. Then the tools are closed and catalysed resin is injected under pressure. Since the system is closed during resin impregnation of the reinforcement, the emission of volatile organic compounds to the air is minimal. In fact, when this process is used with unsaturated polyester resin the level of styrene (the most common monomer used) in the air is often undetectable in the moulding area; this is in complete contrast to contact moulding where, during the moulding/consolidating process, levels of styrene emissions can attain several hundreds of parts per million in the air.

Generally, RTM requires the use of resins which have been modified to meet the requirements of the process; these modifications mainly take the form of rheological and gelation improvements to ensure good fibre wetting and cure, when it is required, as soon as possible after the tool has been filled with resin.

Parts produced by RTM moulding have consistent quality, dimensional stability and two mould quality surfaces. The process is usually cost effective when 100 plus identical components are required and tools, especially the metal lined composite type, can produce thousands of components before refurbishment becomes necessary. The process is used to manufacture small- to medium-sized components with mouldings 5–6 m long can easily be produced using this technique. Tool turnaround is rapid, for example, a delivery-van top around 5 m long can be produced in 25 minutes using this technique – with a cycle time which includes a gelcoating step.

Vacuum-assisted resin injection (VARI)

Vacuum-assisted resin injection is a variant of RTM and is reputed to reduce void content in the component and give better consolidation of the laminate. As with RTM the reinforcement is laid dry into one of the match tools and, when the tools are closed, a partial vacuum is applied to both consolidate the glass pack and remove the air. A separate vacuum arrangement on the mould edges ensures that the moulds are clamped together for the duration of the moulding process. The mould is filled with resin by pump or by gravity feed. Once the mould is filled, the vacuum lines to the mould cavity are clamped off and the component allowed to cure.

As with RTM, the economics of the VARI process requires that reasonable volumes of identical components are manufactured to amortise tooling costs effectively.

Vacuum bagging/vacuum injection

Vacuum bag or vacuum injection is a reduced cost version of other cold match tool processing[30]. In this process only one tool is required because once the dry reinforcement has been placed in the tool followed by, where necessary, other 'films' required for the process, a plastic vacuum bag is used to seal the space between the bag and the tool. The mould is then evacuated and once a seal has been established the vacuum is reduced to about 0.5 bar before catalysed resin is drawn through the reinforcement. Once the tool is full the mould is sealed and curing occurs. Vacuum bagging is the least expensive of the room temperature, closed mould processes because only one tool is required. Although the films used for the process increase the cost compared with the contact moulding process, the increase in productivity, improved quality, and reduced raw materials cost, generally offset the increased costs of the process. In addition, the level of volatile organic compound released to the atmosphere is minimal, ensuring ready compliance with Health and Safety Executive (HSE) and Environmental Pollution (HMIP) legislation without the need for costly extraction and scrubbing equipment.

Vacuum bagging is a very effective method for manufacturing sandwich laminates and for bonding core materials to fibre reinforced skins because the pressure applied is substantial, uniform and easily applied to complex shapes. Just 0.5 bar equates to approximately 5 tonnes over an area of 1 m^2.

13.3.6 Pultrusion

Pultrusion can be used to manufacture extremely strong components with high fibre content by pulling reinforcing fibres through a bath of catalysed resin and then into a heated die for forming and curing[30]. Both open and closed profiles can be produced to fine tolerances with intricate details. Pultruded sections are generally used for structural applications, for example: ladders, railing, bridge sections, I-beams, etc. Hence, there is great potential for pultruded composite materials in building and construction applications.

13.4 Case studies of composites in building and construction

13.4.1 Cladding and roofing – historical applications of FRP in buildings

The literature is littered with examples of composite clad buildings built in the late 1960s and throughout the 1970s, hence, only a few examples of special interest will be discussed here before considering more recent applications[14, 29, 31, 35, 36].

The American Express (Amex) headquarters (Europe) building, Brighton

Much has been written about the American Express building in Brighton with its striking, strong horizontal visual emphasis, which was created through the contrast of blue tinted windows and white GRP cladding (see Fig. 13.5). The specification analysis details all aspects of the building including construction, external finishes, window systems and, of course, the GRP cladding panels[37]. The requirement for a permanently white building was instrumental in the decision to use GRP, since it was considered to be one of the few materials capable of maintaining 'a pristine white colour indefinitely'. The balance in achieving colour fastness whilst maintaining good surface spread of flame resistance was achieved by using a non-fire-retardant marine gelcoat, which had proven durability for marine structures, backed up with a filled fire-retardant resin. At the time, this resulted in the need for compromise over the requirement for a Class Zero building, which required a BS 476 Part 7 Class 1 pass, with a down-rating from a Class 1 system to a Class 2 when a non-fire-retardant gelcoat was used[4,6]. Today, Class 1 can be achieved even when using a non-fire-retardant gelcoat[4,38]. Weathering test data and a full-scale mock-up of the cladding system were necessary before the client was convinced about the new material.

The cladding beams are each 1.3 m deep with the largest span of 12.6 m between reinforced concrete columns. The GRP is structural in that it was required to support the 4400 m² of glass windows. The upper face of the beam also acts as a walkway to allow general maintenance and window cleaning (see Fig. 13.6). The deflection of the largest span was limited to one-thousandth of the span.

Extensive full-scale testing proved the resilience of the fixings, the structural integrity of the GRP to meet the wind loading conditions, and the ability of the composite to resist fire and water penetration.

Figure 13.5 *The American Express building. Reproduced by permission of Gollins Melvin Ward Partnership (Architects) and Peter Hodge & Associates (GRP Consultants).*

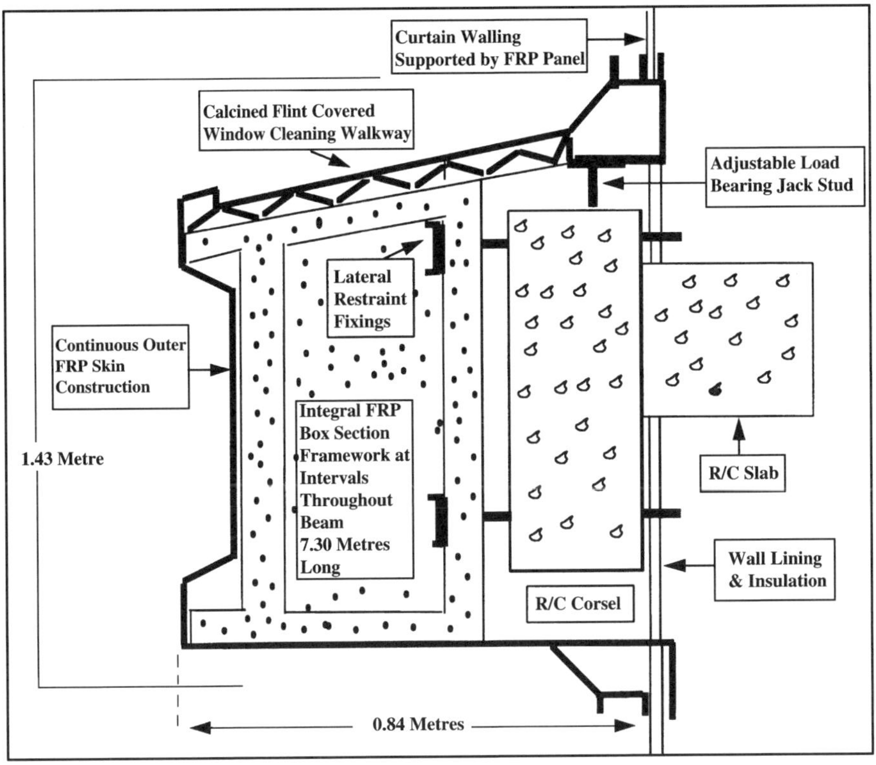

Figure 13.6 *Diagram of details of the beam, showing the walkway used for the American Express building.*

In 1998, the Amex building was inspected 25 years after it was built[39]. There were no signs of structural damage and only very minor laminate surface degradation. This was described as 'clearly illustrating that GRP is indeed an eminently suitable building material, provided that close attention is paid to all the noted parameters'. Even the serious hurricanes of late 1986, which resulted in the loss of the Amex building roof, 'had no adverse effect on the GRP panels nor their fixtures to the building'.

The gelcoated surface has weathered to a semi-gloss finish but with little or no serious yellowing or fading. Whilst there was some evidence of star cracking within the gelcoat in the more highly stressed areas, such as corners, the area affected was very small and the damage considered to have occurred within the first two to three years of installation. These areas were examined but there was no evidence of gelcoat lifting, blistering or other degradation.

The next example is also an encouraging testament to the long-term durability of GRP.

Mondial House, Post Office Telecommunications building, London

At the time of its commissioning in 1975, Mondial House was Europe's largest telephone exchange. The building is situated on the north bank of the River Thames and was subject to strict planning constraints. Since the views from local historical buildings had to be maintained the new building was designed as a wide-based, low-rise, stepped structure with an overall height limit of 46 m.

Cantilevers project from the body of the building (see Fig. 13.7) to give several horizontal bands of GRP cladding between the windows. In addition to providing the desired visual effect, the cladding also encloses service ducts around the perimeter of each floor. Hence, the design allowed for individual panels to be removed from the outside without disturbing neighbouring panels. Handling and jointing were improved with the use of stiffening at regular intervals. The flat faces of panels were fabricated using sandwich construction to provide both stiffness and thermal insulation. The whole building required the use of over 1000 mouldings.

Mondial House was also inspected in 1998 with the same general conclusions as for the Amex Building[39]. It was the durability of both Mondial House and the Amex building which promoted a study in the early 1990s for a material to reclad the concrete stanchions on the Liverpool Roman Catholic Cathedral (see Fig. 13.1). The refurbishment will be discussed in more detail in the next section. The durable non-fire-retardant, marine grade gelcoat, shown so effectively in Fig. 13.8, is backed-up with a very good fire-resistant resin polyester and the complete system meets BS 476 Part 7 Class 1 and the Class 0 requirements of the UK building regulations for non-combustible buildings[4,6,38].

The overall effect of the GRP on the cathedral cladding is shown in Fig. 13.9.

Terminal 2 pedestrian walkway, Heathrow Airport, London

In mid-1975 a rebuilding programme was started to modernise and improve passenger facilities in Heathrow's Terminal 2 building. The new walkway was added to the facade of the existing building to provide better access to and from public transport and car parks[36,40].

Figure 13.7 *Mondial House. Reproduced by permission of Brendal Plastics Anmac Ltd (GRP Fabricators).*

Figure 13.8 *The GRP clad stanchion of the Liverpool Roman Catholic Cathedral. Reproduced by permission of: Viv Wiliams Varley (Architects); Smart Crosby International (GRP Consultants); Minster Composite Products Ltd (GRP Fabricators).*

Figure 13.9 *The renovated Liverpool Roman Catholic Cathedral near completion. Reproduced by permission of: Viv Wiliams Varley (Architects); Smart Crosby International (GRP Consultants); Minster Composite Products Ltd (GRP Fabricators).*

Modular GRP design was employed to negotiate a difficult geometrical path (Figs 13.10 and 13.11). The modules were of sandwich construction and met the insulation test requirements as defined in BS476 Part 8, where exposure to a temperature of 900°C on one side of the panel must not result in a temperature exceeding 160°C on the other side within 30 minutes[41]. An integrity test, a means of

Figure 13.10 *The external view of the Terminal 2 walkway, at London Heathrow Airport. Reproduced by permission of Anmac Ltd (GRP Fabricators).*

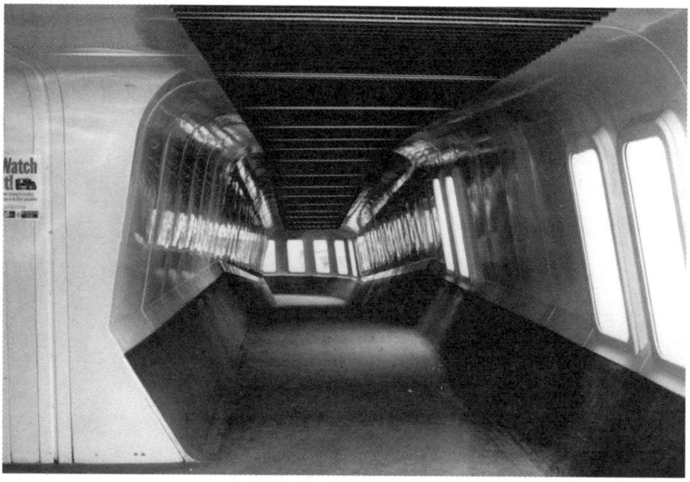

Figure 13.11 *An internal view of the Terminal 2 Walkway at London Heathrow Airport.*

determining whether any cracks or openings caused by extensive heat, will allow flames or hot gases to pass through the panel, was also required. Integrity was determined by holding a cotton wool pad against the side of the panel remote from the 900°C flame and measuring the time to ignition.

The tests, carried out by a Government approved test centre, showed that the panel had a fire insulation time of 48 minutes and an integrity time of 54 minutes. After 60 minutes exposure to the 900°C flame the test was terminated and the panel was considered to still be structurally stable. The modules are bolted to the steel and concrete framework and to each other, using polysulphide sealing compound.

The New Convent Garden flower market at Nine Elms

The roof of the New Covent Garden flower market at Nine Elms (see Fig. 13.12), built in 1973, covers 2.5 acres and used 924 GRP mouldings measuring 3.5 m × 3.5 m. The interior non-translucent surfaces were coated with an intumescent coating[23]. At the time, the roof was the biggest GRP roof that had ever been constructed. The GRP was not used in any significant structural way in this application but it is unlikely that any other material could have performed so many of the functional requirements simultaneously.

Sharjah International Airport

In the mid-1970s the requirement for 'glistening' white domes for the construction of the terminals at Sharjah International airport in the UAE (see Fig. 13.13) made GRP a natural choice. Since the domes were required to be spherical each unit could be cast from the same master mould surface. In the larger domes the entire surface is covered by only seven different panel types. The GRP units are supported on a steel framework which contains the mechanical services and supports the internal domed ceiling.

The Grand Mosque in Bahrain

Again, the unique properties of GRP enabled this unique structure in Bahrain (see Fig. 13.14) to be rapidly constructed. This light, structural dome has exhibited excellent durability under the harshest of UV light conditions.

Manchester City Football Club stand

The GRP roofing at Manchester City Football Club (see Fig. 13.15) was a major breakthrough in both design and size of contract for GRP. The material met both the

Figure 13.12 *The new Covent Garden Roof at Nine Elms in London. Reproduced by permission of: Collins, Melvin Ward and Partners (Architects); Nachschen, Croft & Leggatt (GRP Consultants); Mickelover Transport Ltd (GRP Roof Supplier).*

Figure 13.13 *Sharjah International Airport. Reproduced by permission of: Sir Williams Halcrow and Partners (Architects); NCL Consultancy Engineers (GRP Consultants); Anmac Ltd (GRP Manufacturers).*

Figure 13.14 *The Grand Mosque in Bahrain.*

fire requirements and stringent structural requirements with each barrel vault unit having to resist uplift forces of up to 16 kN. The units were successfully tested with loads of 25 kN.

13.4.2 Cladding and roofing – modern applications of FRP in buildings

Despite the early successes with fibre reinforced plastics in building applications, the material has never become established as a major structural material in the building

Figure 13.15 *The Manchester City Football Club stand, Manchester, UK.*

industry. However, in recent years it is re-emerging as its inherent benefits are, again, being acknowledged, particularly with respect to the production of complex shapes and effective finishes that can add unique characteristics to buildings. Probably the best example of this is the recladding of the stanchions on Liverpool Roman Catholic Cathedral.

Liverpool Roman Catholic Cathedral recladding

The work on Liverpool's Roman Catholic Cathedral has been mentioned previously with respect to the effectiveness of the finish used (see Figs 13.1, 13.8 and 13.9). However, it is worth first explaining the problem with the original cladding on the concrete stanchions and how GRP was chosen over other potential cladding materials. The problem with the original mosaic tile cladding is clearly seen in Fig. 13.16 where the mosaic tiles have either debonded from the mortar 'adhesive' or the mortar 'adhesive' has debonded from the structural concrete. The result was large lumps of material falling from the stanchions creating a dangerous situation. The choice of materials for recladding was short-listed to stainless steel or GRP. Since, stainless steel would require regular cleaning to prevent chloride ion attack from the marine environment in Liverpool, it was considered to be an expensive long-term option with

Figure 13.16 *Liverpool Roman Catholic Cathedral showing damage to the mosaic tiles. Reproduced by permission of: Viv Wiliams Varley (Architects); Smart Crosby International (GRP Consultants); Minster Composite Products Ltd (GRP Fabricators).*

respect to maintenance. Hence, attention turned to the long-term durability of GRP. Proven success in marine structures for over 40 years and the obvious durability of the coatings on the Amex and Mondial House buildings, eventually, resulted in GRP being chosen. Overlapping GRP components not only cover the original, damaged cladding but also prevent further spalling of the mosaic tile layer. The finish on the new cladding is an attractive gelcoat based granite effect, as shown in Figs 13.16 and 13.17. The gelcoat is non-fire-retardant but is backed up with the same type of fire-retardant resin used to produce Paris Disneyworld GRP components resulting in a Class 0/1 cladding system[4,6,38]. The gelcoat has been well proven (on marine structures since 1965) to give excellent gloss and colour retention.

Since GRP is very resistant to water degradation there is no chance of the material becoming damaged and spalling in the same way as the original traditional mortar and mosaic tile based cladding.

The Arabian Towers Hotel in the United Arab Emirates

The offshore structure, shown in Fig. 13.18, is the tallest building in the Middle East. Completed in 1999 it is not only futuristic in design but also in the materials used in

Figure 13.17 *Liverpool Roman Catholic Cathedral stanchion newly clad with GRP. Reproduced by permission of: Viv Wiliams Varley (Architects); Smart Crosby International (GRP Consultants); Minster Composite Products Ltd (GRP Fabricators).*

Figure 13.18 *The Arabian Towers Hotel Dubai, United Arab Emirates.*

its construction. Much of the external and internal cladding is glass reinforced plastic using durable gelcoats backed up with fire-retardant resin based fibre reinforced plastic. Over 40 tonnes of resin were used on the inside of the building alone.

Louvres at Lancaster University

The award winning library extension at Lancaster University has 57 GRP louvres to protect and shield the glazed wall (see Fig. 13.19). Each louvre is 6.2 m long and made in two halves, which have been bonded together with a urethane acrylate based adhesive[18]. Since the site is very exposed the louvres were designed to resist wind speeds of 100 mph – which would result in a load of 2 tonnes on each louvre.

The original internal steel framed GRP clad design was revised and the final design was based on an all-GRP self-supporting structure; this resulted in a 50% weight saving with resultant savings on installation costs, as well as removing corrosion concerns with the original steel frame based design.

Wind break in London

The versatility and low tooling cost of GRP allowed the innovative structure in Fig. 13.20 to be designed to be both functional and artistic. The main purpose of the structure is to act as a windbreak but the ease of manufacture of highly coloured GRP components with long-term maintenance-free characteristics encouraged the design of a very unusual structure, which produced the desired performance characteristics whilst having aesthetic appeal.

Similar innovative design with GRP resulted in its use for the construction of parts of Paris Disneyworld and a UK based company won a £2 million contract to supply fire-resistant GRP components based on well proven resin and gelcoat systems[38].

The next three examples are typical of the decorative uses of GRP.

Figure 13.19 *The louvres protecting the glazed wall at Lancaster University, UK. Reproduced by permission of: CETEC Consultancy Ltd (GRP Consultants).*

Figure 13.20 *The Windbreak in London. Reproduced by permission of: CETEC Consultancy Ltd (GRP Consultants).*

David Lloyd Centre at Milton Keynes

Over 1300 m of radius bull-nosed fascia, columns and feature panels (Fig. 13.21) were manufactured in GRP for this application.

Tesco store, Southend-on-Sea

The project for Tesco at Southend-on-Sea involved the manufacture of 12 m × 6 m high stone-effect GRP columns, high level parapet cladding, brown featured spandrel panels and grey lead-effect lower flashing (see Fig. 13.22).

The lightweight, excellent durability properties of GRP enable inexpensive features to be incorporated on any type of building to enhance its aesthetic appeal in any location.

Cornice on traditional built structures

Traditionally built structures may benefit from additional aesthetic features which would be too costly to achieve using traditional materials (in terms of both material and installation costs) . However, GRP makes such minor, but effective improvements easily affordable, as shown in Fig.13.23.

Internal decorative cladding at the Gateshead Metro Centre

Many of the benefits of GRP are brought to the internal finish of buildings to add character and aesthetic features. The Gateshead Metro Centre (see Fig. 13.24) is a fine

Figure 13.21 *The David Lloyd Centre, Milton Keynes, UK. Reproduced by permission of: Diespeker (GRP) Ltd (GRP Consultants).*

Figure 13.22 *The Tesco store at Southend on Sea, UK. Reproduced by permission of: Diespeker (GRP) Ltd (GRP Consultants).*

example of the use of internal decorative GRP cladding, fascias, columns, planters, bulkheads and water features. Although the material was chosen for its ability to provide the maximum visual appeal, its fire-retardant and vandal-resistant properties were considered equally important in the decision to use GRP.

Figure 13.23 *A cornice on a 'traditional built' structure. Reproduced by permission of: Minster Composite Products Ltd (GRP Consultants).*

Figure 13.24 *The Gateshead Metro Centre, Gateshead, UK.*

Small GRP building components

There are numerous examples of small components manufactured in GRP for building applications. Most are semi-structural – for example, a garage door might require a span of 3.5 m, a maximum sag at the centre point of only 20 mm, and resistance to wind loading in excess of 65 mph. The surface can be made to have a wood-like appearance but unlike wood GRP does not split, rot or require maintenance.

Porches, as shown in Fig. 13.25, also take advantage of the lightweight, low maintenance, high performance characteristics of GRP for enhancing housing.

Figure 13.25 *GRP Porches. Reproduced by permission of: Minster Composite Products Ltd (GRP Consultants).*

Explosion-proof panels

The need to protect buildings from the blast of explosions such that the structure and the contents are not damaged or destroyed has been investigated for many years. Recently, a GRP curtain wall cladding panel (shown in Fig. 13.2) was designed to meet the European EXR2 requirements of 2 bar over pressure resistance without breaking. The panel passed the test with ease and even passed successive blast tests until some debonding of the GRP skin occurred. The detailed design has been well documented in the project reports and the panel described in a data sheet as well as in the literature[42,43,44].

The project involved determining the correct GRP skin design as well as the core characteristics and the choice of resin and surface coating. The manufacturing technique was also fully investigated with resin transfer moulding (RTM) providing the best solution to produce consistent panels.

The surface of the panel (shown in Fig. 13.2) has been coated with a polyester resin granite-effect gelcoat based on a high performance marine grade resin and containing polyester resin based chips. Similar coatings and bulk casting materials are used for kitchen and sanitary-ware applications where a stone effect is desired. This type of granite-effect material is commonly referred to as solid surface material and examples of its internal use in buildings are shown in Figs 13.26 and 13.27[45]. Again,

Figure 13.26 *A granite-effect solid surface kitchen surface.*

Figure 13.27 *A granite-effect solid surface counter.*

the lightweight, durable characteristics of unsaturated polyester resins make it an ideal choice for manufacturing solid surface materials, especially in terms of ease of installation and cost when compared to granite or stone surfaces. The resin based material is also easy to join to provide aesthetically pleasing seamless bonded joints.

13.4.3 Bridges, decking and highways

Bridge construction and protection

In 1981 a proposal was accepted to suspend a floor beneath the girders of steel and concrete bridges to provide inspection and maintenance access, and to enclose the steelwork to protect it from further corrosion. In fact the corrosion rate of steel within such enclosures drops to a negligible level[46,47].

FRP was considered to be the ideal material for the manufacture of such enclosures because of its light weight, high strength and durability. Interlocking pultruded GRP planks were first used to install a permanent enclosure on the A19 Tees Viaduct in the UK. The floor area of 16 000 m^2 contains 250 tonnes of composite material[48,49].

Probably the most talked about composite bridge is the revolutionary, all composite, cable-stayed, suspension footbridge over the River Tay in Scotland – the Aberfeldy Bridge (see Fig. 13.28). The performance of the Aberfeldy bridge has been monitored carefully since its construction in 1992. Details of the bridge are summarised in the SPI brochure, *FRP Composites in Construction Applications*[50]. There are also many other interesting examples, of FRP in construction applications, in this publication with other notable references to:

- 'Eliminating Corrosion at Niagara Falls', where FRP has replaced badly corroded steel, on the US side of the Falls, for handrails, stairways, stands and decking.
- 'Lift Bridge Solves Problem in England', where a new, lightweight bridge was required to cross a canal. The bridge not only had to support the axle loading of heavy goods vehicles, up to 44 tonnes, but had to be easily lifted to allow canal traffic to pass. The Aberfeldy composite system concepts were again applied with great success.
- 'First Structures with Composite Cable Pre-stressing', where carbon fibre based composite cables were used for the manufacture of 12 prestressed piles, which supported two full bays of the pier.
- 'US Navy Test Pier Showcases Composites', where the world's first full-scale, all-composite demonstration pier was put under test at a US Navy's Advanced Waterfront Technology Test Site in California. The deck components were manufactured from isophthalic acid based resin and in some cases vinyl ester resin using glass fibre reinforcement. The high resistance of glass fibre reinforced composites to marine environments made it a natural choice for long-term durability with minimum maintenance.
- 'Composite Grating for Public Pier', where glass fibre reinforced isophthalic acid based polyester resin is used as decking with sufficient strength to support a three-axle truck with a load of 14.5 tonnes per axle.
- 'Eurotunnel "Chunnel" Pushes the Envelope', where within the maze of interconnecting passages over 3500 tonnes of pultruded composites support the 1300 km of electrical cable and hundreds of miles of fibre optics.

Figure 13.28 *The Aberfeldy Footbridge, Scotland, UK. Reproduced by permission of: Scott Bader Company Ltd (Resin Suppliers); Maunsell Structural Plastics (GRP Consultants).*

There are many other examples of the structural use of FRP in bridge construction in the publication *A Look at the World's FRP Composites Bridges* where the evolution of bridge engineering using FRP composites is charted[51]. Since the mid-1970s many bridges in all parts of the world have installed composite materials. Reference 51 cites over 80 examples of the use of composites in bridges.

There are also other examples of composites for civil engineering applications in *Composites for Infrastructure* where design, specification and connection issues are discussed as well as repair[52].

Repair to bridge structures

In recent years many traditionally built structures around the world have been found to be in need of urgent renovation because of long-term deterioration, a reduction in structural integrity, or because of the increased structural loading requirements due to changes in legislation, to improve factors of safety, and increases in vehicle weights.

Since many examples of the use of composites in the repair of structures are cited in the literature, just a few examples will be discussed here. Again, many examples come from the USA and Canada where composites in civil engineering applications, both for new-build and repair, is one of the fastest growing sectors of the composites industry[50,52].

Typical repair applications include:

- 'Carbon fibre strengthens bridges', where cracks in concrete bridges are repaired with carbon fibre reinforced-epoxy resin based patches[50].
- 'Slab strengthening for increased loads', highlights how in South Carolina bridge decks, which could barely carry legal loads, were improved to a higher standard of safety using carbon fibre based laminates which improved the flexural resistance of the decks[52].
- 'Girder Shear Strength Reinforcement', discusses the renovation of approximately 5000 bridges within Canada's Alberta province which had been determined to be deficient in shear strength. Initial testing of carbon fibre epoxy resin based strengthening systems started on one bridge in 1995, took just two weeks to complete and since no heavy equipment was required, traffic disruption was reported to be insignificant[53]. Other work has also been reported on the applications of FRP for repairing Canadian structures[53].
- 'Composites provide timely bridge repair', where a crash damaged prestressed concrete bridge beam was declared unrepairable using bonded steel plates because of the level of damage and twist in the beam. So the choice was either total replacement or an innovative composite based solution; this repair took just 3 nights in comparison with an estimated complete closure for one month if replacement had been the chosen course of action[49].
- Wooden utility poles have, for sometime, been given extended lives by the use of a composite wrap-around system[50]. In recent years, the possibilities of using similar and modified techniques for concrete column renovation have been under serious investigation[52,54,55,56].
- 'Laminate and Wraps' describes how the performance of concrete columns can be significantly enhanced by wrapping with fibre reinforced composites. One particular example involved wrapping 6 columns, on a highway in San Diego County in the USA, with filament wound pre-impregnated carbon fibre tows, in a precise way, so that they would meet the seismic requirements of the area[52].
- 'Column Wrapping' describes how 'advanced fibre reinforced polymer technology is currently being trialled by the Highways Agency to protect bridge columns from "bridge bashing" incidents. Evaluation of the technique is being progressed to confirm whether this process can bring costs down'[54].

It was concluded that although the composite materials which are currently available are expensive, they can be rapidly installed with low levels of resource and disruption – compared with other solutions – which makes the 'total' job cost more attractive. It was also stated that the 'composite solution' provided improved durability and lower maintenance costs compared to other solutions. The final comment was particularly encouraging for composite solutions for infrastructure. 'Wherever disruption costs feature in the selection of techniques for strengthening works, these composite materials are serious contenders and may come to dominate the market'.

13.5 Conclusions

For more than a quarter of a century fibre reinforced plastic has found niche applications in the building and construction industry. The material has been shown to stand the test of time, particularly with respect to its resistance to marine and general environmental degradation, to the extent where it is becoming an increasingly familiar choice for repair to structures where traditional building and construction materials are either failing or require upgrading. However, for all the numerous examples of the use of fibre reinforced composites in the building and construction industry, the volumes of materials used are small in comparison with traditional building materials. No doubt this is primarily a question of education with civil engineers unlikely to be presently educated to the same level of understanding of composite materials as they are with traditional building materials, such as concrete, brick, stone, steel, aluminium and wood. Equally, the composites industry must provide information in the form and in the language required by civil engineers – as codes and standards. Knowledge about composite materials exists in terms of performance durability, test data and case studies, but it is rarely available in the format required to enable ease of understanding and conversion into trusted specifications. Similarly, architects need to be targeted with information about composites to encourage specification of a material, which has so much to offer in terms of freedom of design, high strength-to-weight ratio, excellent durability, minimum maintenance and excellent fire resistance.

Fibre reinforced plastics are a material with a future in the building and construction industry. Their novel characteristics cannot, and will not, be ignored by innovative, creative architects and engineers. However, the suppliers of resins, reinforcing fibres and the converters could do much more to increase the pace of market penetration and in so doing gain rich rewards. Of course, this need not and should not be at the expense of traditional materials since all materials have something to offer and where a synergistic benefit can be achieved by using different materials together, then all those involved will benefit.

References

[1] Parkyn, B. (Ed.), *Glass Reinforced Plastics*, Butterworth, 1970.

[2] Scott Bader Company Limited, *Case Histories of GRP Boats*, Scott Bader, Wollaston, Wellingborough, Northants NN29 7RL, 1973.

[3] British Standards Institution, BS 476:1989, *Fire tests on building materials and structures – Part 6: Method of test for fire propagation products*, London: BSI.

[4] British Standards Institution, BS 476:1989, *Fire tests on building materials and structures – Part 7: Method for classification of the surface spread of flame of products*, London: BSI.

[5] French Surface Spread of Flame and Smoke Standards, Epiradiateur and NFF 16-101

[6] *UK Building Regulations, 1985,* Section B.

[7] *A Strategy for the Investigation of Plastics & Fire*, British Plastics Federation, 1976.

[8] British Standards Institution, BS 4422:1969, *Part 1: Terms associated with fire*, London: BSI.

[9] 'Endurance of Composite Hulls', *Composite Hulls*, May 1988, p. 16.

[10] Batchelor, J., 'Use of Fibre Reinforced Composites in Modern Railway Vehicles', *Materials in Engineering*, Vol. 2, 1981, p. 172.

[11] Roberts, R., 'Design and use of GRP piping in chemical plants: BS 7159, 1899 system design code', Paper 9, *BPF Reinforced Plastics Congress*, 1990, p. 99.

[12] Norwood, L.S. and Taylor, B.R., 'Resin systems for GRP fuel storage tanks', Paper 17, *BPF Composites Congress*, 1992, p. 97.

[13] British Standards Institution, BS 4994:1987, *Design and construction of vessels and tanks in reinforced plastics*, London: BSI.

[14] Leggatt, A.J., *GRP & Buildings*, Institute of Structural Engineers, 27 January 1977.

[15] Hollaway, L., (ed.), *BPF Handbook of Polymer Composites for Engineers*, Woodhead Publishing, 1994.

[16] Scott Bader Company Ltd, *Technical Literature, Building & Construction Product Range and Applications*, Scott Bader Company Ltd, Wollaston, Wellingborough, Northants, NN29 7RL.

[17] Norwood, L.S. and Taylor, D., 'Unsaturated polyester resin in solid surface applications', *BPF Composites Congress 1998*, Session 6, Paper 3.

[18] Scott Bader Company Ltd, *Technical Literature, Crestomer Advantage 10 & Crestomer 1152PA*, Scott Bader Company Ltd, Wollaston, Wellingborough, Northants NN29 7RL.

[19] Smith, C.S., *Design of Marine Structures in Composite Materials*, Elsevier Applied Science, 1990, p. 292.

[20] Norwood, L.S., Nixon, J. and Wood, A., 'Crestomer adhesives for the small and large GRP moulder', *Proc. of the 19th Int SAMPE Europe Conference,* La Défense, Paris, France, April 22–24 1998, p. 51.

[21] Hollaway, L., (Ed.), 'Procedures for designing materials', *BPF Handbook of Polymer Composites for Engineers*, Woodhead Publishing, 1994, Chapter 5, p. 116.

[22] Hollaway, L., (Ed.), 'Procedures for designing materials', *BPF Handbook of Polymer Composites for Engineers*, Woodhead Publishing, 1994, Chapter 6, p. 136.

[23] Scott Bader Company Limited *Technical Literature, Crystic® Fireguard® – 75PA*, Scott Bader Company Limited, Wollaston, Wellingborough, Northants NN29 7RL, UK.

[24] British Plastics Federation, *A Strategy for the Investigation of Plastics and Fire*, BPF, 1976.

[25] Purser, D.A., *Toxicity Assessment of Combustion Products in Fires*, Combustion Toxicology Research, Huntingdon Research Centre, UK.

[26] Nixon, J., 'GRP protects equipment and personnel offshore' *Reinforced Plastics Conference,* December 1992.

[27] Scott Bader Company Ltd, *Technical Literature for Crystic 343A*, Scott Bader Company Ltd, Wollaston, Wellingborough, Northants NN29 7RL, UK.

[28] British Plastics Federation, Fire Committee, *Bath Place*, London: Rivington Street.

[29] Scott Bader Company Limited, *Crystic Polyester Handbook*, Scott Bader Company Limited, Wollaston, Wellingborough, Northants NN29 7RL, UK.

[30] Ball, P., 'Manufacturing Processes', *BPF Handbook of Polymer Composites for Engineers*, (Ed.) Leonard Hollaway, Woodhead Publishing, Chapter 2, p. 73.

[31] Brookes, A. *Cladding in Buildings*, Third edition, London: E. & F.N. Spon, p. 39.

[32] HMSO, *UK Environmental Protection Act 1990*, HMSO, London.

[33] HMSO, UK Statutory Instruments, 1994 No. 1271, The Environmental Protection (Prescribed Processes & Substances etc.) (Amendment) Regulations 1994, HMSO, London, UK.

[34] UK Secretary of State's Guidance – Processes for the Manufacture of Fibre Reinforced Plastics, 1996.

[35] Scott Bader Company Ltd., *Great Crystic Buildings*, Scott Bader Company Ltd., Wollaston, Wellingborough, Northants NN29 7RL.

[36] *The Design & Fabrication of Building Components in Glass Reinforced Polyesters*, Scott Bader Company Ltd., Wollaston, Wellingborough, Northants NN29 7RL.

[37] Norwood, L.S., *Specification Analysis Amex House*, Wollaston, Wellingborough: Scott Bader Company Ltd., Northants NN29 7RL.

[38] Scott Bader Company Ltd., *Technical Data Sheets, Crystic 356PA & Crystic Gelcoat 65PA*, Wollaston, Wellingborough: Scott Bader Company Ltd., Northants NN29 7RL.

[39] Starr, T., 'Get the strength of GRP around you', *Reinforced Plastics*, June 1998, p. 42.

[40] Hollaway, L., (Ed.), 'Designing for fire performance', *BPF Handbook of Polymer Composites for Engineers*, Woodhead Publishing, 1994, Chapter 10, p. 262.

[41] British Standards Institution, BS 476, *Part 8: Fire tests on building materials and structures, Test methods and criteria for the fire resistance of elements of building construction*, London: BSI, (replaced by BS 476 Parts 20–23).

[42] *PERA Blast Panel Report on Project No. P81038*, PERA, Melton Mowbray, Leicestershire.

[43] Fibre Reinforced Polymer Composites for Blast Resistant Cladding, CIRIA.

[44 Alderson, A., 'Fibre reinforced polymer composites for blast resistant cladding', Paper 22, *Composites & Plastics in Construction Conference* 16–18 November 1999, Watford, UK: BRE.

[45] Norwood, L.S., and Taylor, D., 'Unsaturated polyester resin on solid surface applications', *BPF Congress 1998*, Session 6.

[46] 'Enclosures put paid to bridge painting', *New Civil Engineer*, 15 October 1981.

[47] Bishop, R.R., *Transport & Road Research Laboratory, Report 3*,

[48] Head, R., 'GRP walkway membranes for bridge access and protection', *BPF 13th Reinforced Plastics Congress*, 1982.

[49] Richmond, B., 'Cable stayed bridge developments relating to wind effects on vehicles', *International Conference on Cable Stayed Bridges*, Bangkok, 1987.

[50] *FRP Composites in Construction Applications – A Profile in Progress*, The SPI Composites Institute, 355 Lexington Avenue, New York, NY 10017, 1995.

[51] *A Look at the World's FRP Composites Bridges*, MDA (Market Development Alliance), SPI Composites Institute, 355 Lexington Avenue, New York, NY 10017,1998.

[52] *Composites for Infrastructure, Guide for Civil Engineers*, Ray Publishing 4891 Independence St., Suite 270, Wheat Ridge, CO 80033, USA, 1998.
(Website: www.raypubs.com)

[53] Redston, J., 'Canada's infrastructure benefits from FRP', *Reinforced Plastics*, July/August 1999, p. 34.

[54] Haynes, M., 'Column wrapping', *Concrete Engineering International,* July/August 1998, p. 43.

[55] McConnell, V., 'Composites make progress in reinforcing concrete', *Reinforced Plastics*, July/August 1999, p. 40.

[56] Haynes, M. and Denton, S., 'Fibre reinforced polymer composites – Column wrapping', Paper 14, *Composites & Plastics in Construction Conference*, 16–18 November 1999, Watford, UK: BRE.

14 The Durability of Underground Structural Steel Structures

Graham Gedge

14.1 Introduction

This chapter considers the durability of underground structures fabricated from structural steel. In particular, the issues of corrosion, and the prevention of corrosion, in steel components buried permanently in the ground. Durability can also be influenced, or indeed controlled, by issues other than corrosion such as fatigue or fracture considerations, however this chapter concentrates on corrosion.

The chapter outlines the basic reactions that occur during the corrosion of steel in aqueous environments such as the ground. It then considers the important factors of the ground that influence the corrosion rates of buried steel. It is important to appreciate that it is the rate of corrosion (in the context of buried structures this is usually expressed in mm/year) that is the most important in assessing corrosion risk rather than the fact of corrosion itself. The chapter then moves on to describe the difference between uniform corrosion and localised corrosion, particularly pitting, and the importance of each of these to different types of structures.

Having established the above basic concepts, the chapter then discusses how to assess corrosion risks based on published data and on field-testing. In particular, it includes a critical review of the common test methods and why these are often unnecessary and misleading. This naturally leads to a brief outline of the common methods of corrosion protection of steel structures buried in the ground. It is not the intention of this chapter to arm civil engineers with the necessary skills to deal with all corrosion and corrosion related tasks. This is a discipline and a specialist skill in its own right. The aim is to highlight for engineers the important factors that need to be considered in carrying out corrosion assessments and the important considerations that have to be made with respect to the work of specialist corrosion sub-contractors.

14.2 Basic principles of corrosion reactions

14.2.1 General corrosion reactions

For corrosion processes to take place, two simultaneous chemical reactions must occur on the metal surface. For steel these reactions can be expressed as:

$$Fe \rightarrow Fe^{2+} + 2e^- \qquad (1)$$

This is described as the anodic reaction involving the dissolution of metal ions and the generation of electrons.

The second reaction is the cathodic reaction. There are two common reactions that can occur in near neutral pH conditions (pH = 4.5 to 9) the dominant reaction is the reduction of the dissolved oxygen.

$$O_2 + 2H_2O + 2e^- \rightarrow 4OH^- \qquad (2)$$

At lower pH values an alternative cathodic reaction can occur involving hydrogen evolution:

$$2H^+ + 2e^- \rightarrow H_2 \qquad (3)$$

This latter reaction can occur in neutral or alkaline conditions however; the oxygen reduction reaction is the preferred reaction if oxygen is available. When dissolved oxygen is not present, hydrogen reduction can occur but the rate of this reaction at neutral or alkaline pH values is so slow as to be negligible for all practical applications.

The important points to note from these reactions are:

- they must occur simultaneously on the metal surface;
- for corrosion to occur in near neutral pH conditions oxygen must be present and dissolved in an electrolyte;
- the anodic reaction generates electrons that are consumed in the cathodic reaction*;
- that no other chemical species are required or involved in corrosion reactions; and
- that if one reaction is slowed down or prevented then corrosion rates will be retarded or stopped.

*This transfer of electrons is a direct measure of corrosion rates. Corrosion scientists often express corrosion rates in terms of a current density of the quantity of electrons transferred per unit area i.e. amps/m^2.

The rate of reaction (or the corrosion rate) is determined by how a material interacts with the service environment. In the case of steel the reaction rate is normally controlled by the cathodic reaction, that is the ease with which hydrogen evolution or oxygen reduction can occur.

The corrosion products resulting from the corrosion of steel are a complex mix of hydrated iron oxides and hydroxides; commonly referred to as rust.

14.2.2 Different forms of corrosion

Corrosion can manifest itself in many different forms and although these will all involve the previously mentioned reactions, the form that corrosion will take can be very different. These different forms of corrosion can be significant in terms of structural integrity and the methods of protection.

General or uniform corrosion

The most obvious form of corrosion is uniform or general corrosion in which the majority of the surface is affected and corrosion rates over the surface are broadly the same. This form of corrosion is found on steel immersed in seawater or indeed buried in the ground. The rate of corrosion usually stabilises in a period of 6 to 12 months and thereafter remains constant. In many naturally occurring environments, uniform corrosion rates are relatively low, being in the range of 0.025 mm/year to 0.150 mm/year.

This form of corrosion is prevalent on buried steel structures such as piles and pipelines.

As will be seen, uniform corrosion rates in the ground are relatively constant throughout the world and, as the rates are well documented, it is easy to establish corrosion risk based on a uniform loss of section.

Pitting corrosion

In pitting corrosion, damage is localised to small areas of the metal surface but the rate of corrosion is often high and can be in excess of 0.5 mm/year.

Pitting corrosion is rarely of significance to primary structural elements, such as piles, where a large number of small holes in the section would not represent a risk to structural integrity. However, this is not always the case and pitting can be of significance for structures or components subjected to fatigue loading, as there is evidence that pits can act as crack initiation sites. Pitting corrosion may also be of concern to components, such as ground anchors and ties, fabricated from wire strand or bars. In the case of strand, pitting can represent a significant section loss in an individual wire that is inevitably of small diameter. Pitting can also lead to premature failure as pits can act as stress concentrators and initiate crack propagation leading to a brittle type failure; this latter aspect is also relevant to bars used for anchors.

Pitting is also of significance for structures such as tanks and pipelines containing fluids or gases, often under pressure, where pitting through the section could lead to leakage of the product.

Anaerobic corrosion

In conditions of complete oxygen depletion (anaerobic conditions) it is possible for corrosion to occur without either the cathodic reduction of water or hydrogen evolution. Under anaerobic conditions corrosion can occur that involves bacteria which are naturally present in soil and water. This is referred to as microbial induced corrosion (MIC).

MIC is a complex subject that is far from well understood. MIC is rarely a significant problem for structures permanently buried in the ground, the exceptions being components where pitting corrosion cannot be tolerated. For most civil engineering applications, engineers need to be aware of MIC. If it is likely to be an issue, specialist advice should be sought. Engineers need to be aware that in many methods of testing (see Section 14.4.2) for the assessment of ground conditions, with respect to corrosion, a test for specific bacteria associated with MIC must be made. The test is for sulphate reducing bacteria (SRB). Where this test is used, and a positive result is obtained, the engineer should not be alarmed; the nature of the test and the ubiquitous nature of SRB almost guarantee a positive test. This is discussed further in Section 14.4.2. SRB induced corrosion is by far the most common type of MIC encountered in civil engineering and a good review of the subject has been written by Stott[1].

Other forms of corrosion

The reader should also be aware that there are other forms of corrosion including:

- differential concentration cells – corrosion resulting from a difference in concentration of a chemical (typically oxygen) between connected parts of the structure or component (The damage can be rapid and severe as the anode (low concentration area) is often small compared to the cathode.);

- crevice corrosion – a special form of differential concentration cell that affects joined parts with a fine separation (<0.25 mm) can occur between faces of flanged connected pipes;
- stress corrosion cracking (SCC) – is a complex corrosion mechanism involving the simultaneous interaction of a tensile stress, a specific environment and a susceptible material; and
- erosion corrosion – the joint action of flow induced erosion and electrochemical corrosion.

These forms of corrosion are rarely associated with buried steel structures. For further information the reader should consult standard corrosion texts such as those given in the references to this chapter.

14.3 The ground as a corrosive environment

The ground is a complex chemical environment that varies considerably both on a global scale and at any particular given location. Thus one might assume that assessing the risk of corrosion at a given location is equally variable with corrosion rates being particular to a specific location. This is not the general rule and an assessment of conditions can be made with surprisingly little information relating to the actual ground conditions and chemistry. In broad terms the ground can be divided into four classes:

- undisturbed ground conditions of pH 4 to 9;
- disturbed ground conditions of pH 4 to 9;
- acidic grounds with pH<4;
- contaminated land.

14.3.1 Undisturbed ground of pH 4 to 9

It is important to appreciate that in terms of assessing corrosion risks, the term 'undisturbed' relates to the ground not having been disturbed by human intervention for a period of 20 to 30 years, rather than over a geological timescale. These conditions are the most commonly encountered in civil engineering.

The basic chemical reactions discussed in Section 14.2.0 indicate that in these pH conditions, the preferred cathodic reaction is the reduction of oxygen. For corrosion to occur, dissolved oxygen must be present at the surface of the steel. If it is not, then corrosion cannot occur. Although precise data regarding the oxygen concentration of the ground are scarce, it is easy to see that the oxygen concentration in undisturbed ground would be both low and constant with time. A low oxygen concentration is beneficial for ensuring low corrosion rates, or indeed no corrosion at all. However, oxygen concentration per se is not the most important parameter because the oxygen must be dissolved in water and then transported to the steel surface.

If water flow and other mechanical means of transport are assumed not to be present in the ground then transport will occur by diffusion. This is a slow process that is controlled by:

- the degree of water saturation – the higher the saturation the slower the diffusion rate;

- the pore structure of the ground – the larger and more open the pore structure the easier transport becomes;
- the temperature of the ground – higher temperatures give higher diffusion rates; and
- the concentration gradient of the oxygen between the steel surface and the surrounding ground.

Consideration must also be given to what happens at the steel's surface during corrosion. In oxygen-depleted environments, there is an initial short period (6 to 12 months) of high corrosion activity during which the immediately available oxygen is consumed in the corrosion reaction. Because of the relative lack of oxygen in the ground, the corrosion products formed are not the normal mix of oxides and hydroxides referred to as rust. The reaction tends to stop at the formation of magnetite – the conversion of magnetite to other compounds proceeding very slowly with time. The formation of relatively stable magnetite has two benefits:

- magnetite is integral with the steel's surface and is a physical barrier to the transfer of oxygen across the soil/steel interface – thus a crucial step in the diffusion process is blocked;
- oxygen diffusing to the steel will react with the magnetite layer to form higher oxides and hydroxides and is therefore consumed in the layer of corrosion products prior to reaching the steel surface.

The combination of slow diffusion within the ground, and the formation of protective scales on the steel's surface, account for the low corrosion rates of steel components placed in near-neutral pH undisturbed ground. These low corrosion rates have been studied by numerous workers over a period of 50 years and have been confirmed for a wide range of naturally occurring ground conditions. Further details regarding the corrosion rate data are given in Section 14.5.

Oxygen concentration versus depth

It is a reasonable assumption that the oxygen concentration will vary with depth, the highest oxygen concentration being close to finished ground level and quickly decreasing with depth. It therefore follows that one might anticipate higher corrosion rates on steel that is close to ground level.

However, in most applications this is not the case. The effect of corrosion rate with depth was investigated by Romanoff[2] and he concluded that there was no significant variation. Romanoff attributed this to the establishment of a large differential aeration cell between the upper (cathodic) and lower (anodic) parts of the structure in the ground in which the upper part is in an environment of higher oxygen concentration compared to the lower part. However, in general, the relative area of the upper part is small compared to the lower part.

In a differential aeration cell of this type, corrosion damage is spread out over the entire anodic area and is supported by a relatively small cathode. Thus, overall corrosion rates are little altered. Conversely, where the area ratio (has a large cathode and small anode), corrosion damage is concentrated on the anodic area and damage can be both rapid and severe.

Fluctuating water levels

It is well established that some of the highest uniform corrosion rates are sustained in conditions where intermittent wetting and drying of steel occurs i.e. inter-tidal and splash zones on marine structures. It might be expected that a similar situation would occur where groundwater levels, surrounding a buried structure, fluctuate. In general this is not the case.

Where the water source is from a depth, it can be assumed that the oxygen level is depleted and the fluctuation does not alter the basic availability of oxygen to fuel the cathodic reaction. In cases where the water may be fully aerated, e.g. seawater on the reverse side of a sheet pile, the corrosion rate is little altered because the rate of corrosion remains controlled by the diffusion of oxygen to the steel surface. This process is slow and, unlike in open seawater oxygen transport is not assisted by wave and current action and corrosion rates are therefore little affected.

14.3.2 Disturbed ground conditions pH 4.5 to 9

Disturbed ground conditions refer to situations where the ground has been excavated and a steel component placed in the excavation prior to back-filling, as would occur for a pipeline. Disturbing the ground does not alter the corrosion reactions that occur, but it makes the reactions, particularly the transport of oxygen, easier to happen and therefore increases corrosion rates. It is therefore normal to provide corrosion protection using protective coatings and/or cathodic protection for steel structures in these environments.

14.3.3 Acidic ground conditions pH < 4

Naturally occurring ground conditions where the pH < 4 are far less common than near neutral conditions, however, they can occur and when encountered the general approach to assessing the corrosion risks needs to be modified. For acidic conditions, the favoured cathodic reaction will be hydrogen evolution rather than oxygen reduction. For the hydrogen evolution reaction to take place the steel must be in contact with water. In this case, the rate of reaction is controlled by the transfer of charge (electrons) and is, therefore, independent of any transport mechanism through the ground. Corrosion rates can, therefore, be very much higher than in near neutral pH conditions; in general as the pH falls the ease of hydrogen evolution increases and corrosion becomes both more likely and the rate of damage greater. There is little or no published corrosion rate data for naturally occurring acidic ground conditions and, where these conditions are encountered, specialist advice should be obtained regarding the corrosion risk and the methods of dealing with it.

14.3.4 Contaminated land

There is an increasing trend towards the re-use of contaminated land and brownfield sites. Both of these conditions can pose a higher corrosion risk than naturally undisturbed ground. Where such conditions are encountered, a detailed corrosion assessment must be made that is similar to that used for disturbed conditions. In addition, the test programme should identify the presence of any organic contaminants at the site. There is little or no published information on the corrosion risks associated with contaminated land. Each site should be considered as an individual case and specialist advice on corrosion risks and methods of prevention should be obtained.

14.4 Assessing the risk of corrosion

Understanding the basic corrosion reactions, and the ground as a corrosive environment, leads naturally on to the concept of assessing the risk of corrosion in terms of predicting likely corrosion rates at a given site. Such assessments can be divided as follows:

- simplified assessment applicable to all ground conditions; and
- detailed assessment applicable to disturbed, acidic and contaminated sites.

14.4.1 General assessment of all sites

The approach given in this section may be applied to all sites where steel is to be used in the ground. At the end of this assessment it can be decided if additional testing is or is not required.

In Section 14.3 it was shown that the ground conditions have a significant influence on both the corrosion reactions and corrosion rates that occur in the ground. It is important, therefore, to establish whether the ground conditions are disturbed, undisturbed or contaminated. However, the type of ground (sand, clay, chalk, etc.) is, at this stage, unimportant from the point of view of corrosion.

The other important parameter in understanding corrosion in the ground, is the pH of the soil or the groundwater. This can only be assessed by on-site testing. Tests should be carried out at a representative number of locations around the site and at various depths in the ground. The results should be classified as follows:

- pH > 9 indicates alkaline ground conditions where corrosion risk in naturally occurring conditions are low. However, high pH values may indicate a contaminated land site and this must be checked.
- pH $= 4.5$ to 9 indicates near neutral conditions in which corrosion rates are controlled by cathodic reduction of oxygen. Corrosion risk can be assessed using published corrosion rate data.
- pH < 4.5 indicates acidic or contaminated ground conditions where corrosion reactions will be controlled by hydrogen evolution. In acidic conditions specialist advice must be obtained on possible corrosion rates and methods of corrosion prevention.

Having established the ground conditions and the pH, Table 14.1 can be used to decide upon the most appropriate course of action:

Table 14.1 shows that in many instances additional protection is required. However, for most civil engineering applications, such as bearing and sheet piling, the ground will be undisturbed and in the pH range 4.5 to 9. In these conditions no additional testing is needed and the corrosion risk may be tolerable within normal design limits (see Section 14.5).

Table 14.1 General corrosion assessment

Ground condition	**pH**		
	>9	**4.5 to 9**	**<4.5**
Naturally occurring undisturbed	Unless contaminated no corrosion protection is required	Use published corrosion rate data to assess risk and method of dealing with the risk	Additional testing required to establish cause of acidity and method of protection
Naturally occurring disturbed	Additional testing required to establish if protection required	Protection methods required	Additional protection required to establish method of protection
Contaminated	Further investigation required for the identification of contamination and method of protection required		

14.4.2 Detailed assessment of corrosion risk

Where a general assessment of risk has indicated that additional testing is required, it will be necessary to carry out a wide range of tests to identify the factors that may influence corrosion. The scope and scale of such testing is difficult to predict, particularly for the less common conditions of contaminated or acidic grounds. For near neutral pH conditions the situation is somewhat simpler because there are establish sets of tests for such grounds. An example of the range of soil sample testing is given in Table 14.2. This is similar to that given in many national standards and guidance documents.

It should be noted that the full range of tests was originally developed for disturbed ground conditions, typically for pipelines where such testing is wholly appropriate. However, many standard documents make no distinction between undisturbed and disturbed ground. Therefore the standards require the full range of testing irrespective of

Table 14.2 Test for a detailed soil assessment

Property	Measured value	Points
Soil type	Fraction passing 63 :m sieve < 10%	
	Plasticity Index (PI) of fraction	+ 1
	Passing 425 :m < 6	
	Fraction passing 63 :m sieve > 10%	0
	PI of fraction passing 425 :m sieve #	
	Any grade	
	PI of fraction passing 425 :m sieve > 6 but < 15	–1
	Any grading	
	PI of fraction passing 425 :m sieve > 15	–2
	Organic matter. 1.0%	
	or material containing peat, cinder or coke	–3
Resistivity (ohm-cm)	> 10 000	+ 2
	< 10 000 but > 3000	+ 1
	< 3000 but > 1000	–1
	< 1000 but > 100	–3
	< 100	–4
pH of soil	6 < pH < 9	0
	5 < pH < 6	–2
	Less than 5 or more than 9	–5
Soluble sulphates (ppm)	< 200	0
	> 200 but < 500	–1
	> 500 but < 1000	–2
	> 1000	–4
Chloride ion (ppm)	< 50	–0
	> 50 but < 250	–1
	> 250 but < 500	–2
	> 500	–4
Sulphide and hydrogen sulphide	Effect on lead acetate paper:	
	no discolouration	0
	slight to moderate darkening	–2
	rapid blackening	–3

the state of the ground. It is the author's opinion that this misses an important distinction and, as such, much of the testing undertaken is unnecessary. However, if local standards, or utility rules, require such testing for all ground conditions, it must be carried out unless a departure or concession can be agreed with the relevant authority.

The corrosion assessment is usually made by summing the total points for all tests and typically placing the soil in one of the following categories:

- 0 or more = non-aggressive
- –1 to –4 = aggressive
- –5 or less = very aggressive

A problem with assessing the data generated from the type of testing shown in Table 14.2 relates to the cumulative nature of assessing the corrosion risk. It can be seen that the final column of the table assigns a value to the risk for each parameter and that the overall risk is determined as the sum of each test. The difficulty is that, in some respects, the tests are measuring, either directly or indirectly, the same parameter and there is a risk of 'double counting' and over stating the total risk for a given situation.

For example, consider the requirement for testing chloride and sulphate ions and resistivity. Each of these parameters can have an influence on corrosion rates in disturbed grounds. If the ground were identified as having a high chloride content (and/or sulphates) it would be expected to have a low resistivity, this being an indirect measure of the ionic conductivity of the ground. It therefore seems inappropriate to assess the risks, as being +9 and a value of +3 would seem more appropriate.

It is the author's opinion that this double counting is wrong and that if, for example, a low value of resistivity were recorded then it is reasonable to assume that there are chlorides and or sulphates present (and vice versa). Thus the testing regime can be very much simplified. Only where there is concern about particular corrosion problems such as pitting should tests for specific salts need to be carried out and included in the assessment.

The conclusion from the general and detailed assessment will be either that the corrosion risk is negligible in which case no, or minimal, prevention is required or that the ground conditions are aggressive to steel and prevention methods are required.

14.5 Corrosion prevention methods

Where testing of the ground has identified the need for additional protection to prevent corrosion there are a number of options available. Similarly, additional protection may be required for components, such as wire or bar ground anchors, that may be more sensitive to pitting type corrosion even in conditions that are not considered generally aggressive. The options for protection of surfaces exposed to the ground can be summarised as:

- provision of additional section thickness to act as sacrificial corrosion allowance over the life of the structure;
- the use of a sacrificial anode or impressed current cathodic protection (CP); and
- the use of protective coatings and tape wraps.

In some applications, most notably high-pressure hydrocarbon pipelines, it is common practice to use a combination of the above methods.

In practice, corrosion allowance is only practical for undisturbed ground conditions where corrosion rates are expected to be low. Where the ground is disturbed, as is the case for a pipeline placed in a trench, then it is unlikely that a corrosion allowance alone will be adequate. In these cases it is common practice to employ cathodic protection and or protective coatings.

14.5.1 Sacrificial corrosion allowance

The concept of a corrosion allowance is based on knowing what the corrosion rate will be for a particular application. Once the corrosion rate is known in appropriate units (mm/yr) then it is relatively easy to calculate the section that will be lost for a given lifetime. The potential difficulty lies in obtaining reliable data for the particular environment or application under consideration.

There are two ways in which data may be obtained:

- in situ testing in the environment under consideration; and
- using published data.

It should be emphasised that the use of a corrosion allowance is generally only applicable to situations where one is dealing with uniform corrosion damage. For localised corrosion (pitting, etc) a corrosion allowance is unlikely to be practical due to the high corrosion rates and, in this case, alternative measures will be required.

In situ testing

In principle, obtaining *in situ* corrosion rate data is relatively straightforward and can be achieved by the use of weight loss coupons. These are coupons of metal, of accurately known dimensions and weight, that are carefully prepared (in the case of steel this will involve blast cleaning or acid pickling), and then exposed to the environment of interest. Samples are removed at regular intervals, carefully cleaned to remove corrosion products and then weighed. The change in weight (weight loss) can then be used to calculate via Faraday's law the corrosion rate in mm/yr.

This apparently simple technique has a number of disadvantages for civil engineering applications in buried ground:

- corrosion rates in the ground are relatively low and measurable results take a considerable time; and
- initial corrosion rates are often considerably higher than long-term rates. This is because the cleaned surface of the coupon is very active until corrosion scales have formed.

To overcome these problems it is necessary to expose coupons for at least 12 months and preferably longer if reliable data is to be obtained. For most construction projects this is impractical and therefore weight loss data is rarely used.

An alternative method of estimating the in situ corrosion rate is to expose parts of the structure that already exists on or near the site and to measure the section thickness. This can be done using simple hand-held ultrasonic meters. If the original section thickness is known then a section loss can be calculated and provided the age of the structure is also known, a corrosion rate can be calculated. This technique can be particularly useful, for example, in the extension of sheet pile installations. Clearly there are a number of possible errors in such an estimation of corrosion rate using this method, and it is best used to confirm, the assessment of corrosion rate made from published data.

Published data

Given that ground conditions can vary considerably, even over a relatively short distance, it may be natural to assume that the corrosive nature of the ground would also vary, leading to a wide range of possible corrosion rates. Fortunately, this is not found to be the case because the most important factors in determining the corrosion rate of steel in undisturbed, near-neutral pH conditions are the pH and the oxygen concentration. It has been found that these factors, and therefore corrosion rate, are remarkably constant around the world.

The original work in this area was carried out by Romanoff in the US who investigated corrosion rates on a large number of piles of different ages[2]. Subsequent work in Europe, Japan, Scandinavia and Australia have confirmed these original findings and defined more clearly data for estimating corrosion rates of steel placed in undisturbed ground[3,4]. These data have, over time, become accepted to such a degree that upper-bound values for corrosion rates are now often found in national standards relevant to ground engineering. Such data permit the calculation of a conservative estimate of the corrosion rate and therefore the corrosion allowance for a given design life.

Table 14.3 provides typical corrosion rate data for commonly encountered conditions relevant to structures placed in undisturbed ground conditions.

It should be noted that these corrosion rates are for guidance only and data from the relevant national, international or local utility standards should be used in preference to these data where such standards exist. The table includes indicative corrosion rates for conditions other than buried steel, this is because, in many instances, a buried component will have parts in other environments and the corrosion assessment needs to take account of this.

Corrosion allowance is commonly used on:

- sheet piled installations in onshore and marine environments;
- driven tubular and H piles;
- driven piles for jetty and harbour installations (although it should be noted that additional protection may be required for the above water section of piles); and
- piled foundations and pile skirts for offshore platforms.

Table 14.3 Typical corrosion rates in various exposure conditions. The data are expressed as the corrosion rate per face exposed to the relevant condition

Exposure condition	Corrosion rate (mm/yr)
Atmospheric	0.035 to 0.05
Marine splash zone	0.75 to 0.09
Marine immersion	0.035 to 0.05
Undisturbed ground	0.005 to 0.01

14.5.2 Cathodic protection (CP)

It will be recalled, from Section 14.1, that the corrosion process requires two simultaneous electrochemical reactions to occur, if corrosion is to proceed, and that prevention of one reaction will stop corrosion. In CP, the aim is to make the whole of the structure a cathode and therefore eliminate the possibility of the anodic reaction taking place. This can be achieved by supplying electrons to the surface of the metal from an external source.

It is beyond the scope of this chapter to provide detailed information on the design of CP systems, as this is a specialist undertaking. However, civil engineers should be aware of the needs of CP designers and contractors when considering this method of protection and this section aims to provide such an awareness. For more detailed information, the reader should consult standard texts and national standards such as those given in the Refs[5,6] .

Sacrificial anodes

In this case protection is provided by connecting a dissimilar metal to the steel structure. The principle upon which anodes work is that when dissimilar metals are electrically connected in a common electrolyte the more negative will corrode in preference to the other. The most commonly used metals are zinc, aluminium and magnesium. Zinc is preferred in saline mud such as occurs on the sea bed or in marine estuaries, aluminium is preferred in seawater and magnesium is preferred for onshore applications and brackish waters of relatively high resistivity.

In all cases the design of sacrificial anodes needs to consider two factors:

- the weight of anode material required to provide protection for the life of the system; and
- the current output and distribution from individual anodes.

The weight of anode material to protect a given area of steel is calculated from the following formula:

$$\text{Weight / kg} = Cd \times A \times L \times 8760/\ \mathrm{U} \times C$$

Where:
Cd = current density in amps/m^2
A = surface area to be protected
L = life in years
U = anode utilisation factor which is dependent on the anode shape
C = anode's electrochemical capacity in amp.hours/kg

For design, typical values for these variables are given in national standards. When designing sacrificial anodes it is better to use a longer rather than a shorter life. Ten years should be a minimum, otherwise the anode size becomes so small that adequate current output cannot be achieved.

The total anode weight is then divided into sensibly sized individual anodes that can be uniformly distributed over, around or along the structure. Some skill is needed in designing individual anodes as the shape of the anode determines the overall current output. In broad terms, long slender anodes that are placed some distance from the structure have better output than shorter stub anodes or those mounted flush on the structure.

The current output from an individual anode is related to the anode resistance to the ground (or water). This, in turn, is related to the anode shape and the resistivity of the ground. For long slender anodes the resistance to earth can be calculated from the following formula:

$$R = \rho/2\pi L \times \ln(4L/r - 1)$$

where

R = resistance to earth or other electrolyte (ohm)
ρ = resistivity of the ground or water (ohm.cm)
L = anode length (cm)
r = effective radius of the anode (cm) where $r = C/2\pi$ in which C is the anode circumference or, for non-circular anodes, the cross-section periphery

For other anodes shapes, and for those mounted on the structure, other resistance formula are more relevant and the most appropriate can be obtained from design codes and standards.

The current output is then calculated from Ohm's law using the open circuit potential of the anode and the protection potential for the structure; values for these are given in relevant standards or design codes.

The total current available from all the anodes must be greater than the minimum required (current density times surface area) at all stages throughout the design life. As the anode is consumed, the diameter is reduced (the effect on length is much less and is usually ignored), therefore the current output will decrease. It is therefore good practice to check the current output when the diameter has decreased by 60 or 75% to make sure that sufficient current is available at the end of the design life. If this is not the case, then the anode shape needs to be altered or additional anodes used, above the minimum weight requirement, to ensure adequate protection.

In very broad terms, low resistivity conditions (seawater) will result in good current output and the design will be controlled by weight considerations. Conversely, in higher resistivity conditions (>1000 ohm.cm) current output will be less and the design will be determined by current considerations rather than weight. At higher resistivity it may be impractical to use sacrificial anodes and impressed current may be more appropriate.

Sacrificial anodes are commonly used to protect steel offshore structures, ships hulls, and for onshore installations such as pipelines or buried tanks. However, sacrificial anodes are most appropriate for the protection of relatively small areas of steel. A typical sacrificial anode installation is shown in Fig. 14.1.

Impressed current cathodic protection (ICCP)

In ICCP the source of electrons is a DC power supply. This is usually from an AC power supply via a transformer rectifier unit (TRU) to convert the supply to DC, to permanent anodes placed in the ground electrically remote from the structure to be protected.

A typical impressed current installation is shown in Fig. 14.2. This shows the anodes arranged in a horizontal groundbed but it is also possible to place the anodes vertically, in a shallow trench or in a deep borehole.

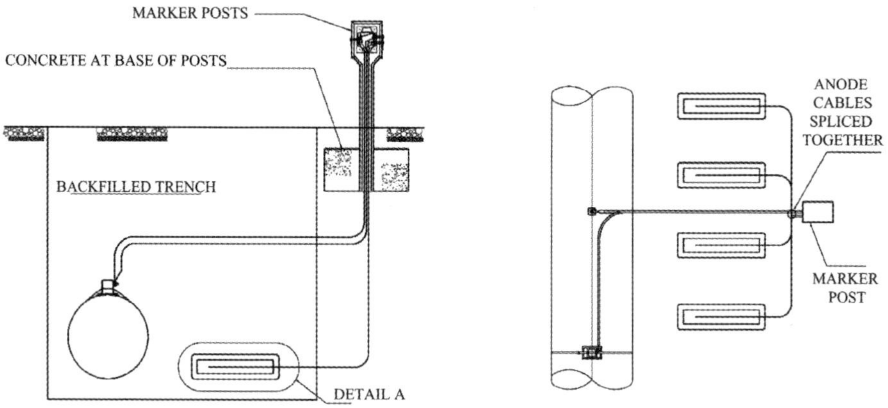

Figure 14.1(a) *Sacrificial anode installation – general arrangements. Note number of anodes required is determined in the detailed design.*

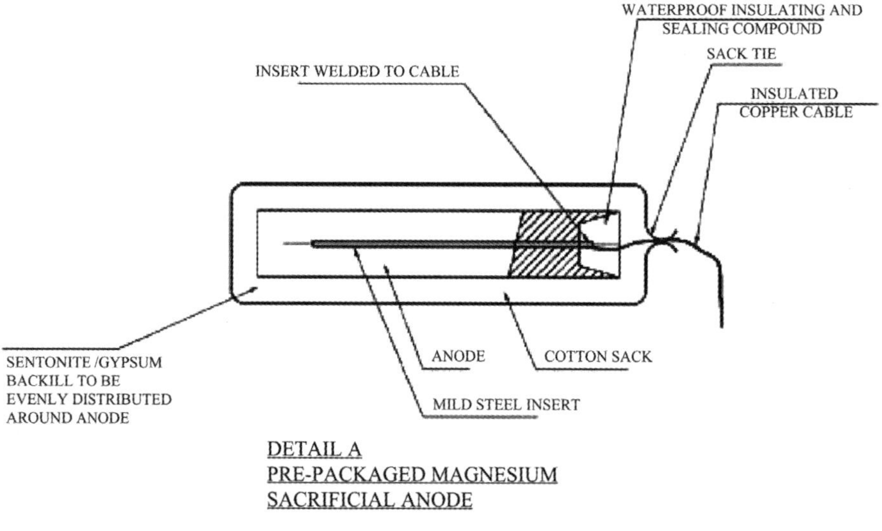

Figure 14.1(b) *Detail of typical sacrificial anode installation.*

The great advantage of ICCP, over sacrificial anodes is that the power output is both much greater and far more flexible. It is therefore possible to protect much greater areas of the structure from a single location, and the system is usually sufficiently flexible to allow for changes in current demand over long periods of time. This flexibility is particularly useful when CP is used in conjunction with protective coatings, as would be the case for a gas pipeline. In this case, the current output from the TRU would be very small, early in the life as modern high quality coatings provide excellent protection in their own right. However, as the coating ages and deteriorates, the current demand will increase and the flexibility that ICCP provides can cope easily with this increase in demand.

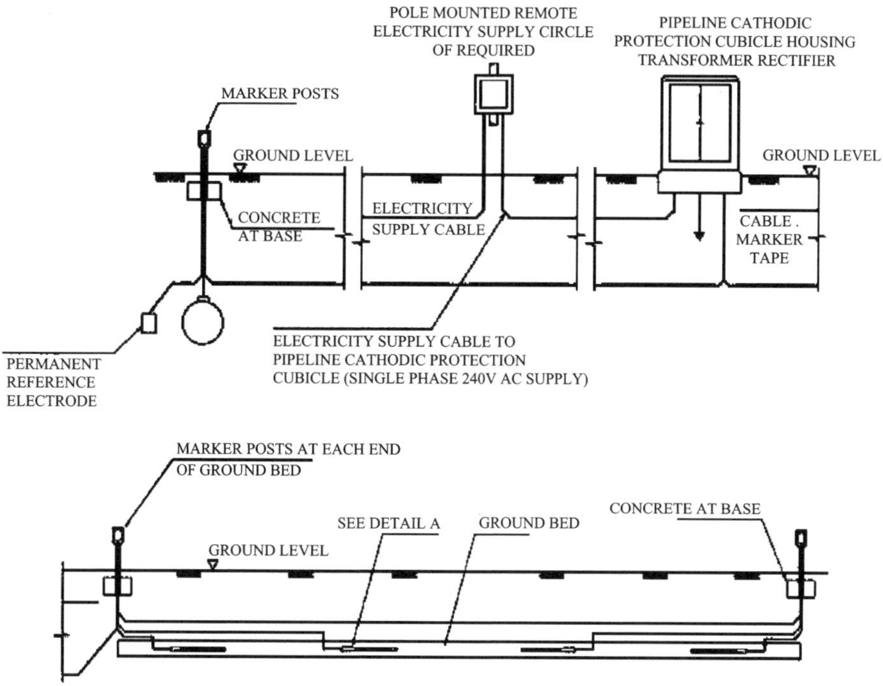

Figure 14.2(a) *Typical impressed current groundbed – general arrangements. Note the number of anodes is indicative and the exact number is calculated in detailed design.*

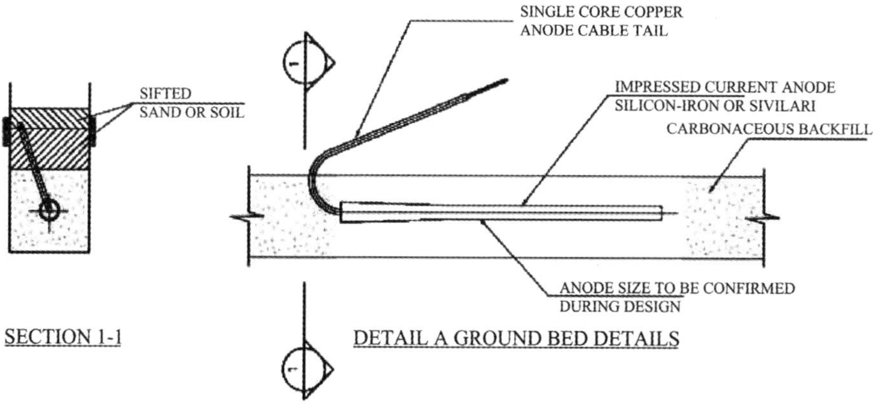

Figure 14.2(b) *Typical detail of impressed current anode.*

With an ICCP system, protection is provided by the passage of electrons onto the structure that then return via the anodes in the groundbed. The design of ICCP systems is a specialist operation that would normally not be undertaken by those directly involved in either the design or construction of buried structures. However, an appreciation of what is required in the design and construction of ICCP systems, by ground engineers, is useful and the following general descriptions, should be helpful in providing some appreciation of the needs of the CP designer/contractor.

The key parts of an ICCP design are shown in Fig. 14.2(a) and it is clear that where this method of protection is to be used the following physical factors need to be taken into account:

- the need for local AC power supply at the proposed groundbed location;
- sufficient land for the installation of the groundbed at a distance (100 to 200 metres) from the structure – it is important that the groundbed is some distance away from the structure to ensure adequate and uniform current distribution to the structure; and
- a secure housing for the TRU.

For these reasons it is desirable and usual, in the case of pipelines, to site ICCP stations at or close to compounds, such as pumping stations or the like.

In designing an ICCP system, it is normal to aim for the total circuit resistance of the installation to be 1 ohm or less, to reduce the power consumption due to the voltage drop associated with current flow through a circuit of known resistance. The key components of this resistance are:

- the resistance of the cables between the TRU and the anode and the TRU and the structure connection;
- the resistance of the anode groundbed to earth; and
- the resistance of the ground.

The contribution of cables to the circuit resistance can be reduced by minimising the cable length and using low resistance cables. In practice the scope is limited. The negative cable will always be a minimum of 100 m long although by careful siting of the TRU, the positive cable can be less than 25 m long. The resistance of the cable can also be reduced by increasing the cross-sectional area, but this must be offset against the increased cost of larger copper cables.

The resistance of the groundbed to earth can be reduced by placing the anodes in a suitably low resistance backfill. It is common to place ICCP anodes in cans filled with coke breeze. The groundbed design also influences the resistance to earth in much the same way that the shape of a sacrificial anode controls the resistance to earth. Most groundbed installations aim to minimise the resistance to earth through the careful design of the orientation, and number of anodes used, in an individual installation. Formulae for the calculation of resistance for different anodes are given in relevant standard and codes of practice.

The resistance of the ground is important as this influences the resistance to earth. High resistivity ground, increases resistance to earth. It is, therefore, normal to site CP stations at locations of low resistivity. Ideal locations are low ground areas that are prone to being wet such as marshes and river flood plains. Locations that are well-drained and/or rocky or high on hills are poor locations for CP stations.

Possible detrimental side effects of CP

There are a number of possible effects that can arise from the use of CP that can have an adverse effect on adjacent structures and vice versa. These effects can be divided into three categories:

- overprotection of the structure resulting in hydrogen generation that can be detrimental to high strength steel components such as bolts, tie bars and wires that are susceptible to hydrogen embrittlement;
- DC stray current interaction/interference with adjacent structures; and
- AC interaction/interference from adjacent structures.

All of these detrimental effects are complex to model and to design against, although if found during service they can be dealt with by reasonably standard mitigation effects.

The first of these effects is potentially serious for buried steel structures incorporating ground anchors, prestressing strand or high strength, bolts although hydrogen embrittlement should not effect ordinary structural steels. The problem arises because, if the CP potential becomes too negative, the cathodic reaction can include a component of hydrogen evolution on the steel surface. Clearly the best way to avoid this is by ensuring the pipeline's CP potential is such that hydrogen evolution cannot occur. However, where CP is providing protection over a large area or distance, the potential close to the CP station may need to be high to achieve effective CP everywhere.

DC interaction occurs where current uses part of an adjacent structure as the return path to the TRU. This will occur only where the alternative path represents a lower resistance than the intended route. The current passes along the adjacent structure for some distance and then returns to its intended route. It is at this point that damage occurs to the adjacent structure. The damage is concentrated in a small area and can be very intense. Clearly this type of stray current effect can happen both to a structure with CP due to current flowing, for example, from a DC rail system onto a pipeline, or vice versa.

This type of problem can be avoided by ensuring that all preferred electrical pathways are of the lowest possible resistance. Guidance on testing and mitigation against DC stray current effects is provided in most codes and standards such as BS 7361 [5].

The effects of AC interference are less common, although incidents do seem to be on the increase, and are usually associated with buried structures close to, and parallel with, high voltage overhead power lines. In this case, the problem is related to induced currents flowing on the buried structure arising from the proximity of the power line. Where it occurs, AC induced corrosion can result in extremely high local current density and associated high corrosion rates.

AC interference can be detected by simple surveys, or through the installation of monitoring coupons and test posts at susceptible locations. AC interference is a relatively new phenomenon and there are no standard or accepted methods of combating its effect. Each incidence needs to be dealt with individually. Some guidance is given in the references.

Both DC and AC effects can cause serious corrosion problems. They can also represent a significant hazard to those working on structures, or otherwise coming into contact with them. It is important, therefore, that the issues of stray current corrosion are dealt with during the design, construction and operation of the buried structures.

14.5.3 Protective coatings

For corrosion prevention of buried structures, protective coatings need to be of high quality and be capable of providing corrosion protection for long, maintenance free, periods of time. In the past, many materials have been used for corrosion prevention but, in recent times, the preferred materials have been limited to:

- fusion bonded epoxy (FBE), mainly for buried pipelines;
- Glass Flake Epoxy (GFE) and Glass Flake Polyester (GFP), mainly used for piles that will be exposed to either immersed or atmospheric conditions after driving, and
- PVC-backed bitumen tape wraps.

All the coatings listed above are characterised by the following:

- unlike most coating specifications, they do not require the application of a corrosion inhibiting primer;
- very high film thickness in the range 0.5 mm to 2 mm;
- preventing corrosion by acting as a physical barrier between the steel and the environment; and
- a high degree of impermeability to water and/or ions.

To some extent, the type of coating will be controlled by the relevant utility, owner's standards, or steel producers specifications. For example, in the UK buried high pressure gas transmission lines are controlled by Transco's specifications and these effectively limit the choice of coating to FBE.

The quality of the coating system is determined largely by the quality of surface preparation, coating application, the quality of repairs and making good on site. The most appropriate surface preparation is abrasive blast cleaning to ISO8501-1 with coating operations following immediately after cleaning[7]. For structures that are to be buried in the ground, the application should take place under workshop conditions. In the case of FBE it is impractical to do this outside shop conditions.

Repairs at joints and damage should be made using the same methods of preparation, and materials that were used in the original shop application. The exception to this is FBE coatings where special liquid formulations should be used. It is important that site-coating quality is comparable to that of shop application if premature failure of the coating is to be avoided. Common experience with all coating specifications is that the first areas to fail are those at site repairs where a lower quality of application has been permitted.

Where steel is to be placed in a pipe trench, or driven into the ground, there is often some, justifiable, concern that the protective coating will suffer excessive damage during back-filling or driving. In the case of back-filling a pipe trench, the risk of damage is limited by the use of controlled fill and the general robust nature of the FBE type coatings. The situation with respect to driven piles is less clear and does, to a degree, depend upon the ground into which the pile is driven. Soft clay and sand type soils should not cause excessive damage, whereas unknown fill material, and ground with rocks, may be more problematic. However, it is unusual for driven piles to require protection on the below ground section of the pile. Coating is usually required above ground and this is unlikely to be damaged during driving.

14.6 Conclusions

This section has outlined mechanisms and processes involved in the corrosion of structural steel items in both disturbed and undisturbed ground conditions. It has been shown that in most cases it is relatively easy to assess the risk of corrosion and identify whether additional preventive measures are required to achieve a given design life. Often it will be found that long-term corrosion rates can be tolerated within normal design limits or by the provision of relatively simple and cost effective prevention strategies.

It is only in special conditions, acidic or contaminated ground or where the ground is intentionally disturbed, that specific corrosion prevention systems maybe required. Where such provisions are necessary, there are established methods of cathodic protection and protective coatings that the engineer can use with confidence to ensure that corrosion does not lead to premature failure.

References

[1] Proctor, R.M. (Ed.), 'Proceedings of the international conference to mark the 20th anniversary of the UMIST Corrosion and Protection Centre', *Corrosion Science*, Pergammon Press, Vol. 35, pp. 667–73.

[2] Schwerdtfeger, X. and Romanoff, Y., *Underground Corrosion of Steel Piling 1962 to 1971*, National Bureau of Standards, NBS Monograph 127, March 1972.

[3] British Steel, *The Corrosion and Protection of Steel Piling in Temperate Climates*, British Steel, October 1992.

[4] Karl Tungesvik, *Investigations of corrosion rates of steel piles in Norwegian marine sediments*, Tungesvik, Oslo: Norges Geotekniske Institutt, 1976.

[5] British Standards Institution, BS 7361:1991, *Part 1: Cathodic protection: Code of practice for land and marine applications*, London: BSI.

[6] Shrier, L.L., Jarman, R.A. and Burnstein, G.T. (Eds) *Corrosion and Corrosion Control*, 3rd edition, Oxford: Butterworth-Heinemann, 1993.

[7] British Standards Institution, ISO 8501-1, *Preparation of steel substrates before application of paints and related products. Part A1: Specification of rust grades and preparation grades of uncoated steel substrates and of steel substrates after overall removal of previous coatings*, London: BSI.

15 Durability and the Standard and Code Making Bodies

David Doran

15.1 Introduction

The need to consider durability was recognised by the British Standards Institution (BSI) in BSCP 3 Chapter IX (1950). This code introduced the concept of 'Designed life of a building or its component parts or installations'. This code remained in place until 1992 when, after several abortive attempts, BS 7543: *Guide to the durability of buildings and building elements products and components* was introduced. In January 1995, Taywood Engineering stated in a confidential report that 'there was little awareness of the guide and even less application'.

The task has now been taken up by the International Standards Organization (ISO) which, with support from BSI and others, is producing a new document entitled *Buildings – Service Life Planning*. The Standard is in eight parts as indicated below. It should be noted that the commentary on this document is up-to-date at the time of writing but may not represent the final outcome of the work of the ISO.

1. General principles
2. Service life prediction procedures
3. Performance audit and reviews
4. Data requirements
5. Maintenance and life cycle costing
6. Life cycle assessment
7. Performance evaluation and feedback of service life data from existing construction works
8. Reference service life

The UK, through BSI, is responsible for drafting 'General principles'[1] and 'Performance audit and reviews'[3]. Fig. 15.1 indicates, diagrammatically, the scope of the ISO standard and the interaction between the parts of the document. The titles of some of the parts differ somewhat from the list shown above because the work has progressed, but, at the time of writing, the diagram had not been updated.

This standard will employ concepts such as life cycle costing and a factorial method for estimating the service life of buildings and components. The factorial method is based upon work by the Architectural Institute of Japan (*Principal guide for service life planning of buildings*).

The highly regarded, but now defunct, Greater London Council included some guidance in its London Building (Constructional) Bylaws 1972–1979 made under the

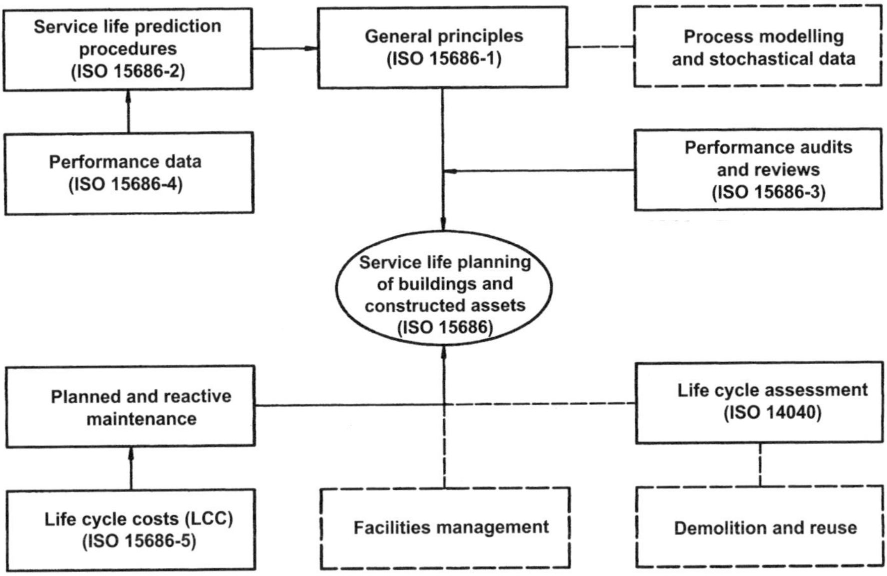

Figure 15.1 *Scope of ISO 15686.*

London Building Acts 1930–1939. Part IV of that document dealt with materials of construction and contained the statement:

> *The purpose of this part of these by-laws is to ensure that material used in the construction of a building is of adequate strength and durability and is of a suitable nature and quality for the purpose for which it is used.*

Detailed advice (mainly related to British Standards) was then given for the following materials:

- aggregates (including sand);
- cements;
- water;
- concrete;
- structural metal and reinforcement for concrete;
- lime for mortar or plaster;
- stone;
- bricks and blocks;
- materials for damp-proofing;
- slates;
- roofing tiles;
- asbestos cement slates and sheet;
- steel and aluminium roofing slates;
- lathing and plastering;
- slab or sheet plastering;

- structural timber; and
- resin bonded wood chipboard.

These regulations were, of course, only applicable in the City of London and in the, then, inner London Boroughs of Westminster, Camden, Islington, Hackney, Tower Hamlets, Greenwich, Lewisham, Southwark, Lambeth, Wandsworth, Hammersmith, and Kensington & Chelsea. Because of the respect in which these by-laws were held they were frequently invoked by other local authorities throughout the country. However in these, out of London, areas they were not enforceable by law. Although these by-laws majored on materials they did not specifically deal with questions of the interaction between materials and the importance of good detailing in achieving the required durability. However, all work had to be 'to the satisfaction of the District Surveyor'. It is worth noting that Dr Alistair Paterson, in his 1984 presidential address to the Institution of Structural Engineers indicated that 59% of construction faults could be attributed to poor detailing.

This chapter traces the involvement of Standard and Code making bodies and others (e.g. The International Union of Testing and Research Laboratories for materials and Structures) (RILEM) and the Housing Association Property Mutual (HAPM) etc. in the quest to codify matters of durability.

15.2 British Code of Practice: CP3 Chapter IX (1950) *Durability*

Chapter IX was part of a comprehensive work entitled *Code of basic data for the design of buildings*. It was published in ten parts and dealt with such topics as lighting, thermal insulation, fire precautions and loading. Chapter IX remained in use until replaced by BS 7543: *Guide to durability of buildings, products and components* in 1992.

In spite of many subsequent attempts the definition of durability given then has probably not been surpassed. It was:

> *The quality of maintaining a satisfactory appearance and satisfactory performance of required functions.*

The code recognised that not all buildings required the same design life and categorised them as follows:

(a) designed to have a life of 100 years;
(b) designed to have a life of 40 years;
(c) designed to have a life of 10 years; and
(d) designed to have a life of less than 10 years.

It was suggested that foundations, structural walls, floors and roofs should normally have a satisfactory life at least equivalent to the designed life of the building. Furthermore that non-structural components such as finishes and decorations need not be expected to be as durable as structural components.

Chapter IX recognised that durability was inextricably linked to maintenance and highlighted the need for periodic renewal of jointing materials. It also called for provision of adequate access to items in possible need of maintenance.

A bold innovation was to list (possibly for the first time) potential causes of deterioration of buildings and their installations. These included:

- atmospheric and climatic action;
- wetting and drying effects;
- soil and ground water action;
- rodent, insect, bacterial, fungal and plant action;
- water supply;
- electrolytic action;
- contact or association of incompatible materials;
- specific chemical action or chemical changes in materials;
- wear;
- impact and vibration;
- action of cleaning and cleansing agents;
- action of domestic or industrial gases; and
- accidental causes, including fire, lightning and floods.

Each of these headings was elaborated by an explanatory paragraph giving state-of-the-art advice. For example, it was stated that damp oak and other timbers might have a corrosive action on metals, particularly, lead, iron or steel. These notes were backed-up by informative appendices dealing with such topics as atmospheric and climatic action; wetting and drying effects; soil and groundwater action; rodent, insect, bacterial, fungal and plant action; electrolytic action; association of incompatible materials and chemical action.

Guidance was also given for the following;

- the susceptibility of building materials to deterioration;
- the effect of design upon the durability of materials;
- the classification of water supplies in relation to their effect upon metals;
- the classification of groundwaters and soils in relation to their effect upon
- concrete; and
- the classification of atmospheric pollution conditions.

It can therefore be seen that Chapter IX, for its time, embraced some quite advanced thinking.

15.3 BS 7543: 1992/2003 *Guide to durability of buildings and building elements, products and components*

The next major attempt to codify durability saw the emergence of BS 7543: 1992. This appeared after a very long gestation period which saw one consultant's draft rejected following public consultation and the production of a new document which, after representations, was down-graded from code status to that of a guide. In 1998 the guide was amended to recognise inter alia that the campaign for better durability was now being headed by ISO.

BS 7543 contained the following sections:

1. general;
2. expression of durability; and
3. appendices.

Significant definitions of the subject were defined thus:

Durability – Ability of a building and its parts to perform its required function over a period of time and under the influence of agents.

Durability limit – Point at which loss of performance leads to the end of the service life.

Service life – Actual period of time during which no excessive expenditure is required on operation, maintenance or repair of a component or construction (as recorded in use).

Required service life – Service life to meet users' requirements (e.g. as stated in the client's brief for a project or in a performance specification).

Predicted service life – Service life predicted from recorded performance or accelerated tests (e.g. as stated by the manufacturer or in a European Technical Approval (ETA)).

Design life – Period of use intended by the designer (e.g. as stated by designer to the client to support specification decisions).

It will be seen that European regulatory bodies such as ETA were beginning to have an influence on durability in general and the UK in particular.

Categories of design life were defined as follows:

- temporary – up to 10 years for non-permanent buildings, etc;
- short life – minimum period 10 years for temporary classrooms, etc,
- medium life – minimum period 30 years for most industrial buildings, etc,
- normal life – minimum period 60 years for new housing, etc; and
- long life – minimum period 100 years for civic buildings, etc.

It may not have escaped the reader's notice that the design life recommendation for housing very closely approximates to the normal local authority loan pay-back period!

Further advice was given for the design life of components or assemblies under the headings of:

- replaceable;
- maintainable; and
- lifelong.

The code also introduced the concept of durability prediction in which the following was highlighted:

- experience in the use of traditional materials embodied in codes of practice, text books and trade association publications;
- certificates assessing the performance of products issued by the British Board of Agrément (BBA) and other authoritative bodies;
- publications from the Building Research Establishment (BRE) and other research bodies; and
- predictions of service life of products provided by their manufacturers. Where possible such predictions should include details of the method of assessment, the variability of test results, the assumed conditions of use, and maintenance requirements.

It was expected that the designer would take into account site conditions, expected quality of workmanship, and expected maintenance levels. A statement on design life should be modified where:

- the usage of parts of the building cannot be foreseen;
- untried products are to be incorporated; and
- exceptionally large assemblies of components are involved.

Maintenance was categorised under the following headings:

- repair only;
- scheduled maintenance plus repair; and
- condition based maintenance plus repair.

In the UK, it is rare for maintenance to be carried out on a systematic basis. The motto (derived from the USA) seems to be 'if it aint busted don't mend it!'

In order to assist in the use of BS 7543 a decision tree was provided to guide the user towards service life prediction. Although fulsome in its detail it was, in retrospect, somewhat complex to use.

The code also attempted to deal with the effects of failure under the following headings:

A. danger to life (or injury);
B. risk of injury;
C. danger to health;
D. costly repair;
E. costly because repeated;
F. interruption of building use;
G. security compromised; and
H. no exceptional problems.

Examples of each category were given, highlighting the penalties of each type of failure. Perhaps the most dramatic of these was the sudden collapse of a structure under category A.

Appendices to the code include the following:

A. Design life data sheet
B. Agents that cause deterioration

- Temperature variations
- Radiation
  - Solar radiation
  - Infrared radiation
  - Ultraviolet radiation
  - Thermal radiation
- Water (including water vapour)
- Normal air constituents
  - Oxygen
  - Sea spray
- Air contaminants
  - Gases that form acids
  - Particulates
- Freeze–thaw considerations
  - Conditions for frost damage
  - Prediction of potential frost damage
  - Requirements for concrete
  - Requirements for render
  - Frost heave
- Wind
- Biological factors
  - Bacteria
  - Insects
  - Fungal and insect attack in timber
  - Rodents and birds
  - Surface growths
  - Plants and trees
- Stress factors
  - Sustained stresses
  - Intermittent stresses
- Chemical incompatibility factors
  - Incompatible materials
  - Leaching and leachates
  - Solvents
  - Contaminated land
  - Expansive materials
- Use factors
  - Normal wear and tear
  - Abuse by user

C. Examples of premature deterioration.

This appendix tabulates 18 examples under the following headings:

- material or component;
- action causing deterioration;
- effect of deterioration;

- service life;
- design life; and
- explanatory notes and references.

The example dealing with slabs supported on fill (i.e. made up ground) gives an action as 'dead load acting on a compressible material'; an effect as 'compaction of fill leading to movement cracks in slab'. The service life is quoted as '2 to 20 years'; the design life as '60 years' and recommendations that fill depths 'be limited to 600 mm'.

The Code terminates with a list of 55 useful references together with a further listing of BSI publications.

A great deal of hard work had gone into the preparation of this guide. It was disappointing, therefore, that in 1995 Taywood Engineering should produce a report which indicated that the document was little known and little used. It did, however, make reference to the work that HAPM (Housing Association Property Mutual) were doing through their publication of a Component Life Manual.

Subsequent discussions led to the setting up of an *ad hoc* group based around Taywood Engineering who set out to improve the approach of industry to this thorny problem. The group was later subsumed into a formal sub-committee (B/500/3: Durability) of the BSI which would complement and advise an international committee in the production of an ISO Guide to the design life of structures. The chairman of the *ad hoc* committee was Dr Roger Browne of Taywood who was elected the first chairman of B/500/3 (a BSI Committee) and succeeded on his retirement by Dr Hywel Davies then of the BRE (Building Research Establishment). To Dr Browne must go the credit for driving this work forward and keeping the UK in the forefront of the initiative.

15.4 *ISO Guide to the Design Life of Structures*

Gradually the work of independent, disparate groups from around the world was brought together by their common interests. The need for durable and sustainable structures was recognised. It is not easy to see what was the catalyst for such unanimity, although the work is now coordinated by the ISO committee known as TC59/SC3/WG9. In a paper delivered to an audience at BRE in 1996 Dr Roger Browne made an impassioned plea for the members of RILEM to support the initiative. It would seem that his words have been heeded.

As mentioned above the ISO Guide is being produced in eight parts which are discussed in more detail below.

15.4.1 Part 1 – General principles

The guide is organised under the following headings:

1. Scope
2. Normative references
3. Definitions
4. Symbols and abbreviated terms
5. Process of service life planning
6. Service life planning – steps in the design process
7. Service life forecasting

8. Service life prediction based on exposure data
9. The factor method for estimating service life
10. Financial and environmental costs over time, and
11. Obsolecence, flexibility and re-use.

Annex A Typical financial costs of buildings over time (in UK and USA)
Annex B Alphabetical glossary of terms
Annex C Example of critical property assessment of alternative specifications
Annex D Agents affecting the service life of building materials and components
Annex E Examples of requirements
Annex F Method for estimating service life of components using factors to represent agents
Annex G Worked examples of factorial estimates
Bibliography

For commentary on some of the above topics see below.

The Scope (1) includes the phrase that the standard describes the principles and procedures that apply to design when planning the service life of components and assemblies, and the buildings of which they form a part. It goes on to caution that consideration must be given to local conditions. Within a country it is obvious that conditions will be more severe in windy coastal areas than in interior, calm secluded spots. It may, however, be less obvious to an inexperienced designer in say Greece that conditions off the north east coast of Canada may be much more severe than off the north east cost of Greece.

The Guide is meant to be applicable to both new-build and refurbishment work and it stresses the need to rectify existing defects when considering refurbishment. Although written specifically for building structures the document will be of considerable interest and help to those in charge of civil engineering structures.

Of particular interest in the Definitions (3) are the following:

Durability Capability – of a building or its parts to perform its required function over a specified period of time under the influence of the agents anticipated in service. (Durability is not a property of a material or component, though the term is sometimes erroneously used as such.)

Service life – Period of time after installation during which a building or its parts meet or exceed the performance requirements.

Estimated service life – Reference service forecast for a building or parts of a building for use as a basis for estimating service life.

Predicted service life – Service life predicted from recorded performance over time e.g. as found in service life models or ageing tests.

Failure – Loss of ability of a building or its parts to perform a specified function.

Life cycle cost – Total cost of a building or its parts throughout its life, including the costs of planning, design, acquisition, operations, maintenance and disposal, less any residual value.

Process of service life planning (5) gives an overview of the issues which should be considered to plan and help to ensure the building's service life.

Service life planning – steps in the design process (6) is dealt with under the following headings:

The brief

This imposes upon the client the need to state clearly the client's overall objectives in sponsoring a building. Many clients do not have the expertise to estimate conditions such as design life with any accuracy and will need considerable guidance from the design team. Equally, the design team will need to proceed with care if untried materials or construction techniques are used in the project. The design team must not be guilty of trying to guarantee what they cannot deliver.

Conceptual and initial design

This includes a warning that professional judgement must be used to check whether the client's aspirations can be achieved within the project restraints of budget, time, performance, maintenance requirements and site-specific issues.

Detailed design

This section suggests that it is unlikely that components will be tailor-made to suit the project and that the design process will be an iterative one of proposing a component, checking its predicted performance against the brief and amending selections.

Specification

This introduces the concepts of value engineering plus life cycle costing and the thought that specification should include measurable/auditable performance criteria. In the selection of materials there is a heavy emphasis on the need for test data from manufacturers. In doing so it accepts that a specifier must maintain an acceptable balance between materials whose performance is well known and innovative materials which may achieve better performance but may lack service life data.

The section carries a strong 'health warning' that allowances must be made in circumstances where the anticipated level of site workmanship is likely to fall below that expected by the manufacturer.

Environment characterisation

Emphasis is given to the environmental uniqueness of each site and the way in which this characteristic can vary within a site. For example although the natural environment may be mild, the use to which a building may be put may impose severe conditions on the materials. This is particularly so in some industrial structures where injurious chemicals are processed.

Initial cost estimates

Although emphasis is laid on the need to work within the project budget, a warning is given that reducing initial costs by specifying inferior materials can lead to additional maintenance requirements thus increasing life cycle costs.

Maintenance plan

This is considered under the following headings:

- cyclical maintenance (e.g. regular redecoration);
- reactive condition-based maintenance (e.g. repairs of defective performance); and
- major refurbishment.

In the light of the poor record of maintenance by many building owners it is perhaps too much to expect that the suggested schedule of planned dates for major refurbishment and replacement of defective parts will be prepared! However, an owner will now be left in no doubt about the potential consequences of his lack of action.

Performance requirements and acceptability

This highlights the need to differentiate between permanent, replaceable or maintainable construction. A table is provided to show suggested minimum design lives for components. As an example, whereas an inaccessible or structural component should match the design life of the building service installations and external works would normally have a design life of not more than 25 years.

Service life forecasting

This section is dealt with under the following headings:

- objective of forecasting;
- precision and reliability;
- use of data for forecasts;
- taking account of variability and reliability;
- use of forecasts;
- issues that can affect forecasting;
- relevant issues;
- agents relevant to degradation (these include mechanical, electromagnetic, thermal; chemical and biological);
- the effect of dose and intensity variations;
- the effects of agents in combination;
- relevant data;
- data recorded over time;
- measurement of degradation; and
- comparison between exposure data and other evidence.

Service life prediction based on exposure data

This section is dealt with under the following headings:

- use of predictions based on exposure data;
- general approach;
- defining the problem;
- preparation;
- pre-testing;
- exposure and evaluation;
- long-term exposure;

- short-term exposure;
- accelerated exposure;
- simulated exposure;
- assessing the data from exposure; and
- interpretation and reporting.

The factor method for estimating service life

In this method, based on the Japanese *Principal Guide for Service Life Planning of Buildings* the Estimated Service Life (ESL) is defined by the equation:

$$ESL = RSL \times A \times B \times C \times D \times E \times F \times G$$

Where

RSL = Reference Service Life
A = quality of component: (e.g. as lower qualities within BS 5977: value 1.0)
B = design level: (e.g. installed in brick wall: value1.0)
C = work execution level: (e.g. no repair of damaged coatings: value 1.0)
D = environment (e.g. indoor environment: value 0.8)
E = outdoor environment (e.g. industrial pollution: value 0.8)
F = in-use conditions: (e.g. appropriate use only: value 1.0)
G = maintenance level: (e.g. none: value 1.0)

The numerical value of the factors varies (usually, but not always, within the range 0.8 to 1.2) The higher values (1.2 etc) indicate the best possible situation.

This is one of the most controversial sections of the guide. In the absence of other data from manufacturers and definitive case studies it is suggested that the factorial method be used. It is dealt with under the headings set out below.

The method is based on work done in Japan by the Architectural Institute. The objective is to assess accurately the service life of components by having regard to the specific conditions in which the component is used. For example, a concrete component that is coated with a render and placed in an indoor environment would be expected to have a longer service life than a similar component without render in a hostile external environment. The Japanese have provided a series of factors to be used in these assessments. The qualitative factors (environment, workmanship, maintenance etc.) are based on research; others are conjectural. Consideration will also need to be given to regional variations to these factors to take account of differing environments.

The ISO Standard gives a number of examples of the method for predicting the service life of building elements and components. These include:

- structural elements for reinforced concrete buildings,
- painted elements of steel framed buildings,
- exposed asphalt waterproofing systems,
- external wall tiling for reinforced concrete building, and
- metal pipework.

The Guide also suggests that the number of (modifying) factors to be considered may vary for differing materials and components.

Financial and environment costs over time

This section introduces life cycle assessment (LCA), a cradle-to-grave concept which goes way beyond considerations of building costs. It is a structured attempt to consider matters of sustainability and recycling without which the non-renewable resources of the planet will become extinct. LCA also includes consideration of embodied energy (i.e. the relative amounts of energy to make or create different types of construction materials). Within this overall framework advice is also given on life cycle costing (LCC) (also known as whole-life costing or through-life costing). Definitions of LCC abound; one of the best (thought to come from an American source) is:

> *The Life Cycle Cost of an item is the sum of all funds expended in support of the item from its conception and fabrication through its operation to the end of its useful life.*

LCC is dealt with in more detail in Part 5 of this Standard.

Obsolescence, flexibility and reuse

This is considered under the following headings:

- obsolescence;
- types of obsolescence;
- minimising obsolescence;
- future use of building; and
- demolition and reuse.

Obsolescence is defined as the inability to satisfy changing requirements and should not be confused with the need to replace due to defective performance. Obsolescence may be functional, technological or economic. Economic obsolescence may occur because maintenance has become unreasonably costly or disruptive. Refurbishment and upgrading are the major strategies to counter obsolescence.

Part 1 is completed by a series of annexes as follows:

A. typical financial costs of buildings over time (in UK and USA);
B. alphabetical glossary of terms;
C. example of critical property assessment of alternative specifications;
D. agents affecting the service life of building materials and components;
E. examples of requirements,
   1. Examples of functional critical properties
   2. Examples of economic critical properties
F. method for estimating service life of components using factors to represent agents, and
G. worked examples of factorial estimates. These are for a steel lintel, a softwood window and fibre-based cement slates.

The guide also includes a bibliography containing references from ISO, RILEM (International Union of Laboratories & Experts in Construction, Materials, Systems & Structures) and other UK, Japanese, Canadian, American and European authorities.

15.4.2 Part 2 – Service Life Prediction Principles (SLPP)

This part of the Guide is considered under the following headings.

1. scope;
2. normative references;
3. terms and conditions;
4. abbreviated terms;
5. general (brief description of SLP);
6. methodological framework;
7. critical review; and
8. reporting.

Annex A gives guidance on the process of SLP.

The scope of the document is quoted here in full:

This International Standard describes a procedure that facilitates service life predictions of building components. General framework, principles and requirements for conducting and reporting such studies are given. This international Standard does not describe the techniques of service life prediction of building components in detail.

15.4.3 Part 3 – Performance audit and review

This part of the Guide was produced by the UK and is considered under the following headings:

1. scope;
2. conformance;
3. normative references;
4. terms and conditions;
5. symbols and abbreviated terms;
6. overview and purpose of the audit;
7. general auditing issues;
8. audit of the design brief;
9. audit of the detailed design;
10. audit of the implementation stages;
11. audit of the operating and life care instructions; and
12. audits during the life of the building.

Annex A deals with the selection of components, assemblies and systems for auditing.

Annex B indicates a sample audit pro forma.

The scope is reproduced here in full:

> *This part of ISO 15686 is concerned with ensuring the effective implementation of service life planning. It describes the approach and procedures to be applied to the briefing, design, construction, and where required, the lifecare management and disposal of a building to provide a reasonable assurance that the measures necessary to achieve a reasonable performance over time will be implemented. It is intended primarily for the auditing of building fabric and mechanical and electrical services installations of new building projects, but much of it is equally applicable to non-building structures and to the refurbishment of existing buildings. This standard does not deal with the cost implications of service life planning or in any detail with the broader issue of sustainability (i.e. embodied energy, decommissioning and site restoration) A service life performance audit is not concerned with early failures (within the first year after installation) that are caused by faulty design, manufacture, handling or installation.*

This part gives very strict guidelines for a durability review of the design, construction and post construction stage. There is a strong hint that this audit should be carried out independently. This independence may not be readily accepted by all parties.

15.4.4 Part 4 – Data requirements

This part includes more than 50 types of data required to make design life calculations. Among the data to be considered will be:

- airborne pollutants & salinity;
- rainfall & humidity;
- degradation mechanism;
- earthquake hazard;
- flood risk;
- temperature range;
- global radiation;
- user requirements;
- UV irradiance; and
- wind.

15.4.5 Part 5 – Maintenance and Life Cycle Costing

At the time of writing this part was still being drafted.

15.4.6 Part 6 – Procedures for considering environmental impacts

This was issued in 2004.

15.4.7 Part 7 – Performance evaluation and feedback of service life data from existing construction works

To be drafted.

15.4.8 Part 8 – Reference service life

To be drafted.

15.5 Commentary

This Guide, when completed, will represent the most comprehensive work on the subject. Its success will depend upon a number of factors including the following.

The supply of a great deal of data. For example the results of long-term exposure tests. Long-term performance can now be predicted in some areas by accelerated testing. The British Cement Association (BCA) now confidently predict the long-term performance of certain types of concrete by carrying out accelerated testing at 38°C rather than at normal ambient temperatures. This has been made possible by comparing the results of accelerated tests with those on concrete samples exposed to atmospheric conditions over many years.

Although these methods will become more sophisticated, it must be remembered that some materials may only be available for relatively short periods and will go out of production before reliable long-term tests can be assessed. A particularly difficult area is in dealing with polymers (plastics) where formulations change very regularly. A plastics manufacturer may say that a formulation has remained constant for a number of years and not be aware of the way in which the quality of the raw materials has changed.

The accuracy of a life cycle cost exercise may be affected badly by inaccurate forecasting of interest rates. Equally the exercise may fail to recognise the state of the commercial market in predicting the cost of replacing items.

Life cycle assessment is gaining popularity as the need for sustainability advances. However, the way in which demolition material can be assessed may change. It is fashionable to assume that, with advancing technology, better ways of handling these materials will arise. This may, however, need to be balanced by environmental demands.

There may be a strong move in favour of independent audits of durability. However, will clients be prepared to pay for such independence or will they expect their 'design and construct' team to take care of this aspect?

Such issues suggest a cautious approach to a most complex subject.

15.6 Final thoughts

BSI, ISO, RILEM, HAPM and others have done well to identify many of the factors that affect durability; the Architectural Institute of Japan has indicated a way forward by introducing the Factorial Method. However, the complexity of the proposed methodology is enormous.

Additional experience will increase the data required to manipulate this methodology. However, it is less certain whether this additional experience will lead to simplification. Much will depend on whether it will be possible to distil a few simple guidelines from this experience and if these guidelines can be successfully built into the changing culture of the construction industry.

Acknowledgements

In compiling this chapter I have drawn freely on my own experience and from information in a number of publications. In particular I would like to pay tribute to BSI and my colleagues on the BSI Committee B/500/3 dealing with Durability. A bibliography at the end of this chapter attempts to cover relevant literature and grateful thanks are accorded to those concerned.

Bibliography

[1] Amato, A. and Eaton, K., *A Comparative Environmental Life Cycle Assessment of Modern Office Buildings*, Ascot, UK: SCI, 1998.

[2] Architectural Institute of Japan (AIJ), *Principal Guide for Service Life Planning of Buildings,* Japan: AIJ, 1993.

[3] British Standards Institution, BSCP CP3:1950, *Code of functional requirements of buildings: Durability,* Chapter IX, London: BSI.

[4] British Standards Institution, BS 7543:1992, *Guide to: Durability of buildings and building elements, products and components,* London: BSI.

[5] Bull, J.W., (Ed.) *Life Cycle Costing for Construction,* Glasgow: Blackie Academic & Professional, 1993.

[6] British Standards Institution, ISO 15686-1:2000, *Buildings and constructed assets – Service life planning, Part 1: General principles*, London: BSI.

[7] British Standards Institution, ISO 15686-2:2001, *Buildings and constructed assets – Service life planning, Part 2: Service life prediction procedures*, London: BSI.

[8] British Standards Institution, ISO 15686-3:2002, *Buildings and constructed assets – Service life planning, Part 3: Performance audits and reviews*, London: BSI.

[9] British Standards Institution, ISO 15686-6:2004, *Buildings and constructed assets – Service life planning, Part 6: Procedures for considering environmental impacts*, London: BSI.
Note: Other parts of this standard which were under preparation at the time of writing have also been referred to in the text.

[10] London County Council, London Building Acts 1930–1939, Constructional By-laws, London: LCC, 1965.
Note**:** This is one of several similar publications by the LCC/GLC.

[11] Paterson, A.C., 'Presidential Address', *The Structural Engineer*, 62A[11], London: Institution of Structural Engineers, 1984.

[12] Somerville, G., (Ed.) *The Design Life of Structures*, Glasgow: Blackie, 1992.

16 The Legal Aspects of Durability

Neil le Roux

16.1 Introduction

This chapter deals with the legal issues which arise from the reasonable expectation that materials and structures produced by the engineering and construction industries should achieve a satisfactory level of durability. The various dictionary definitions of 'durability' include the ability to resist wear or decay or attack by the elements over a long period of time. Although 'durability' is recognised as a term of art within the construction industry, its closest equivalent in construction law is the statutory requirement under the Sale of Goods Act 1979, as amended by the Supply of Goods and Services Act 1982 and the Sale and Supply of Goods Act 1994, that all goods and materials supplied shall be of 'satisfactory quality'. It is usually implied that the contractor warrants that the materials used and the completed product will be reasonably fit for the intended purpose. It follows that if the product is not likely to achieve satisfactory durability it will be unfit for its intended purpose and materials lacking durability will not satisfy the test of 'satisfactory quality'. It is not surprising therefore that under the 1994 Act 'durability' is specifically referred to as one of the required aspects of quality.

The construction industry differs from other industries which produce consumer products because the end product of a construction project is almost always unique. Each site presents its own problems to be solved by the construction team. Because of the large number of parties involved, often with conflicting interests, even a project of modest size is likely to produce complex issues of management and co-ordination. Construction law is therefore concerned not simply with a specialised area of law, but includes risk assessment and control and the principles of management. Errors in these areas are just as likely to produce disputes as errors of design and workmanship.

It should be the aim of all contributors to a construction project to deliver a quality product which is 'right first time' (i.e. without wasteful redesign or reconstruction) which is delivered on time and to budget, and with zero defects (Egan 1998). The Latham Report (1994) identified eight features which clients would normally wish for in a project including: fitness for purpose, worthwhile guarantees, and satisfactory durability. Research at that time showed that commercial clients were dissatisfied with the performance provided by the construction industry in each of those three areas. More recently Egan (1998) concluded that the industry as a whole was under-achieving and that too many of the industry's clients were dissatisfied with its overall performance.

Construction law is a large and complex area. In this short chapter, an attempt will be made to consider some of the elements of construction law which are most likely to have an effect upon the attainment of satisfactory durability of the construction product and those elements which will apply when it is not achieved. For a more general study of this extensive subject the reader is urged to consider the works which are referred to in the bibliography. On no account should the contents of this chapter be regarded as a substitute for obtaining and relying upon the advice of a properly qualified construction lawyer.

16.2 Procurement methods and risk assessment

Every construction project involves risks which are unavoidable. These risks may include the physical conditions on site, mal-administration, delays, physical damage to property, injury to persons, insolvency and disputes.

Having perceived the need for a construction project, the employer ought to carry out a detailed assessment of the benefits, risks and financial constraints associated with the particular project. The employer will then have to decide which party is in the best position to control or minimise the particular risk and to transfer those risks which the employer is not prepared to accept. If the contractor is required to accept a risk which is difficult to assess, or is beyond their control, this will inevitably be reflected in an inflated tender price.

The selection of contract, or procurement method, ought to be determined by the extent to which the employer wishes to exercise direct control over the contract or prefers to accept little or no risk. The effect of risk on the different forms of contract which are in current use is discussed below. The achievement of satisfactory durability ought to be a common aim whatever form of procurement is adopted. Satisfactory durability is less likely to be achieved if the employer selects an inappropriate form of contract. The wrong form of contract will probably have a direct effect upon the performance of the construction team and may result in defective design, materials and workmanship.

There are currently three main types of construction contract in general use in the private sector. In each case the contractor undertakes to supply work and materials for the erection of a building or structure for the benefit of the employer. The type of contract which is selected dictates whether the detailed design is supplied in whole or in part by the employer or by the contractor.

16.2.1 Traditional contracting

Traditional or general contracting is still used on the majority of projects and is the one with which the industry is most familiar. In the building industry the most commonly used standard form (Parris 2002) is the Private Edition of JCT 98 (Joint Contracts Tribunal Standard Form of Building Contract 1998 Edition). [The equivalent standard form contract in civil engineering is ICE 7 (Institution of Civil Engineers Conditions of Contract 7th Edition 1999).] JCT 98 succeeded JCT 80 and incorporated many of the recommendations of the Latham Report (Latham 1994) to coincide with the passing of the Housing Grants, Construction and Regeneration Act 1996 on 1 May 1998. It is this standard form and its predecessors which will be the focus of comment in this chapter. Under JCT 98 the employer engages a contractor to build what the employer's architect has designed. Other professionals may be engaged by the employer including structural engineers, services engineers and quantity surveyors. The task of the quantity surveyor is to be responsible for the planning and control of the financial aspects of the project and it is the responsibility of the architect (or sometimes a separately appointed project manager) to lead the design team and to co-ordinate and manage the connection between design and production. Certain parts of the work are carried out by specialist sub-contractors nominated by the employer. There is therefore a clear separation between the design and the construction of the project.

Latham (1994) saw the lack of co-ordination between design and construction as being the cause of many of the problems in the construction industry. Where defects occur the employer is often left with a dispute between the designers and the contractors as to whether the problem is one of design or of workmanship. Contractors will often blame late or inappropriate design information for delay in completion of construction. The traditional form of contract is therefore not seen as appropriate where the employer's priority is the timely completion of a technically advanced or highly complex building.

16.2.2 Design-and-build

The problems created by the separation of design from construction led to the emergence of design-and-build contracting in various forms. These contracts are sometimes called turnkey or design-and-construct contracts. Under this type of contract the contractor accepts responsibility for both design and construction. The extent of the design obligation will depend upon the terms used in the particular contract which may be *ad hoc*. The best known standard form for design-and-build work is the JCT Standard Form Contract With Contractor's Design (WCD 98) of which the structure and wording is very similar to JCT 98 (Chappell & Powell-Smith). Under WCD 98 a statement of the 'Employer's Requirements' is submitted to the contractor who responds with the 'Contractor's Proposals'. The 'Employer's Requirements' and the 'Contractor's Proposals' become, in effect, the specification and no 'Bills of Quantity' (see below under contract documents) is therefore required under WCD 98. When agreement is reached the contractor becomes responsible for undertaking the design work contained in the 'Contractor's Proposals'. The contractor is also responsible, in accordance with the requirements and proposals, for constructing the building, and for co-ordinating and integrating the project.

The most important advantage of design-and-build is that responsibility for the project rests entirely with the contractor (Murdoch and Hughes 1996). Unlike traditional contracting there is no scope for dispute over the otherwise contentious issue of whether defective work arises from design or workmanship, or if it is the responsibility of the employer's designer, the main contractor or a sub-contractor.

Design-and-build is suitable where the building is not highly complex and standard processes and components can be used. Design-and-build is not suitable where the employer anticipates the need for variations during the contract period.

16.2.3 PFI

The Private Finance Initiative (PFI) is a way of funding capital projects, such as hospitals and schools, in the public sector. The contractor accepts responsibility for the financing, design, construction and operation of the project which is then leased back to the public sector body over a fixed period. Thus the cost to the public sector is spread over a number of years instead of having to be financed as a lump sum. One of the perceived advantages of PFI is that maintenance of the facility is outsourced and remains the responsibility of the contractor. Thus it is in the interest of the contractor to ensure the durability of the structure as the risk of defects remains their responsibility for the duration of the project which is likely to be for many years.

16.2.4 Construction management

Construction management is a form of contract in which the employer (rather than the contractor) enters into a series of direct contracts with the specialist and trade contractors. The employer engages the 'construction manager' whose responsibility is to manage and co-ordinate the various contractors.

This form of contract was developed by employers who wanted to have 'hands on' involvement in the project. It is appropriate to technically complex projects and requires the employer to take an active rôle in the management process. The construction manager replaces the architect by having responsibility for supervision and management while the architect, with the assistance of the other professional designers, produces a specification to which the trade contractors' work will have to conform. Much of the design is provided by the specialist contractors who, together with the trade contractors, will be directly responsible to the employer for any breach of contract.

The advantage of construction management is that complex projects can be brought swiftly to completion with scope for variations throughout the process. The work is let under different contracts as the project develops and there is therefore little certainty about the price of the project until it is completed. As the employer accepts much of the risk, there should be a corresponding benefit in cost saving against, for example, a design-and-build contract where the price is certain.

16.3 Contract documents

16.3.1 Agreement and Conditions of Contract

Although there are many forms of construction contract, they will usually contain an agreement, detailed conditions of contract, a specification, bills of quantity, a set of drawings and other incidental documents. The 'Agreement' records the general obligations of the parties such as the contractor's obligation to carry out the work and the employer's obligation to pay the contract price. The 'Agreement' incorporates the much more detailed clauses which are contained in the 'Conditions of Contract'. The 'Conditions of Contract' (which are often in standard form) are intended to provide the framework for the efficient control and administration of the work.

16.3.2 Drawings

The drawings are generally produced by the employer's design team and the aim is to convey the designer's ideas and intentions to the contractor. Under a traditional contract it is usual for the contract drawings to be followed by more detailed drawings. The detailed design drawings show how the different parts of the building interact. This can be a complex area of design and problems often arise where different specialist teams undertake different parts of the design. As a consequence, problems may not be identified until construction of that element of the design begins. The lack of early planning and co-ordination frequently gives rise to construction disputes (Uff 1999).

16.3.3 Specification

The specification defines the materials and products to be used, the standard of work required, any performance requirements and the conditions under which the work is to be executed. To ensure that the specification is accurate it should be prepared in a systematic and methodical manner. Specifications often describe the work to be undertaken in great detail, but sometimes may comprise a 'performance specification' requiring the details to be provided by the contractor. This carries with it a risk to the employer as it may be in conflict with the conditions of contract which do not impose a design obligation upon the contractor.

Under a traditional JCT 98 Contract 'Without Quantities', the specification is the overriding description of the quality and quantity of the work which is required. Under a design-and-build contract, the specification is set out in detail in the 'Employer's Requirements' and in the 'Contractor's Proposals'.

16.3.4 Bills of Quantity

'Bills of Quantity' define the whole of the contractor's obligations for both quality and quantity of work (Murdoch and Hughes 1996). Their principal use under the JCT form of contract is for assessing interim payments by approximate measure under a lump sum contract. Where there are bills of quantity, the specification will not be a separate document but will be incorporated in the bills.

Under engineering contracts (e.g. ICE 7) the quantity of work described in the bills is only an estimate of what is required of the contractor. At the end of the contract the work actually carried out is re-measured for the purpose of the final payment. The quality of the work is subject to any tests which the engineer may require at the contractor's expense. Rules for the standardisation of the descriptions of work are set out in a separate document known under the JCT contract as the relevant 'Standard Method of Measurement'.

16.3.5 Priority of documents

Discrepancies can arise because the contract documents may comprise: conditions of contract, the specification, bills of quantity, drawings, and other items. This makes it necessary to decide which document is to have precedence and how any inconsistency may be resolved.

Both the JCT 98 and ICE 7 contracts require the contractor to refer to the contract administrator to issue instructions to resolve any discrepancy. If this causes delay or disruption, the contractor will be entitled to an extension of time and to compensation for loss and expense. If the instructions constitute a variation, the contractor will be paid for the extra work.

The contract administrator should apply the rules which govern the interpretation of contracts subject to any express term to the contrary in the contract itself. Under the common law rules, written words take precedence over typed words, and typed words over printed words. The principle is that the parties' true intentions are more likely to be reflected in a document such as the bills of quantity than standard printed conditions of contract.

Many of the standard forms of contract (including JCT 98) provide, however, that nothing in the contract bills shall override the contract conditions. This has led to numerous cases in which the obvious intention of the parties (as described in the bills) has been overridden. An example of this is the case of *English Industrial Estates Corporation v George Wimpey & Co (1973) I Lloyd's Rep*. Under an earlier JCT form of contract, the employer took possession and occupied part of a factory as provided in the contract bills. The contract conditions, however, laid down a different procedure for taking possession. The part of the works which had been taken over was destroyed by fire. The Court of Appeal held that, despite the provision in the contract bills, the contractor remained responsible to the employer under the contract conditions for the part which had been destroyed even though the employer had taken possession of it.

16.4 Contractual relationships

The essentials of any binding contract are:

- two or more parties;
- an intention by them to create a legal relationship;
- an agreement; and
- consideration unless the contract is by deed.

'Consideration' is an exchange of obligations such as payment of the contract price by the employer in exchange for the works being carried out by the contractor. Legally binding contracts may be written or oral or may be part-written and part-oral. The contractual obligations, or terms, may be express terms or may be terms which are not referred to in the contract, but are implied by law.

16.4.1 Privity of contract

A fundamental principle of contract law is the doctrine of 'privity of contract', i.e. that only the parties to a contract can take the benefits of that contract or be subject to its burdens or obligations. As an example, a third party (such as an employer) cannot take advantage of the contractual duties which the sub-contractor owes to the main contractor. Similarly the main contractor cannot enforce contractual duties which are owed by the architect to the employer, as no contractual relationship exists between the main contractor and the architect. Tort, by contrast, is a common law remedy which may be available in the absence of any private agreement and to a much wider class of persons. Liability, however, will usually be restricted to damages for physical injury to persons or 'other property'. Damages for economic loss arising from defects in the building itself are excluded – *Murphy v Brentwood District Council (1991) 1 AC 398*.

16.4.2 Inequity caused by law of privity

It is often the case that by the time defects are discovered in a commercial building the property has passed into the hands of a subsequent owner or tenant. Under the doctrine of caveat emptor (let the buyer beware), the current owner or tenant, unless they have obtained collateral warranties, will have no cause of action in contract against the previous owner or landlord. Nor is there a remedy in tort. In *D & F Estates Ltd v Church Commissioners for England (1989) 1 AC 177* it was held that the tenants of a flat were not entitled to recover the cost of renewing plasterwork from the main contractors with whom they had no contractual relationship. It was held that unless the defective plasterwork had caused personal injury or damage to property other than the structure itself, the loss was 'pure economic loss' and irrecoverable in tort.

The problem of finding a building owner a remedy to recover substantial damages for breach of contract after he has parted with the property has long troubled judges. It is not unusual for an employer to commission a building for sale or lease to a third party. If the third party later discovers defects in the building he will be prevented from claiming damages from the contractor in the absence of a collateral warranty. Contractual damages are compensatory in nature and, under the general rule, the employer cannot recover substantial damages as he has suffered no loss. In the absence of any remedy, the damages sustained by the third party would fall into a 'legal black hole'. To get around this problem the House of Lords held in *Linden Gardens Trust Ltd v Lanesta Sludge Disposals Ltd; St Martin's Properties Corporation v Sir Alfred McAlpine & Sons (1994) 1 AC 95* that the employer could sue a contractor on behalf of the third party, thereby creating an exception to the general rule.

In *Alfred McAlpine Construction Ltd v Panatown Ltd (2000) AC* a contract was entered into between a contractor and an employer for the construction of a building upon the land of a third party who would own the building. The House of Lords held that the employer was entitled to claim substantial damages from the contractor for defects in the building, only if the third party who had actually suffered the loss had no direct remedy against the contractor. In this case the third party had a direct remedy against the contractor by virtue of a duty of care deed and the employer was therefore entitled to only nominal damages.

16.4.3 Law reform on privity

To address the unsatisfactory state of the law on privity of contract the Contracts (Rights of Third Parties) Act, 1999 was passed on 11 May 2000. This gives third parties the right to enforce a contract term if the contract expressly so provides, or if it appears that the contracting parties intended a term to be enforceable by a third party, and that the term grants a benefit in favour of the third party.

The third party will not have to provide any benefit (consideration) to the contracting parties before being able to enforce the benefits under the contract. The third party will be put in the same position as the original contracting party and will be subject to any defences (such as limitation) which would have been available against the original contracting party.

The third party must be identified expressly in the contract, by name, but may be described as a member of a class or as answering to a particular description. The party need not be in existence when the contract is entered into. This means that financial institutions, purchasers and tenants may be granted rights under the contract. Instead of insisting upon collateral warranties, lawyers may require that benefits be conferred upon their clients in the original construction contract itself. Construction lawyers, however, have generally adopted a more conservative approach and advise their clients that it would be safer to continue to rely upon collateral warranties.

There are restrictions in the Act preventing the original contracting parties from altering the third party's entitlement by later cancellation or variation of the contract without the third party's consent. Without these restrictions this would defeat the object of avoiding the need for collateral warranties. If the contracting parties choose to avoid the entire effect of the Act (and many do) they may do so simply by providing in the contract that no benefits are intended to be conferred upon any third party. The Act does not deal with the practical problem of how a third party is to find out if any benefits have been conferred upon him under the contract. He has no statutory right to the information. Consequently a third party who has not seen the contract will have to incur wasteful legal costs to obtain the information in the absence of co-operation from the contracting parties.

16.4.4 Collateral warranties

The decline in tort as a remedy for the recovery of economic loss since the mid-1980s resulted in the growth of collateral warranties as a routine requirement by end users who would not otherwise be in a direct contractual relationship with those responsible for the design and construction of the project.

The construction contract between employer and main contractor is central to a construction project. There are, however, many more parties involved on a project, resulting in a complex arrangement of contractual relationships. Thus end users such as prospective purchasers, tenants or financial institutions funding the project require contractual rights against the designers and contractors in the form of collateral warranties.

The intention of a collateral warranty is to give to the subsequent owner or tenant the same contractual rights that the employer/developer would have had against the warrantor had there been no change in ownership.

The following terms are often included in the various collateral warranty forms to restrict the liability of the warrantor:

- that the rights under the warranty may not be assigned (e.g. to a second purchaser or sub-tenant);
- that the warrantor shall be entitled to raise any defence which could have been raised against the employer;
- that liability is limited to the cost of necessary remedial works and that consequential losses (such as loss of profit) are excluded;

- that if other parties are jointly liable, the warrantor shall be liable for no more than the share of responsibility which a court would allocate against him if all other potential defendants were sued. (This is intended to cover the position where other parties responsible for the design and construction have not also given collateral warranties.)

As giving a collateral warranty increases the risk of being sued, it is essential that consultants make suitable arrangements with their professional indemnity (PI) insurers before entering into collateral warranties. Failure to do so could result in subsequent claims being repudiated by the insurers or the policy itself being avoided. Some insurers provide a collateral warranty advisory service in order to ensure consistency in the terms which are undertaken by their insured.

16.5 The contractor's obligations

The contractor's obligation is to supply work and materials for the erection of a building or other structure for the benefit of the employer. While the contractor may undertake the contractual obligation of providing the detailed design, it is more usual that the design work is carried out by the architect or engineer engaged by the employer.

The contractor's obligations are essentially to comply with the terms of the contract documents which specify the work that is to be done and the time within which it is to be completed. The contractor is required to comply with instructions issued pursuant to the contract by the employer's agent, usually the architect or engineer.

In the absence of a contrary provision in the contract, the contractor is required to provide all elements which are necessary to complete the project irrespective of what is contained in the specification. In *Williams v Fitzmaurice (1858) 3 H&N. 844* the contractor claimed extra payment for the floorboards because they had been unintentionally omitted from the specification of the fixed sum contract. It was held that the contractor was required to provide the whole of the material necessary for the completion of the work to make the house habitable and that flooring was not an extra.

The employer gives no implied warranty of the suitability or nature of the site and the risk of adverse site conditions rests with the contractor. In *Sharpe v Sao Paulo Brazilian Railway Company (1873) LR 8 Ch. App. 597* the contractor entered into a fixed sum contract to construct a railway between two stations in Brazil. The employer provided engineer's plans, but these proved to be useless. The contractor had to excavate twice as much as expected. The contractor claimed extra payment, but it was held that, since the employer had given no warranty about the accuracy of the plans, the contractor was not entitled to extra payment.

In another instance the court would not find any implied warranty on the part of the employer that the demolition and replacement of Blackfriars Bridge could be built according to the engineer's design. In *Thorn v London Corporation (1876) 1 App. Cas. 120* the contractor was held not to be entitled to recover the delay and extra expense incurred as a result of caissons, designed by the engineer, proving to be useless.

Of course, under a modern form of building contract such as JCT 98, where the contract states that the bills of quantity contain an exhaustive definition of the works, the contractor's obligation under the contract is simply to provide what is described in the bills. If the contract bills description is inadequate, the employer will be required to pay for any extra work undertaken by the contractor in completing the building.

16.5.1 Fitness for purpose and quality of materials

It is fundamental to the achievement of the durability of a building or structure that it is built in a workmanlike manner using materials of satisfactory quality which are fit for their required purpose. It is also necessary that the completed building or structure is fit for its intended purpose. The courts have readily implied warranties in respect of these obligations in construction contracts.

The common law, as created by the judges, is overlain by the implied warranties created by the Sale of Goods Act 1979, the Supply of Goods and Services Act 1982 and the Sale and Supply of Goods Act 1994. These Acts substantially codified the common law concerning implied terms of quality and fitness for purpose. The contractor, therefore, has an implied obligation to carry out the work in a workmanlike manner using materials of 'satisfactory quality'. If the materials have been specified in the contract then they will need to be of the kind described and without defects. The contractor further undertakes to ensure that both the workmanship and materials used on the project will be reasonably fit for the purpose for which they are intended, unless the obligation is excluded by the circumstances of the contract.

The term 'satisfactory quality' is defined under the 1994 Act as being that attained by goods which 'a reasonable person would regard as satisfactory, taking account of any description of the goods, the price (if relevant) and all the other relevant circumstances'.

The implied condition that goods supplied under the contract are reasonably fit for the purpose for which they are intended does not apply where the employer does not rely on the contractor's skill and judgment, or where such reliance would be unreasonable. The implied condition is limited to situations where the employer has expressly or impliedly communicated to the contractor any particular purpose for which the materials are being acquired.

In *Young & Marten Ltd v McManus Childs Ltd (1969) 1 AC 454* the House of Lords held a sub-contractor liable to the main contractor for defective roof tiles even though the make and type of tiles (Somerset 13) had been selected by the employer by nominating the specialist supplier. The tiles, which were otherwise suitable, came from a defective batch and became useless within a few years. The House of Lords held that the sub-contractor was liable for breach of the implied warranty that the materials supplied were of good quality. There was, however, no implied warranty as to their fitness for purpose because the tiles were not chosen by the sub-contractor whose skill and judgment had not been relied upon for that purpose.

By contrast, in *Gloucestershire County Council v Richardson (1969) 1 AC 480*, the House of Lords held that the implied duty as to fitness had been excluded. The contractor had been required to obtain concrete columns from a supplier nominated by the employer. The columns had latent defects, but the contractor was not liable because the employer knew that under the terms of the sub-contract the supplier excluded liability for defects. The position in Young & Marten was different because the sub-contractor was not prevented from seeking redress 'down the chain' from the supplier of the tiles.

The two House of Lords' cases illustrate the need to consider each particular contract to ascertain the degree of reliance placed by the employer on the contractor.

In *University of Warwick v Sir Robert McAlpine (1988) 42 BLR 1*, the contractor was instructed by the employer's architect to order a specified type of epoxy resin from an identified sub-contractor for repair work to cladding. The contractor raised objection to the use of the epoxy resin although it had been tested by the employer's professional team. The remedial works failed and Garland J held that the contractor was not liable since the employer had not relied upon the contractor's skill and judgment.

Under the 1979 Act, goods were of 'merchantable quality' if they were suitable for one of a number of purposes for which goods of that general type are used. In *Aswan Engineering Ltd v Lupdine Ltd (1987) 1 WLR 1* the Court of Appeal held that heavy-duty plastic pails which were suitable for the intended purpose of transporting their contents on board ship, but which perished when stacked in excessive heat on a quayside in the Middle East, were of merchantable quality. That case may have been decided differently under the 1994 Act which now provides that goods should be fit 'for all the purposes for which goods of the kind in question are commonly supplied'.

In *Rotherham Metropolitan Borough Council v Frank Haslam Milan & Co Ltd and M J Gleeson (Northern) Ltd (1996) 78 BLR 1* the Court of Appeal had to decide if steel slag used by the contractor as hardcore fill was of merchantable quality and reasonably fit for the intended purpose. Under a JCT 63 (With Quantities) Contract, the employer specified in detail the required properties of hardcore fill to be used in construction and the contract provided for approval of the fill by the employer's design team following inspection and testing. Evidence was accepted that, although the steel slag was of merchantable quality for other purposes, because of its expansive properties it was unsuitable for use as fill material beneath the foundations and floor slabs, which were consequently damaged. The expansive properties of steel slag were not known by the contractor and the design team, but were known within the industry at that time (1979). It was held that there had been no breach by the contractor of the implied term to use materials of merchantable quality and that the normal implied term as to fitness for purpose had been excluded in the absence of reliance by the employer on the contractor's skill and knowledge.

16.6 Duties of the design consultant and contract administrator

16.6.1 Design

Under the traditional forms of contract (e.g. JCT 98 or ICE 7) the overall responsibility for the design, supervision and administration of the carrying out of the works rests with the architect or engineer (referred to here as 'the designer'). The designer's duties will usually be set out in his terms of engagement with the employer either under the standard conditions of engagement produced by the relevant professional governing body (e.g. RIBA or ACE) of which the designer is a member, or some variation of those standard conditions. The basic duty of the designer is to prepare skilful and economic designs and thereafter to supervise and administer the carrying out of the work on site in the best interests of the employer (Uff 1999).

In common with all professionals, architects and engineers do not warrant or guarantee that they will achieve the results that their client desires. This is in contrast to the contractor's express or implied warranty that the building or structure will be fit for the intended purpose. The standard which an architect or engineer is generally expected to achieve in discharging their duties to the employer (both in contract and in tort) is the reasonable skill, care and diligence of an ordinary competent and skilled architect or engineer (Jackson & Powell 2003).

While an architect or engineer will generally be expected to exercise no more than reasonable care and skill, there may be particular circumstances which require them to take special steps to discharge their duty.

In *Greaves & Co Contractors Ltd v Baynham Meikle & Partners (1975) 2 Lloyd's Rep. 325 CA*, a structural engineer was employed by a contractor under a design-and-build contract to construct a warehouse. The engineer knew that the first floor of the warehouse was to be used for the storage of oil drums to be stacked and moved by forklift trucks. The engineer's design failed to make sufficient allowance for vibrations caused by the forklift trucks and the floor cracked and was unfit for its required purpose. The contractor was therefore liable to the client for the cost of replacement and sued the engineer claiming that it was an implied term of the engineer's engagement that the design would be fit for its intended purpose. The Court of Appeal held that although the engineer had not been guilty of negligence, he was nevertheless liable to the contractor for failure of the design. It was held that it was the common intention of the contractor and the engineer that the design of the warehouse would be fit for the required purpose which gave rise to a term implied as a matter of fact.

The case has not, however, been followed and a number of other reported cases have reaffirmed the standard duty to be one of reasonable skill and care.

In *Hawkins v Chrysler (UK) Ltd and Burne Associates (1986) 38 BLR 36* the architects had designed, specified and supervised the installation of showers which caused injury to an employee who slipped on wet tiles. The Court of Appeal held that there was no reason to imply any warranty that the materials selected by the architects would be fit for their intended purpose and that the standard test of reasonable care and skill would apply.

The question of whether the designer's conduct falls within established practice is ultimately a question to be decided by the judge. The designer will be judged by the standards expected of a competent designer at the material time that the work was done and not with the benefit of hindsight and with knowledge gained within the profession subsequently. This is known as the 'state of the art' defence. (*Perry v Tendring District Council; Thurbon v Same (1984) 30 BLR 118*). The standard of competence will almost invariably be a matter of expert evidence and opinion (*Warboys v Acme Investments (1969) 4 BLR 133 at 139 (CA).*)

In project-related services the designer is under a duty to the employer to produce an economic design. Indeed, it is common for employers to insist upon cost reductions in schemes which are presented to them by designers. The designer may therefore be tempted or driven to the use of new, untried and cheaper materials and construction methods. If there is some element of risk involved in their use, the designer is under a duty to warn the client of the possible consequences (*Independent Broadcasting Authority v EMI Electronics and BICC Construction (1080) 14 BLR 1 HL*). Indeed in certain circumstances, if the risk appears too great, the designer should decline to proceed with the project. Similarly, if the task required by the client is beyond their competence, designers must decline the appointment or advise the client to seek other specialist advice.

In *Moresk Cleaners v Hicks (1966) 2 Lloyd's Rep. 338*, the architect, without the employer's express authority, delegated to the contractor the design of a reinforced concrete structure on a sloping site. The design proved defective and the architect contended that they had implied authority, as the employer's agent, to employ the contractor. The court disagreed and held that the architect was liable. Modern construction technology, however, has meant that delegation of the design of numerous specified parts of the project is inevitable and may, therefore, be implied from the circumstances of the case. Consequently, in *Merton London Borough Council v Lowe (1982) 18 BLR 130,* CA, an architect was held to have been entitled to rely upon the expertise of a specialist proprietary ceiling sub-contractor, where details of the design were kept as a trade secret by the sub-contractor.

Where an architect, with the employer's authority, instructs a specialist (such as a structural engineer) to undertake an element of the design (e.g. the foundations) their responsibility will be to co-ordinate and direct the expert's work. However, if any danger or problem arises, of which an architect of ordinary competence ought reasonably to be aware, the architect has a duty to warn the employer and may not rely blindly upon the specialist (*Investors in Industry Ltd v South Bedfordshire DC (1986) QB 1034*). In *Chesham Properties Ltd v Bucknall Austin Project Management Services (1996) CILL 1189* it was held that in the circumstances of that case the architects were obliged to report to the employer on any deficiencies on the part of the engineer and the quantity surveyor in carrying out their duties. They were not, however, held to be under a duty to report to the client upon their own deficiencies.

Where the designer assumes a supervisory duty under the terms of engagement (as is often the case) their responsibility in respect of design continues until the project is completed, usually by the issue of a final certificate (*Brickfield Properties v Newton (1971) 3 All England Reports 328*).

The traditional form of building contract creates a separation between design and construction. This separation was much criticised by Egan (1998) on the grounds that it leads to wasted time and effort being spent on site during the construction process in trying to make designs work in practice.

The co-ordination and integration of the different elements of the overall design of a project is a complex task. The design team may include an architect, structural engineer, electrical services engineer, heating and ventilating services engineer, public health engineering consultant, landscape architect and interior designer. They in turn may have further specialists working for them. Contractors, sub-contractors and sub-sub-contractors carrying out installation work are also likely to have design responsibilities. The distinction between design and workmanship is often finely drawn and many design decisions will not be specified in the contract bills.

Latham (1994) emphasised that, irrespective of the chosen procurement route, effective management of the design process is crucial to the success of the project. Latham recommended the appointment of a lead manager and co-ordination between consultants and specialist contractors including detailed checklists of design requirements in their appointment documents which should also be set out in the main contract documentation.

16.6.2 Inspection and supervision

It is essential to the achievement of durability that the work which is carried out on site is executed by the contractor properly in accordance with the contract documents. Where an architect or engineer is appointed by the employer as a contract administrator, certificates will have to be issued by them as the work proceeds entitling the contractor to proportionate stage payments of the contract price. As the contract administrator is under a general duty to the employer to exercise reasonable skill and care, they will be under a duty to satisfy themselves that the work which they are certifying has been carried out properly and in accordance with the contract documents. The architect or engineer is therefore required to visit the site at appropriate intervals to inspect the progress and quality of the works and to determine that they are being executed in accordance with the contract documents. The duty to exercise skill and care means that the architect or engineer is required to detect such defects and bad workmanship as would a reasonably competent architect or engineer – *Imperial College of Science & Technology v Norman & Dawbarn (1987) 8 Con LR 107*.

The contract administrator's conditions of appointment will not usually envisage that they will be required to make frequent or constant inspections. Where this is likely to be the case, it is appropriate that a clerk of works or a resident architect be appointed. In *East Ham Corporation v Sunley (Bernard) & Sons Ltd (1966) AC 406*, the House of Lords acknowledged the limited nature of the architect's obligation to inspect and examine the works and recognised that some defects may escape their notice.

It is important that the contract administrator inspects the principal parts of the work before they are hidden from view. If necessary they should require the contractor to give notice before that part of the work is carried out. In the classic case of *Jameson v Simon (1899) IF Court of Session 1211*, the architect made weekly visits but failed to inspect the 'bottoming' before the laying of cement floors of a house under construction. The 'bottoming' contained rubbish including wood which caused dry rot and the architect was held liable.

Under the standard JCT 98 Contract the contract administrator has the power to carry out tests of any materials and to order that work which has been covered up be opened up for inspection. If the work is found to be defective the costs will fall on the contractor, but they will be paid for the work if no defects are found. If the work is defective, however, the defects will have to be made good at the contractor's expense. Furthermore, once defects have been found, the contract administrator may require further tests, the cost of which will have to be borne by the contractor whether or not further defects are discovered.

The requirement that the contractor shall execute the work on site properly in accordance with the contract documents means that the contractor has a duty to supervise their own work force on site to ensure that the work is carried out with due care and skill. The contract administrator's rôle is therefore secondary to that of the contractor. The contract administrator's duty is to protect the employer's interests in seeing that the work is executed by the contractor in accordance with the contract documents.

This duty places the contract administrator in a vulnerable position where the contract is not properly executed by the contractor. It is not possible for the contract administrator to detect every defect, but in practice if defects are discovered after completion of the building, the likelihood is that the contract administrator will be sued together with the contractor. If, as is often the case, the contractor has by then gone out of business or is insolvent, or if the contract administrator has issued a final certificate (see below under contract administration), the contract administrator (who will be assumed to have professional indemnity insurance) may be the only remaining target for the aggrieved owner. Because of the readiness of the courts to hold architects liable under such circumstances, the architect's governing body (RIBA) has progressively amended the architect's terms of appointment to distance the architect's duty from 'supervision' to 'site visits' and 'inspections'. In practice, however, this has made little difference to the position of the architect when sued.

16.6.3 Quality assurance

In view of the increasing use of design and construct and construction management contracts there has been an increasing emphasis upon contractors improving their level of self-supervision. Contractors have for some time been able to become accredited with the British Standards Authority under BS 5750. Quality assurance certification under BS 5750 is intended to improve quality by requiring that it be in 'conformance to requirements'. There is a documented procedure intended to ensure that materials and workmanship supplied are as close as possible to what is specified. Many doubts have been expressed, however, about whether producing what is specified will ensure that the contractor produces a good, high quality building (Murdoch and Hughes 1996). Specified standards do not necessarily equate to good standards. Ultimately quality is more likely to be achieved by selecting the right team to carry out the project in a spirit of co-operation than by confrontation and conflict.

16.6.4 Contract administration

Apart from designing the project the architect or engineer (under a traditional contract) will often be appointed to undertake the rôle of contract administrator. Under other forms of contract this rôle may be undertaken by a contractor or project manager. The duties of the contract administrator will usually involve supervision and inspection (see above) to ensure that the work complies with the contract documents; authorising variations under the terms of any variation clause in the contract; advising the employer on contractual matters such as insurance and the effects of inflation; giving instructions to the contractor as provided under the contract, and issuing certificates to record some event or decision of the contract administrator under the terms of the contract.

The three main types of certificate are interim certificates, final certificates and certificates recording an event under the contract. These may be certificates of practical completion, certificates of completion of making-good defects and certificates of non-completion. The most significant of these certificates in relation to the issue of durability is likely to be the final certificate because of the effect that the issuing of a final certificate may have upon the rights of the parties in respect of latent defects.

The party given the responsibility for issuing certificates (usually the architect or engineer) is under a duty to act impartially between the employer and the contractor while at the same time exercising due care and skill as agent for the employer

(*Sutcliffe v Thackrah (1974) AC 727*). Although the contract administrator will owe a duty of care to the employer, it appears that no such duty is owed to the contractor who will have to pursue other remedies under the contract (such as adjudication or arbitration) if aggrieved by the architect's or engineer's certificate (*Pacific Associates Inc v Baxter (1989) 2 All ER 159*).

Under earlier editions of the JCT standard form of contract (e.g. JCT 63) the architect's final certificate was a statement that the whole of the works were complete in all respects in accordance with the contract. A final certificate became conclusive evidence of the matters stipulated in it in any proceedings provided neither party had commenced adjudication, litigation or arbitration before the issue of the certificate or within 28 days of it being issued.

Under JCT 80 (which preceded JCT 98) the contract provided that the quality of materials and standards of workmanship were to be to the reasonable satisfaction of the architect. In *Crown Estate Commissioners v John Mowlem & Co Ltd (1994) 70 BLR 1* the Court of Appeal held that the final certificate under JCT 80 was conclusive that the architect was satisfied with the quality and standards of all materials, goods and workmanship. The effect of the decision was to make it practically impossible for an employer to bring an action for latent defects against the contractor after the final certificate had been issued.

To overcome this unsatisfactory situation the contract forms were amended concerning the conclusive aspect of the architect's 'satisfaction'. Under JCT 98 the final certificate is conclusive on the architect's satisfaction only if the architect has specifically stated in the bills of quantities or specification that some item of goods, materials or workmanship is to be to their satisfaction or approval (Parris 2002).

Defects will often not materialise within 28 days of completion and may remain latent for many years thereafter. This is particularly so where durability is concerned. It follows that if a final certificate has been issued the contractor may have an absolute defence and the party most likely to be sued will be the architect or engineer who issued the final certificate. Although the wording of JCT 80 and final certificates issued thereunder have since been amended to avoid this consequence, it remains to be seen how the courts will interpret final certificates in the future.

16.7 Contractor's design obligations

Under the traditional form of construction contract, responsibility for design usually rests with the employer's design team while the contractor has responsibility to construct the work strictly in accordance with the contract documents. While the architect or engineer responsible for the design is obliged only to exercise the care and skill of a reasonably competent architect or engineer, the contractor warrants expressly or impliedly that the work done and materials supplied will be fit for their intended purpose. The traditional distinction between design and construction is removed under modern design-and-build or 'package' contracts where the contractor accepts responsibility for both design and construction. This means that if the building is not fit for its intended purpose the employer will not have to prove negligent design or workmanship to succeed in holding the contractor liable.

In practice, the employer's design team will usually produce the 'Employer's Requirements' and the employer will then enter into a contract with the contractor to include the 'Contractor's Proposals' comprising the detailed design. Often the design team's appointments will be novated by the employer to the contractor. This has the effect of transferring the employer's responsibility for design to the contractor who thereby undertakes the more onerous responsibility of ensuring that the building is constructed fit for its intended purpose.

If the contract limits the contractor's responsibility for design to 'a reasonable care and skill' requirement, difficulties can be created where elements of the design (e.g. those provided by specialist sub-contractors) are on the basis of fitness for purpose or in accordance with a performance specification. Thus the standard of design required of the sub-contractor by the main contractor may be higher than the standard required of the main contractor by the employer who, because of the principle of privity of contract, will be unable to seek redress against the sub-contractor on the basis of fitness for purpose.

The courts, however, will readily imply a fitness for purpose obligation where the contractor is responsible for both design and construction. In *Viking Grain Storage Ltd v T H White Installations Ltd & Anor (1985) 33 BLR 103* an experienced specialist contractor entered into a package deal of design, execution and management for the construction of a grain storage and drying installation on the employer's land. The employer alleged defects concerning the steel superstructure, the cladding and the ground of the site. The judge held that the contractor could be held strictly liable for each of the defects, because each would constitute a breach of the overall implied obligation that the 'finished product' should be reasonably fit for its intended purpose.

Where it is clear that an employer did not rely upon the contractor in respect of the design, the implied term of fitness for purpose will not arise. In the Irish case *Norta Wallpapers (Ireland) v Sisk & Sons (Dublin) Ltd (1978) IR 114*, a specialist sub-contractor nominated by the employer supplied and erected the roof of a factory. It was held that no warranty of fitness for purpose could be implied against the main contractor since the main contractor had no alternative but to accept and adopt the sub-contractor's design.

16.8 Legal remedies for defective buildings

16.8.1 Contractual remedies

This section deals only with remedies for latent defects which occur after completion of a project because these are the type of defects most likely to arise from problems concerning durability. Where the defect arises from a breach of contract by the contractor (e.g. bad workmanship) or by the designer (e.g. lack of reasonable skill and care) the employer may sue in court or commence arbitration proceedings for breach of contract. This is subject to the important proviso that the breach of contract or professional duty did not occur before expiry of the relevant limitation period (six years for a simple contract and twelve years for a contract executed as a deed). For the position of a subsequent owner who has no contractual relationship with the contractor or designer see Section 16.4 Law Reform on Privity.

16.8.2 Damages and remoteness

Damages for breach of contract are intended to place the innocent party in the same position (so far as money can do it) as they would have been in had the contract been properly performed. They are not intended to return the parties to the same position as if the contract had never been made. Not all losses flowing from the breach of contract are recoverable, some being described as 'too remote'.

The classic case dealing with damages for breach of contract is *Hadley v Baxendale (1854) 9 EX 341* in which Baron Alderson described two categories of recoverable loss. Under the 'first limb' the innocent party is entitled to recover losses for breach of the contract which naturally arise in the usual course of events, of which the parties are deemed to be aware. Under the 'second limb' the innocent party is entitled to recover additional or special consequences which were within the contemplation of both parties (i.e. foreseeable by them) at the time they made their contract as the probable result of the breach.

In *Hadley v Baxendale* the owners of a mill in Gloucester contracted with common carriers to deliver a broken crankshaft to an engineer in Greenwich. The defendants delayed delivery for several days during which time the mill was at a standstill. The plaintiffs were held not entitled to recover their loss of profit because the defendants did not know that the plaintiffs had no spare crankshaft and that the mill could not operate until a new crankshaft had been installed. The rule in *Hadley v Baxendale* was restated in *Victoria Laundry (Windsor) Ltd v Newman Industries Ltd (1949) 2 KB 528.* The plaintiff required a new boiler to undertake work of an exceptionally profitable nature. The work was lost when the boiler was not delivered on time. Because the defendant was unaware of the new work at the time the contract was entered into, it was held that the plaintiff could recover no more than the usual profit for that type of work. The special (second limb) profit was therefore too remote to be recoverable.

The claimant will have to prove that the damage he suffered was caused by the defendant's breach of contract and not by some other factor. In *Quinn v Burch Bros (Builders) (1966) 2 QB 370* an independent plastering sub-contractor suffered injury when a makeshift trestle which he had erected collapsed. Under the contract the employer was obliged to provide necessary equipment, but had failed to supply the sub-contractor with a suitable step ladder. The Court of Appeal held that the accident had been caused by the sub-contractor's use of the trestle and not by the employer's breach of contract in not providing suitable equipment.

16.8.3 Measure of damages

The measure of damages awarded for breach of contract is usually the actual cost of repairs although the claimant may be compensated for the diminution in value of the property where it would be wholly uneconomic and/or unreasonable for the claimant to insist upon repair or replacement of the defective building.

In *Harbutts Plasticine v Wayne Tank & Pump Co (1970) 1 QB 447* the plaintiffs' factory had burnt down as a result of the defendants' breach of contract. The factory was old and before the fire the plaintiffs had already decided to build a new factory. The Court of Appeal dismissed the so called 'betterment' argument and held that the plaintiffs were entitled to the cost of rebuilding even though this meant that they got new for old.

In *Ruxley Electronics v Forsyth (1996) AC 344* the defendant refused to pay for a swimming pool which the plaintiff contractor had built to a depth of 6 ft instead of 7 ft 6 ins. Although this did not affect the value of the property nor the use of the swimming pool, the defendant counterclaimed £21,560 being the estimated cost of rebuilding the swimming pool. The House of Lords held that the cost of rebuilding would be disproportionate to the benefit and upheld the trial judge's award to the employer of only £2,500 for loss of amenity. The sum awarded was in line with damages usually awarded for 'distress and inconvenience' such as may be recoverable for a frustrated holiday.

The date on which the cost of repairs should be assessed has been held to be not the date of breach, but the date when the repairs ought reasonably to have been carried out. In *Dodd Properties (Kent) v Canterbury City Council (1980) 1 All ER 928* the plaintiffs delayed carrying out repairs to their garage premises for ten years during a period of high inflation because they were impecunious and liability was disputed by the defendants. It was held that, in the circumstances, the defendants had acted reasonably and they were awarded damages of £30,000 being the current cost of repair even though this was nearly three times what the repairs would have cost ten years previously.

16.8.4 Mitigation of loss

Under the principle of restitution the plaintiff is entitled to be put in the same position as if the contract had been properly performed. Consequently the claimant will not be entitled to recover as damages any loss which he could reasonably have avoided. The guiding principle is one of reasonableness. In *William Tompkinson & Sons Ltd v Parochial Church Council of St Michael in the Hamlet (1990) 6 Const LJ 814* an employer unreasonably refused the contractor permission to return to the site to carry out remedial work during the defects liability period. The employer was therefore held not entitled to recover the additional cost of employing another contractor to do the work.

Once remedial work has been carried out, the court will be reluctant to infer that the employer has failed to mitigate the loss, provided the employer has acted reasonably. In *Board of Governors of the Hospitals for Sick Children and Another v McLaughlin & Harvey plc (1987) 19 Con LR 25* there was a dispute between experts for the respective parties before costly remedial piling work was carried out. Although the employer's expert was heavily criticised, Judge John Newey, QC held that the employer had acted reasonably in relying upon the advice of an eminent expert concerning the remedial works. If the expert had acted negligently, however, the chain of causation would have been broken, giving rise to a claim against the expert.

16.8.5 Contributory negligence

Contributory negligence which occurs at the time of breach ought not to be confused with mitigation of loss which concerns events after the breach. If the employer is partly at fault, the recoverable damages may be reduced by the court if the claim is brought in tort – Law Reform (Contributory Negligence) Act 1945. In *Forsikringsaktisselskapet Vesta v Butcher (No.1) (1989) AC 852* it was held that contributory negligence would apply if the plaintiff could have brought the action in tort (i.e. negligence) even though the claim was for breach of contract. Contributory negligence will not apply, however, where the breach of contract does not correspond to a common law duty or depend upon negligence.

16.8.6 Remedies under the Defective Premises Act 1972

The Defective Premises Act 1972, imposes a duty to build dwellings properly so that they are fit for habitation. Any person taking on work for, or in connection with, the provision of a dwelling (whether the dwelling is provided by the erection or by the conversion or enlargement of an existing building) owes a duty (if the dwelling is provided to the order of any person) to that person and to every person who acquires an interest (whether legal or equitable) in the dwelling.

The duty imposed under the Act is to ensure that the work which is taken on is done in a workmanlike or, as the case may be, professional manner, using proper materials and so that as regards that work the dwelling will be fit for habitation when completed.

If the person who carries out the work for another does so properly and in accordance with instructions given by that other person, he shall be treated as having discharged his duty under the Act, except where he owes a duty to warn the other person of any defects in the instructions and fails to discharge that duty to warn. A person (e.g. the owner) shall not be regarded as having given instructions merely because he has agreed to the work being done in a specified manner, with specified materials or to a specified design.

The cause of action for breach of duty under the Act is deemed to have accrued at the time when the dwelling was completed. If remedial work is later carried out, however, the cause of action in respect of the remedial work is deemed to have accrued at the time when the further work was completed. Any action under the Act needs to be commenced **within six years of completion** and the Act renders null and void any term of an agreement which purports to exclude or restrict the operation of its provisions.

When the Act was introduced it did not apply to houses covered by the NHBC scheme which affords purchasers of newly built homes a ten year guarantee against any defects in the property. The Act therefore became a dead letter because most new houses were constructed under the NHBC scheme. In any event, by the date of its introduction, an equivalent, and in some cases better, remedy could be obtained in the courts under the tort of negligence. The current NHBC scheme ('Buildmark') which came into force in 1988 is not an 'approved scheme' under the Act. It follows that the provisions of the Act apply to all dwellings completed after that date. Thus an owner may have protection under the NHBC scheme as well as under the Act. It is regrettable that the period adopted by the legislature for the bringing of claims is six years instead of the ten year period (as used under the parallel NHBC scheme) since it is generally accepted that the great majority of defects will manifest themselves within ten years.

16.8.7 Breach of building regulations

Section 38 of The Building Act 1984 provides that, subject to contrary provision in the regulations, breach of a duty imposed by the building regulations is actionable if it causes damage.

Section 38 of the Act does not yet apply, however, and it must be assumed therefore that it was the intention of Parliament that failure to comply with the building regulations does not, of itself, give rise to a civil action for breach of statutory duty. This was the view of Judge John Newey, QC in *Perry v Tendring District Council; Thurbon v Same (1984) 30 BLR 118.*

16.8.8 Remedies at common law

Difficult legal problems can arise where no contractual or statutory remedy is available to the party who suffers loss as a result of latent defects. An employer may find that his cause of action is 'statute barred' if the defect does not manifest itself within six years of the breach of contract. Where an employer sells the property to another, the subsequent owner will have no contractual remedy in the absence of effective collateral warranties or deeds of assignment. The subsequent owner will have no cause of action against the employer under the principle of caveat emptor (let the buyer beware). The claimant may therefore be forced to seek a common law remedy in tort.

The claimant in an action for the tort of negligence must show: (1) that the defendant owed them a duty of care; (2) that there was a breach of that duty; and (3) that the breach caused recoverable damage. In the leading case of *Donoghue v Stevenson (1932) AC 562* Lord Atkin held that the consumer of a bottle of ginger beer could sue the manufacturer for illness caused by the discovery of a decomposing snail in the empty bottle. The principle was extended in *Dutton v Bognor Regis Urban District Council (1972) 1 QB 373* where Lord Denning MR held that the council was liable to a second purchaser of a house built on a rubbish tip and that the builder, if sued, would also have been liable in tort. Dutton, which was approved by the House of Lords in *Anns v Merton London Borough Council (1978) AC 728* opened the floodgates to numerous claims including many which had been thought to be stale because of the less generous limitation period allowed in contract.

Following a period of retrenchment in subsequent cases the House of Lords held in *D & F Estates Ltd v Church Commissioners for England (1989) 1 AC 177* that a builder's liability in tort is limited to defects which cause either injury to persons or physical damage to property other than the building itself (e.g. where a defective garage roof falls on the occupier's car). Damage to the building itself is regarded as pure economic loss and is therefore irrecoverable. Although *Anns* was criticised it was not until *Murphy v Brentwood District Council (1991) 1 AC 398* came before the House of Lords that *Anns* was declared to have been wrongly decided. In *Murphy* it was suggested that there were two possible exceptions which might entitle a subsequent owner to recover damages for remedial works against the contractor or designer. The first exception is where the defect in the building creates a danger to an adjoining property and the subsequent owner carries out repairs to avert that danger. The second exception is where one part of the building causes damage to another part (the so-called 'complex structure' theory).

The examples given are of a central heating boiler exploding and damaging a house or a defective electrical installation malfunctioning and setting fire to the house. It was stated by Lord Bridge that the position would be different in the case of defective foundations causing cracking because once the cracks appeared the structure as a whole would be seen to be defective and the nature of the defect would be known.

16.9 Dispute resolution

Dispute resolution concerns the various processes by which the parties to a dispute may resolve their differences. Civil procedure is the body of rules of court applicable to the litigation process leading ultimately to an enforceable judgment. All other forms of dispute resolution are alternative (ADR) to litigation and include arbitration, adjudication and mediation. Because arbitration often follows a very similar procedure to litigation in construction cases it has regrettably become synonymous with litigation in the view of many practitioners and is not seen as a form of ADR. Under the Housing Grants, Construction and Regeneration Act 1996 a new statutory form of adjudication has been created in relation to defined construction contracts to provide rapid temporarily binding determination of construction disputes. As in the case of arbitration, any award is enforceable only by the court.

16.9.1 Litigation

Litigation is an adversarial process by which parties have their dispute determined by the court. Because the process was all too frequently used by defendants as a device to delay or frustrate the payment of legitimate claims, the rules governing court procedure underwent fundamental change in the form of the Civil Procedure Rules (CPR) which were introduced in 1999. The CPR are intended to give litigants cheaper and faster access to justice. Whereas litigation was previously claimant-driven, the courts are now encouraged and empowered to intervene to control the speed and content of litigation and to encourage the parties to co-operate at all stages of litigation and to settle cases wherever possible.

Most construction disputes will be brought in the Technology & Construction Court (the TCC) whose judges have much experience in dealing with cases of technical complexity. Before suing, however, the parties are required to follow a 'Pre-action Protocol', intended to promote the early exchange of information so that the claim can be fully investigated and, if possible, be resolved without the need for litigation. If the dispute proceeds to litigation, the claimant begins the process by issuing and serving a claim form to which the defendant must file a defence or an acknowledgement of service. The defendant may serve a counterclaim. The basic facts upon which the parties rely should be contained in these 'statements of case' which should be accompanied by any key documents. At an early stage there will be a case management conference (CMC), at which the court will seek to narrow the issues and dictate how the litigation is to be conducted. The court can strike out a claim if it discloses no reasonable grounds for bringing or defending the claim or for failure to comply with any rule, practice direction or court order.

There will usually follow disclosure and inspection of relevant documents. Directions will be given concerning without prejudice meetings between experts appointed by the parties and where appropriate the court will require the appointment of a single joint expert. Witness statements, each supported by a signed ‘statement of truth’, must be exchanged between the parties before commencement of the trial. At the trial the parties will present their cases and witnesses, including experts, may be cross-examined by the other side. When all the evidence has been heard, the judge is obliged to give a judgment which, in the absence of an appeal, may be enforced by the successful party.

16.9.2 Arbitration

The arbitration of construction disputes is a private process, supported by the Arbitration Act 1996, by which the parties choose an independent arbitrator to whom they refer their dispute for a final binding decision by him. The essential requirements for an arbitration are:

- a dispute;
- an agreement to arbitrate (as contained in many standard form contracts); and
- a submission or reference to arbitration.

Section 1 of the Arbitration Act 1996 provides that the object of arbitration is to obtain the fair resolution of disputes by an impartial tribunal without unnecessary delay or expense. The parties should be free to agree how their disputes are resolved, subject only to safeguards as are necessary in the public interest. The arbitrator is required to act fairly and impartially as between the parties, giving each party a reasonable opportunity of putting their case and dealing with that of their opponent. They are required to adopt procedures suitable to the circumstances of the particular case, avoiding unnecessary delay or expense so as to provide a fair means for the resolution of the dispute. Subject to any written agreement between the parties concerning procedural and evidential matters (such as the arbitration agreement or relevant rules) the arbitrator has power to decide all procedural and evidential matters. Effective and experienced arbitrators are therefore able to adopt a more flexible approach to dispute resolution and to have regard to the convenience of the parties concerning timetable, venue and procedure.

Where the contract between the parties requires (as construction contracts frequently do) that their dispute shall be referred to arbitration, the court will be bound under s. 9 of the Arbitration Act 1996 to stay any court case on the application of the defendant unless they have shown an intention to defend the case rather than to choose arbitration.

An important difficulty in construction arbitration is that the arbitrator has no power under the Act to order the consolidation of proceedings or for there to be concurrent hearings. Because construction projects frequently involve a number of parties under different contracts, disputes frequently involve a number of parties. Unless the various parties to the dispute (e.g. employer, consultants, contractor and sub-contractors) consent to multi-party arbitration (and they may refuse for tactical reasons) a number of separate arbitrations may result, leading to additional expense and the possibility of inconsistent findings by different arbitrators. Consensual multi-party arbitration may take the form of consolidation or concurrent hearings. Consolidated arbitrations become a single arbitration in which one arbitrator makes a single award in respect of all the parties. In the case of concurrent or back-to-back arbitrations separate arbitrations take place but, because the same arbitrator is involved, the findings are likely to be consistent. The process is nevertheless cumbersome and costly.

The arbitrator is required to act fairly and impartially and may be removed by the court under section 24 if substantial injustice has been caused to the applicant. Where there has been serious irregularity the court may remit the award, set it aside or declare it to be of no effect.

The arbitrator is required to award costs on the general principle that costs should follow the event. Recoverable costs shall be determined on the basis that there shall be allowed a reasonable amount in respect of all costs reasonably incurred.

Leave to appeal against the arbitrator's award on a question of law will be granted by the court only in very limited circumstances. Unlike the judgment of a court, an arbitration award cannot by itself be enforced against the unsuccessful party. The successful party has to obtain leave of the High Court to enforce the award as a judgment, or they have to bring an action on the award and seek summary judgment. If the arbitrator has exceeded their jurisdiction (e.g. by making an award on issues not contained in the reference to arbitration) the unsuccessful party may object to the court against enforcement of the award.

16.9.3 Alternative Dispute Resolution (ADR)

The essential features of alternative dispute resolution (ADR) which distinguish it from either litigation or arbitration, are that it is:

- voluntary; and
- non-binding on the parties.

Of the various types of ADR the most important is mediation. Mediation involves the parties voluntarily engaging an experienced and neutral third party to assist them in achieving an acceptable settlement agreement. It is the task of the mediator to explore with the parties the strengths and weaknesses of their respective positions. The parties will usually meet in open session to present brief submissions to the mediator. Thereafter the mediator will establish and maintain lines of communication and seek to find common ground between the parties through a number of confidential and separate meetings with the parties or with both parties present as may be appropriate. It is essential that the parties at the mediation have authority to conclude a binding settlement and the whole process is intended to take place on the same day.

Under the CPR the court has a duty to encourage and facilitate the use of ADR in appropriate cases. The parties will be required to consider the use of ADR in all cases. The costs of the process are not usually recoverable.

The focus of ADR is not upon legal rights (as is the case in litigation and arbitration), but upon the commercial interests of the parties. Since most court cases are settled before they get to trial (because of the huge costs, unpredictability and risks involved) ADR provides a practical and sensible way of achieving a settlement and avoiding the costs of litigation. ADR shares with arbitration the advantage of privacy which may be a significant factor where professional or commercial reputations are at stake.

16.9.4 Adjudication

One of the key recommendations of the Latham Report (1994) was that adjudication should be the normal method of dispute resolution under construction contracts. This recommendation has been adopted under s. 108 of the Housing Grants Construction and Regeneration Act 1996. Under section 108 a party to a construction contract (as defined) has the right to refer a dispute arising under the contract for adjudication. The construction contract must be in writing and entered into after 1 May 1998. Construction operations are widely defined under the Act but do not apply to residential occupiers.

The adjudicator must act impartially, avoid unnecessary expense and is required to reach a decision within 28 days of referral of the dispute. There is no provision under the Act for the adjudicator to make an award for costs. The adjudicator is entitled to take the initiative in ascertaining the facts and the law and the adjudicator's decision is binding until the dispute is finally determined by legal proceedings, by arbitration or by agreement. Judges of the TCC have been robust in summarily enforcing adjudicators' decisions under an abbreviated court procedure.

Adjudication, which overrides both litigation and arbitration, provides a swift and effective remedy against a party to a construction contract who refuses to make payment. It is now widely used, but is not intended for complex cases involving latent defects when durability is likely to be an issue. Latent defects usually take a long time to become apparent and it remains to be seen if disputes over defects will increasingly be referred to adjudication.

16.10 Limitation

Limitation is particularly relevant to the issue of durability because, by its nature, the lack of durability is apt to reveal itself long after the project has been completed. Limitation can be defined as the period of time specified by statute (or sometimes agreed by contract) within which legal proceedings must be commenced once a cause of action has accrued. Proceedings brought after the expiry of this period are usually described as 'time barred' or 'statute barred'.

The significance of the expiry of the limitation period is that in the majority of cases, it provides the defendant with a complete defence to the claim. The claimant's remedy (i.e. to bring an action in court or to pursue an arbitration) is extinguished but not the right of action.

The relevant limitation periods are calculated from the date when the cause of action accrued. This event differs between claims for breach of contract, claims arising in the tort of negligence and claims for breach of statutory duty under the Defective Premises Act 1972.

The current limitation periods relevant to construction claims under the Limitation Act 1980, as amended by the Latent Damage Act 1986, are as follows:

- claims in contract: 6 years from date of breach of contract (12 years if contract was executed as a deed);

- claims in tort of negligence: 6 years from occurrence of relevant damage, whether or not damage is discoverable (*Pirelli General Cable Works v Fabre (Oscar) & Partners (1983) 2 AC 1*);
- claims in negligence may additionally be brought within 3 years of the damage first becoming reasonably discoverable;
- no claim for tortious negligence may be brought after the expiry of 15 years from the negligent act (the 'long stop');
- claims under the Defective Premises Act 1972 must be brought within 6 years of the date of completion of the dwelling or the remedial works to the dwelling.
  The three year discoverability rule (s. 14A of the Limitation Act 1980) does not apply to claims brought under DPA 1972;
- claims under the Civil Liability (Contribution) Act 1978 for indemnity or contribution by one party against another who is liable for the same damage must be brought within 2 years of the judgment or settlement;
- no limitation applies where there has been fraud or 'deliberate concealment' of the relevant act (s. 32 of the Limitation Act 1980) until the claimant has discovered the fraud or concealment or could have done so with reasonable diligence.

The deliberate commission of a breach of duty which is unlikely to be discovered for some time amounts to 'deliberate concealment' of the facts concerning the breach of duty. An inadvertent, accidental or unintended breach of duty of which the perpetrator was not aware is not a 'deliberate commission of a breach of duty' – *Cave v Robinson Jarvis & Rolf (2002) AC*. Fraud in this sense means more than outright dishonesty and includes 'unconscionable' behaviour (e.g. not caring if the action is right or wrong). In the case of *Kijowski v New Capital Properties Ltd (1987) 15 Con LR 1* a contractor was held to have been guilty of deliberate concealment where he knowingly covered up defective foundations.

The case of *Pirelli* concerned cracks in a chimney which occurred before 1971. The writ was issued more than six years later in 1978. Experts agreed that the cracks were not discoverable until 1972. They were not in fact discovered until 1977. The House of Lords held that the claim was time barred. The fact that the damage was not reasonably discoverable at the time the cause of action accrued was irrelevant. To mitigate the harshness of this régime Parliament enacted the Latent Damage Act 1986 which introduced the secondary three year period beginning from the date when the claimant could reasonably have discovered the damage (s. 14A of the Limitation Act 1980, as amended by the Latent Damage Act 1986). This secondary period does not apply to claims in contract (*Societé Commerciale de Reassurance v ERAS International Limited (1992) 2 All ER 82*).

16.10.1 Concurrent liability in contract and in tort

The limitation periods applicable in cases of breach of contract are generally less favourable than those in tort. For this reason, claimants whose actions are time barred in contract will often frame their cause of action in tort to take advantage of the more generous limitation period. In *Henderson v Merrett Syndicates Ltd (1995) 2 AC 145* the House of Lords determined that Lloyd's Names, whether they had direct contracts with the managing agents or not, could bring parallel concurrent causes of action against the managing agents in tort, thereby entitling the Names to rely upon s.14A of the Limitation Act 1980.

Concurrent claims in tort against professionals or quasi-professionals for economic loss are based upon the principle in *Hedley Byrne & Co v Heller & Partners (1964) AC 465*, which is not applicable to contractors and sub-contractors. It follows that claims arising from latent defects may often be statute barred against contractors and sub-contractors (e.g. for bad workmanship or quality of materials not to required stand-ard) but not statute barred against the construction consultant (e.g. an architect for having failed adequately to 'supervise' the contractor). Moreover construction consultants are invariably insured for claims against professional negligence while contractors are not insured against defective materials or workmanship. Under the principle of joint liability the full burden of a claim for latent defects in a building may fall upon the construction consultant for not having noticed defects rather than the contractor or sub-contractor whose poor workmanship or materials was the cause of the defects.

16.10.2 Arbitration clause and limitation

Construction contracts and professional terms of engagement often contain an arbitration clause, i.e. that any dispute between the contracting parties will be referred to an arbitrator for his decision. If one of the parties is sued by the other in court this is a technical breach of the arbitration clause. The defendant may therefore apply to have the action stayed so that the matter may be referred to arbitration. The court then has no discretion but to order the action to be stayed (s. 9 Arbitration Act 1996). If the defendant shows an intention not to seek arbitration, but to defend the action (e.g. by taking a step in the action) the right to claim a stay in the action will be lost. Once the action has been stayed the claimant's only remedy will be to refer the dispute to an arbitrator provided, however, that the claim is not yet time barred. It follows that if a claimant prefers to pursue a claim in court rather than by way of arbitration and the claim is soon to become time barred, it is essential that notice to refer the matter to arbitration is given at the same time as court proceedings are commenced to protect the claim from becoming both stayed in court and time barred against arbitration.

16.11 Insurance

Although the various risks associated with a construction project may have been allocated between the respective parties under the construction contract, the party responsible under the contract for a loss which has occurred, may dispute liability or be financially unable to meet the loss. Building contracts therefore invariably contain provisions concerning insurance arrangements to cover loss and damage. These provisions, however, will cover generally only loss and damage which has arisen during the period of construction.

Most construction contracts will require two different types of overlapping insurance cover as follows:

- a material damage policy:
- a liability damage policy.

A material damage policy covers loss or damage to the contract works (e.g. through fire, storm, flood, etc.). This type of policy is often taken out in the joint names of the contractor and the employer. A liability insurance policy covers claims made against the contractor and/or employer by third parties for damage to persons or property arising from the works. This requirement will usually be covered by the contractor's 'public liability' (PL) and the 'employer's liability' (EL) policies. These claims are dealt with under the policy in existence when the incident giving rise to liability occurs.

Construction consultants (such as architects and engineers) will invariably have professional indemnity (PI) insurance to cover them against claims for professional negligence. PI policies are typically renewable on an annual basis and have a limit of indemnity which may be in respect of each claim notified or the aggregate of claims arising during the year. PI policies are written on a 'claims made' basis and cover only claims which are made against the insured professional during the period of insurance (usually 12 months) irrespective of when the negligent 'act, error or omission' occurred. It follows that professionals are required to make arrangements to cover potential claims long after completion of construction and even after their retirement or death. It is also essential that the consultant notifies any claim during the year when it is first made otherwise the policy will have lapsed and the claim will not be covered. Consultants also have a duty to disclose to their insurers any material facts (e.g. a potential claim) before the renewal of their policy otherwise their insurers may avoid the policy on the grounds of the non-disclosure of material facts.

16.11.1 Project insurance

On massive projects (e.g. airport developments) the employer may often take out a single project policy to avoid gaps in insurance cover arising from the various overlapping insurance policies which would otherwise be in place. By securing a single co-ordinated insurance package the employer can avoid litigation associated with the difficulties of apportioning liability between the numerous parties to a project.

The project insurance policy is a single combined construction 'all risks' and third party policy of insurance which indemnifies the employer, the main contractor or management contractor and all other contractors and sub-contractors. Although professional consultants may be included under the policy, they will be covered only in respect of claims arising before practical completion of the project. Consultants will therefore not be covered in respect of claims arising from latent defects which are discovered after completion of the project.

16.11.2 Latent defect insurance

Defects causing a lack of durability in commercial buildings are most likely to be latent and the resultant damage may not be discoverable until long after completion of the building. With the exception of PI insurance held by the professional consultants, the insurance arrangements under standard building contracts will be of no assistance to the owner or occupier of a commercial building in the case of latent defects, as standard insurance provision operates only to cover risks which arise during the period of construction.

The unfortunate employer is therefore faced with the prospect of arbitration or litigation to establish liability against those parties who were responsible for the defects. He will have to prove causation of any loss long after the negligent acts complained of. Because of the complex nature of construction litigation the costs of pursuing a claim for latent defects are likely to be substantial. Under the principle of 'joint and several liability' the employer may have a choice of several parties from each of whom the employer's losses may be recovered in full. This works to the disadvantage of construction consultants and their insurers because the other parties (contractors and sub-contractors) will be uninsured and the employer need sue only the party with the deepest pockets. The construction consultant may also be vulnerable to a concurrent claim in tort where claims in contract against the contractors and sub-contractors are already statute barred.

In addition to the inherent risk and cost of litigation the employer may find that the professional consultant has failed to maintain a sufficiently high limit of indemnity under his PI policy or the claim may not be covered under the policy on grounds of late notification or non-disclosure. If the professional consultant has become insolvent, the client may sue the professional indemnity insurers direct under the Third Parties (Rights Against Insurers) Act 1930, but only after a judgment has been obtained against the insolvent construction consultant.

To address these problems a government committee was set up and, in 1988 produced the BUILD (Building Users' Insurance against Latent Defects) Report. The report concluded that the current arrangements for securing redress for latent defects did not serve satisfactorily the interests of either clients or 'producers' (i.e. consultants and contractors). The committee recommended that instead, redress should be obtained by means of a non-cancellable first party material damage insurance policy which would ensure the repair of defects and damage covered by the policy without proof of fault. This BUILD policy would be negotiated by the developer or building owner at the preliminary design stage. The committee envisaged that BUILD policies would cover initially only the most potentially troublesome elements: the structure (including foundations) and the weathershield. During its currency, the policy would be transferable to successive owners and whole-building tenants. The committee considered it essential that provisions be incorporated to provide protection for a period of 10 years from date of practical completion, that insurers' rights of subrogation against the producers be waived and that risk assessment and verification be undertaken by independent consultants appointed by the insurer.

The construction and insurance industries have been slow to take up the BUILD recommendations, the principal reluctance on the part of insurers being the waiver of their rights of subrogation against consultants; and on the part of developers, the requirement for a duplicate professional team to provide a technical audit in view of the associated costs and delays which this would involve.

In 1994 Latham recommended legislation for compulsory latent defects insurance for 10 years from practical completion for all future new commercial, retail and industrial building work on the basis defined in the BUILD Report. Parliament, however, was reluctant to impose continental style compulsory insurance on property owners and these recommendations were not incorporated in the Housing Grants, Construction and Regeneration Act 1996. Latent defect insurance is now generally available for smaller commercial developments and premiums have halved since the Latham recommendations.

References

Chappell, David and Powell-Smith, Vincent (1999) *The JCT Design and Build Contract*, Second Edition, Blackwell Science Ltd.

Egan, Sir John (1968) Rethinking Construction: *The Report of the Construction Task Force*, HMSO, London.

Jackson, Rupert M, QC and Powell, John L, QC (2003) *Jackson and Powell on Professional Negligence*, Fourth Edition, Sweet and Maxwell, London.

Latham, Sir Michael (1994) *Constructing the Team, Final Report of the Government/Industry Review of Procurement and Contractual Arrangements in the UK Construction Industry*, HMSO, London.

Murdoch, John and Hughes, Will (1996) *Construction Contracts: Law and Management*, Second Edition, E. & F. Spon, London.

Chappell, David (2002) *Parris's Standard Form of Building Contract: JCT 98*, Third Edition, Blackwell Science Ltd.

Uff, John, QC (1999) *Construction Law*, Seventh Edition, Sweet and Maxwell, London.

Appendix 1

Table of cases

Alfred McAlpine Construction Ltd v Panatown Ltd (2000) AC.
Anns v Merton London Borough Council (1978) AC 728.
Aswan Engineering Ltd v Lupdine Ltd (1987) 1 WLR 1.
Board of Governors of the Hospitals for Sick Children and Another v McLaughlin & Harvey plc (1987) 19 Con LR 25.
Brickfield Properties v Newton (1971) 3 All England Reports 328.
Cave v Robinson Jarvis & Rolf (2002) AC.
Chesham Properties Ltd v Bucknall Austin Project Management Services (1996) CILL 1189.
Crown Estates Commissioners v John Mowlem & Co Ltd (1994) 70 BLR 1.
Dodd Properties (Kent) v Canterbury City Council (1980) 1 All ER 928.
Donoghue v Stevenson (1932) AC 562.
Dutton v Bognor Regis Urban District Council (1972) 1 QB 373.
D & F Estates Ltd v Church Commissioners for England (1989) 1 AC177.
East Ham Corporation v Sunley (Bernard) & Sons Ltd (1966) AC 406.
English Industrial Estates Corporation v George Wimpey & Co (1973) I Lloyd's Rep.
Forsikringsaktisselskapet Vesta v Butcher (No.1) (1989) AC 852.
Gloucestershire County Council v Richardson (1969) 1 AC 480.
Greaves & Co Contractors Ltd v Baynham Meikle & Partners (1975) 2 Lloyd's Rep. 325 CA.
Hadley v Baxendale (1854) 9 EX 341.
Harbutts Plasticine v Wayne Tank & Pump Co (1970) 1 QB 447.
Hawkins v Chrysler (UK) Ltd and Burne Associates (1968)38 BLR 36.
Henderson v Merrett Syndicates Ltd (1995) 2 AC 145.
Hedley Byrne & Co v Heller & Partners (1964) AC 465.
Imperial College of Science & Technology v Norman & Dawbarn (1987) 8 Con LR 107.
Independent Broadcasting Authority v EMI Electronics and BICC Construction (1080) 14 BLR 1 HL.
Investors in Industry Ltd v South Bedfordshire DC (1986) QB 1034.
Kijowski v New Capital Properties Ltd (1987) 15 Con LR 1.
Jameson v Simon (1899) IF Court of Session 1211.
Linden Gardens Trust Ltd v Lanesta Sludge Disposals; St Martin's
Properties Corporation v Sir Alfred McAlpine & Sons (1994) 1 AC.
Merton London Borough Council v Lowe (1982) 18 BLR 130, CA.
Moresk Cleaners v Hicks (1966) 2 Lloyd's Rep. 338.
Murphy v Brentwood District Council (1991) 1 AC 398.
Norta Wallpapers (Ireland) v Sisk & Sons (Dublin) Ltd (1978) IR 114
Pacific Associates Inc v Baxter (1989) 2 All ER 159.
Perry v Tendring District Council; Thurbon v Same (1984) 30 BLR 118.
Pirelli General Cable Works v Faber (Oscar) & Partners (1983) 2 AC 1.
Quinn v Burch Bros (Builders) (1966) 2 QB 370.
Rotherham Metropolitan Borough Council v Frank Haslam Milan & Co Ltd and M. J. Gleeson (Northern) Ltd (1996) 78 BLR.
Ruxley Electronics v Forsyth (1996) AC 344.
Sharpe v Sao Paulo Brazilian Railway Company (1873) LR 8 Ch. App.597.
Societé Commerciale de Reassurance v ERAS International Limited (1992) 2 All ER 82.

Sutcliffe v Thackrah (1974) AC 727.
Thorn v London Corporation (1876) 1 App. Cas. 120.
University of Warwick v Sir Robert McAlpine (1988) 42 BLR 1.
Victoria Laundry (Windsor) Ltd v Newman Industries Ltd (1949) 2 KB 528.
Viking Grain Storage Ltd v T H White Installations Ltd & Anor (1985) 33 BLR 103.
Warboys v Acme Investments (1969) 4 BLR 133 at 139 (CA).
Williams v Fitzmaurice (1858) 3 H&N. 844.
William Tompkinson & Sons Ltd v Parochial Church Council of St Michael in the Hamlet (1990) 6 Const LJ 814.
Young & Marten Ltd v McManus Childs Ltd (1969) 1 AC 454.

Appendix 2

Table of statutes

Arbitration Act 1996.
Building Act 1984.
Civil Liability (Contribution) Act 1978.
Contracts (Rights of Third Parties) Act 1999.
Defective Premises Act 1972.
Housing Grants, Construction and Regeneration Act 1996.
Law Reform (Contributory Negligence) Act 1945.
Latent Damage Act 1986.
Limitation Act 1980.
Sale and Supply of Goods Act 1994.
Sale of Goods Act 1979.
Supply of Goods and Services Act 1982.
Third Parties (Rights Against Insurers) Act 1930.

Index